Introduction to the Strong Interactions

INTRODUCTION TO THE STRONG INTERACTIONS

Nathan W. Dean
Iowa State University

GORDON AND BREACH SCIENCE PUBLISHERS
New York London Paris

Copyright © 1976 by

Gordon and Breach Science Publishers Inc.
1 Park Avenue
New York, N.Y.10016

Editorial office for the United Kingdom

Gordon and Breach Science Publishers Ltd.
42 William IV Street
London W.C.2.

Editorial office for France

Gordon & Breach
7–9 rue Emile Dubois
Paris 75014

Library of Congress catalog card number 73-85292. ISBN 0 677 02750 8.
 Printed in Great Britain.

For Mary

Preface

The strong interactions, which provide the force holding together the nucleus, pose one of the most perplexing problems of contemporary physics. In times past, we had the proton and the neutron as "fundamental" particles, interacting through the exchange of a meson. Now any list of fundamental particles contains scores of entries running to several pages, and gets longer with each new edition. Indeed, there appears a real possibility that we are going to find an endless variety of "elementary particles", if that term can meaningfully be applied to an object belonging to so large a set. The profusion of particles is further complicated by the complexity of their interactions. It is a difficult field; its experiments require a great amount of ingenuity and an even greater amount of money, and its theories confound the trusted and familiar techniques of quantum mechanics. It is a glamorous field, for the same reasons, and occupies the research efforts of a large number of physicists and physics graduate students. While progress is certainly being made, we cannot yet claim more than a very limited knowledge of strong processes, and our understanding of them remains practically nil.

Although we may not at present have any satisfactory theory of strong interactions, however, we do have a well-developed phenomenology which helps to classify particles and correlate observations. It consists primarily of three areas – unitary symmetry, the analyticity of the scattering amplitude, and the Regge pole theory. This book is intended to provide an introduction to that phenomenology by setting forth the basic features of these three topics.

The unitary symmetry scheme furnishes a periodic table by means of which we are able to bring order into the list of particles. It arises as an extension of the concept of isospin, which postulates the Lie group SU(2) as a symmetry of the strongly interacting particles, to include strangeness as an additional conserved quantity. The resulting group is SU(3), and it is remarkably successful. Surprisingly, however, its classification cannot rest on a basis of known particles; it requires instead a new sub-unit, the "quark", which is a building block of all particles but has not yet been discovered experimentally. The addition of spin to the SU(3) quantum numbers produces a still larger symmetry group, SU(6), which accounts correctly for the spins of the particles classified by SU(3). Both of these symmetries are only approximate, and we attempt in Part I of the book to show some of their failures as well as their successes.

The second area stems from the assumption that the "smoothness" of nature will be reflected in the analytic structure of the scattering amplitude. The methods of the theory of analytic functions of a complex variable then become applicable to the strong interactions, and we can guess the singularities from a knowledge of the physical effects which produce them. Poles of the scattering amplitude are shown to correspond to bound states and resonances, and branch points can be identified through unitarity as resulting from intermediate states. Knowing the singularity structure of the amplitude, we can continue it from one region to another. In a relativistic kinematics this continuation relates the scattering of particles with that of antiparticles in "crossing symmetry". By putting all of these ideas together, we obtain both exact results, such as dispersion relations, and new approximation techniques known as "bootstraps".

Finally, there is Regge's transformation into the complex angular momentum plane. This esoteric mathematical step leads to two unexpectedly disparate insights. First, it provides an elegant unification of families of bound states and resonances. A Regge trajectory produces sets of particles with identical quantum numbers except that their spins differ by two units. Second, it suggests an eminently sensible parametrization of high energy scattering data. This second aspect of the Regge theory has been used extensively in fitting a number of different reactions; it has also led to more general developments.

We attempt to present a treatment of these three topics at a level accessible to students who have completed a year's course in quantum mechanics. Although quantum field theory is essential to any rigorous development of particle theory, to postpone the introduction of these important ideas until that rather ponderous mathematical apparatus is mastered seems to us an unnecessary delay. Most of our arguments are therefore based on simple potential theory. Some familiarity with the standard mathematical techniques of theoretical physics has been assumed, in particular a basic knowledge of the theory of functions of a complex variable.

The concept of this book, and its organization, originated in a course given to graduate students at Vanderbilt University, where the beginning chapters were written. The Aspen Center for Physics provided a stimulating atmosphere for writing further chapters, and the finishing touches were applied at Iowa State University. The encouragement and cooperation of friends and colleagues in all three locations were essential to this task and are gratefully acknowledged. In essence, this book is a distillation of my own conceptions about particle physics, and all who have played a part in their formulation are, in some sense, my coauthors. They should not, of course, be held responsible. Finally, I should like to express my appreciation to my wife, who typed each chapter while I was writing the following one.

N.W.D.

Ames, Iowa

Contents

Preface vii

Chapter 1 INTRODUCTION 1
Describing Interactions 2
The Transition Amplitude 6
Quantum Numbers 12

PART I – GROUP THEORY

Chapter 2 THE ROTATION GROUP 19
The Rotation Group 19
Tensor Operators 26
Lie Groups 28

Chapter 3 THE ISOSPIN GROUP SU(2) 34
Applications of Isosymmetry 41

Chapter 4 SU(3) 47
Generating the Strange Particles 48
The SU(3) Algebra 50
Generators and Casimir Invariants 55

Chapter 5 PARTICLE CLASSIFICATION IN SU(3) 59
The Sakata Model 65
Meson Classification in the Sakata Model 70
Baryon Classification in SU(3) 71
Quarks 74

Chapter 6 CONSEQUENCES OF SU(3) 76
Baryon Mass Relations in SU(3) 76
Meson Mass Relations 83
Electromagnetic Properties 84
Three-Particle Vertices 87
Two-Body Scattering 92

Chapter 7 ADDING SPIN 97
SU(4) 97
SU(6) 103
SU(6) Assignments 108
Consequences of SU(6) 112

Chapter 8 THE INDEPENDENT QUARK MODEL 116
The Low-Energy Quark Model 117
Additivity 121

APPENDIX A
A1 Irreducible Representations of SU(N) 128
A2 Table of the Hadrons 134

Reading List for Group Theory 138

PART II – ANALYTICITY

Chapter 9 SINGULARITIES IN POTENTIAL THEORY . . . 143
Scattering by a Central Potential 144
The Square Well 150
Bound State Poles 151
Resonance Poles 153

Chapter 10 THE COMPLEX k-PLANE 163
The Jost Function 165
Analytic Properties of the Jost Function 168
The Yukawa Potential 175
Singularities of $S_\ell(k)$ 178

Chapter 11 THE COMPLEX ENERGY PLANE 181
Analyticity of $A_\ell(E)$ 182
The Forward Scattering Amplitude 188
Relativistic Scattering and the Left-Hand Cut 192

Chapter 12 UNITARITY AND ANALYTICITY 203
The S-Matrix 203
Applications of Unitarity 206
Unitarity and Relativistic Scattering 212
πN and NN Dispersion Relations 218

Chapter 13 RELATIVISTIC SCATTERING THEORY . . . 224
Relativistic Description of Two-Particle Scattering 225
Crossing and Physical Regions 227
Unequal Masses 233
Singularities in the stu Plane 238
The Mandelstam Representation 246

Chapter 14 FURTHER TOPICS 253
Partial Wave Dispersion Relations 253
The N/D Method 256
Bootstraps 260
Phase Considerations from Crossing Symmetry 262

Reading List for Analyticity 265

PART III – REGGE THEORY

Chapter 15 COMPLEX ANGULAR MOMENTUM 269
The Sommerfeld–Watson Transformation 273
Properties of a(λ, k) in Potential Theory 278
Exchange Forces and Signature 282
Factorization 286
Comparison with Experiment 288
Nucleon Resonances 290

Chapter 16 RELATIVISTIC REGGE POLES 293
Regge Poles from Dispersion Relations 293
Signature in Relativistic Regge Theory 296
Crossed-Channel Regge Poles 299
The Regge Amplitude 302
Line Reversal and the Pomeranchuk Theorem 307
Daughters 311
Ghosts, Sense, and Nonsense 315
Conspiracy and Evasion 317

Chapter 17 REGGE PHENOMENOLOGY 319
Pion–Nucleon Total Cross sections 320
Charge Exchange and the ρ Trajectory 323
Elastic Differential Cross Sections and Polarizations 329
Regge Parametrization of Other Processes 332

Chapter 18 REGGE CUTS 340
Problems at Higher Energies 340
Regge Cuts 343

Chapter 19 DUALITY 352
Finite Energy Sum Rules 353
A Dual Amplitude 358
Dual Bootstraps and the Veneziano Amplitude 361

APPENDIX B SPIN COMPLICATIONS 365
Polarization 366
Partial Wave Analysis 370

Reading list for Regge Theory 373

Postscript 374

Index 375

Chapter 1

INTRODUCTION

There was a time when one could speak meaningfully of "fundamental particle physics". Although our knowledge of the basic interactions of matter was extremely limited, the existence of a small number of elementary particles interacting in a limited number of ways seemed clear on experimental as well as philosophical grounds.

At first, of course, only electromagnetism and gravitation were observed, and Einstein tried unsuccessfully to combine these two phenomena in a "unified field theory". Probing more deeply into the nuclear structure of the atom, however, revealed the necessity for two additional forces. The first of these, now said to arise from the *strong interactions*, was required to account for the binding of protons and neutrons in the nucleus, while the other explained beta decay as a result of the *weak interactions.* Neither of these is directly observed in the macroscopic world because they act only over exceedingly small distances.

Esthetically, a theory involving four fundamental forces is certainly less satisfactory than a completely unified one; but that does not mean nature cannot choose such a theory. In any case, our current understanding of physics accepts these four basic interactions as distinct and independent, and classifies elementary particles by the way in which they participate in each one. Thus the mass of a particle measures how strongly the gravitational interaction affects it, and the charge similarly reflects the electromagnetic interaction. To determine the strengths of the short-range forces is more difficult experimentally, but leads to the conclusions indicated by calling them "strong" and "weak".

If this description of nature is to be called fundamental, however, it should involve only a small number of basic particles upon which the four forces act. All charged particles interact electromagnetically, of course, and all with nonzero mass feel gravitation. But there are, for example, only four particles (and their four antiparticles) which participate in the weak interactions but not in the strong. These four are the electron, the muon, and their two associated neutrinos. The first two have much smaller masses than the proton, and the neutrinos are apparently massless; for this reason these particles are called *leptons*, from the Greek for "light".

The strongly interacting particles were originally separated correspondingly into *mesons* (medium) and *baryons* (heavy), but since both of these classes are involved in the same interaction they are now often jointly called *hadrons*. Over the past two decades, the hadrons have proliferated to such an extent that they call into serious question the fundamentality of particle physics. To the proton and neutron were added first the π mesons, or pions; then the heavier strange mesons and baryons; then, with the interpretation of resonance effects as particles of extremely short lifetime, a bewildering profusion of new states. Particles and almost-particles abound. To call four forces fundamental may be reasonable, but to call a hundred different particles "elementary" stretches the imagination to the breaking point.

The strong interactions thus pose an intriguing conceptual problem not dissimilar to that which led from the periodic table to atomic structure. Constructing models which can account for the diversity of phenomena observed has therefore become the principal concern of a large number of theoretical physicists. The difficulties encountered may be separated roughly into two parts: first, finding classification schemes which will group particles into larger and larger multiplets, and second, understanding how they interact with each other. Substantial progress has been made with the former by means of the mathematical techniques of Lie group theory. The latter, however, is a much more difficult nut to crack, because the methods of perturbation theory cannot be used reliably; the interaction is so strong that the series is divergent. Radical departures are therefore required from the procedures that have proved successful in quantum electrodynamics.

In this book we shall explore these two topics and introduce the methodology which has developed in strong interaction physics for treating them. First, however, a basic vocabulary must be developed; this initial chapter will therefore be devoted to a review of some of the basic ideas of describing interactions and labeling particles.

Describing Interactions

All of the ideas we have about how particles interact in the microscopic domain are based on quantum mechanics. Processes in which particles are created or destroyed, or drastically change their identities, should be more properly treated by means of quantum field theory; but no one has been able to make field theory work very well for solving problems in the strong interactions. It is nevertheless an important tool, despite its quantitative inefficacy, because it can be used to show that the amplitudes describing these processes have certain very basic properties. The non-relativistic Schrödinger equation is certainly not a correct description of particle physics, but it also possesses most of these properties; in fact, they are much more easily abstracted from it than from field theory. We shall therefore work in this simpler theoretic framework, which, although strictly not applicable to the problems we wish to study, makes the basic ideas easier to assimilate.

Since physical quantities are generally not dimensionless, the first step in any calculation should be choosing a set of units. There are in nature three fundamental dimensionalities – length, time, and mass. Choosing units for these three (e.g. the mks or cgs system) determines the units of all other quantities. For a given problem, the most convenient "natural" units may be chosen so that frequently occurring parameters have unit values. There are, however, two dimensionalities in which nature herself has chosen fundamental values. One of these is a velocity, namely the speed of light c; the other is angular momentum, which is quantized in units of $\hbar$. Let us therefore choose our system of units so that "$c = \hbar = 1$", i.e. so that velocities are measured as fractions of c, and angular momenta as multiples of $\hbar$. These choices impose two constraints on our system of units; specifically, the unit of length is c times the unit of time and $\hbar/c$ divided by the unit of mass. We still may choose one of the three units for convenience. The quantity which has been chosen, for better or worse, is the electron volt ($1\ \text{eV} = 1.60 \times 10^{-19}$ joule); energies in particle physics are measured in terms of $1\ \text{MeV} = 10^6\ \text{eV}$ or $1\ \text{GeV} = 10^9\ \text{eV}$. (The latter is sometimes called a BeV, for billion electron volts; but in Europe 1 billion = 10^{12}, and this notation is gradually yielding to GeV to avoid confusion.) Since $E = mc^2$, this means that masses are measured in GeV/c^2, or for brevity simply in GeV – units of $\hbar$ and c may be neglected by this convention. Lengths are then measured as multiples of $\hbar c/(1\ \text{GeV}) = 1.97 \times 10^{-16}$ meter, and times as multiples of $\hbar/(1\ \text{GeV}) = 6.58 \times 10^{-25}$ second. The values of a number of important physical quantities in this sytem of units are given in Table I-A.

TABLE 1-A Some common quantities expressed in units of $\hbar$, c, and eV

Quantity	Dimensions	Unit	Value
Mass	eV/c^2	1 kg	5.61×10^{29} MeV/c^2
		1 amu	931 MeV/c^2
		proton mass	938.3 MeV/c^2
		$\pi^\pm$ mass	139.6 MeV/c^2
		electron mass	511.0 keV/c^2
Length	$\hbar c$/eV	1 meter	$5.07 \times 10^{15}\ \hbar c$/GeV
		Bohr radius of electron	$2.68 \times 10^{11}\ \hbar c$/GeV
		Compton radius of pion	7.41 $\hbar c$/GeV
Area	$(\hbar c/\text{eV})^2$	1 millibarn (10^{-27} cm^2)	2.57 $(\hbar c/\text{GeV})^2$
Charge	$(\hbar c)^{1/2}$	e	0.0854 $(\hbar c)^{1/2}$ (= $(\hbar c/137)^{1/2}$)

Interactions produce transitions, and that is why we study them. The physical quantity of fundamental importance is the transition rate $w_{\beta\alpha} = (d/dt)p_{\beta\alpha}$, where $p_{\beta\alpha}$ is the probability that a system originally in the state labeled by α will be observed in another state β. It is given by the "Golden Rule" of time-dependent perturbation theory, in the above units, as

$$w_{\beta\alpha} = 2\pi|\mathscr{T}_{\beta\alpha}|^2\delta(E_\alpha - E_\beta), \tag{1.1}$$

where $\mathscr{T}_{\beta\alpha}$ is the transition amplitude between the two states.

How we interpret (1.1) in terms of experimentally significant quantities depends on the transition involved. The two principal situations which occur involve the decay of an unstable particle or the scattering of one particle from another (or from a potential representing such an interaction). In the former case, the initial state α is a single-particle state, while the final state will contain two (or more) particles moving apart with an appropriate kinetic energy. The total decay rate w_α is then the sum of the rates for all final states β, limited by the delta-function in (1.1) to those having the same energy as the initial decaying state. If there are $\rho(E_\beta)$ states with energy E_β, then

$$w_\alpha = \sum_\beta w_{\beta\alpha} = \int dE_\beta \rho(E_\beta) 2\pi|\mathscr{T}_{\beta\alpha}|^2\delta(E_\alpha - E_\beta)$$

$$= 2\pi|\mathscr{T}_{\beta\alpha}|^2\rho(E_\alpha), \tag{1.2}$$

with $\mathscr{T}_{\beta\alpha}$ referring only to final states having $E_\beta = E_\alpha$. In general, (1.2) represents a constant decay rate, and the probability of observing the state α may therefore be obtained from

$$dp_\alpha = (-w_\alpha dt)p_\alpha$$

yielding

$$p_\alpha = p_\alpha(0)e^{-w_\alpha t}, \tag{1.3}$$

leading us to define the lifetime τ_α as

$$\tau_\alpha = 1/w_\alpha. \tag{1.4}$$

(A much more detailed discussion of resonance decay will be given later.)

The second method of studying interactions looks at the scattering of an incident beam by a target particle. This class of interactions includes not only elastic scattering through some angle, as often represented by non-relativistic potential theory, but also "inelastic" processes, in which the interaction changes the identities of the incident and target particles, or in which additional particles may be created. Theoretical interpretation of multiparticle production processes would not be expected to progress until two-particle scattering is well understood, and our comprehension of them is still quite rudimentary. Thus our attention will be confined in this book to two-body scattering. For the moment, let us consider only elastic scattering of particles having mass m.

It is convenient to work here in the center of momentum system, where the momenta of the two final state particles may be denoted by k and −k. Let us neglect temporarily the other quantum numbers of the particle and use the standard normalization of one particle per unit volume. Then the density of states in momentum space is $1/(2\pi)^3$, and the transition rate to states specified by a momentum lying within the differential element d^3k_β is given by

$$d^3w_{\beta a} = 2\pi|\mathscr{T}_{\beta a}|^2\delta(E_\beta - E_a)\,\frac{d^3k_\beta}{(2\pi)^3}\,. \tag{1.5}$$

Since $E_\beta = k_\beta^2/2m$, the delta function limits the magnitude of $\mathbf{k}_\beta$ but not its direction, and the transition rate for scattering into the differential solid angle element $d\Omega_\beta$ is

$$\begin{aligned} d^2w_{\beta a} &= \frac{1}{(2\pi)^2}\left[\int |\mathscr{T}_{\beta a}|^2\delta\left(\frac{k_\beta^2 - k_a^2}{2m}\right)k_\beta^2 dk_\beta\right]d\Omega_\beta \\ &= \frac{1}{(2\pi)^2}\,mk_a|\mathscr{T}_{\beta a}|^2 d\Omega_\beta, \end{aligned} \tag{1.6}$$

with $\mathscr{T}_{\beta a}$ referring only to energy-conserving final states. This transition rate still depends on the normalization of the initial state a, i.e. on the incident flux of particles; with one particle per unit volume moving at a speed $v_a = k_a/m$, the flux is simply v_a particles per second per unit area. The transition rate for unit incident flux defines the differential cross section,

$$\begin{aligned} d\sigma_{\beta a} &= d^2w_{\beta a}/v_a \\ &= \left(\frac{m}{2\pi}\right)^2|\mathscr{T}_{\beta a}|^2 d\Omega_\beta, \end{aligned} \tag{1.7}$$

since $k_\beta = k_a \equiv k$ because of energy conservation. The transition amplitude in this case is a function of k and of the scattering angle θ defined by

$$\mathbf{k}_\alpha \cdot \mathbf{k}_\beta = k_\alpha k_\beta \cos\theta = k^2 \cos\theta,$$

and writing $m\,\mathcal{T}_{\beta\alpha}/2\pi = f(k, \theta)$ yields the familiar result

$$\frac{d\sigma}{d\Omega} = |f(k, \theta)|^2. \tag{1.8}$$

The generalizations of (1.8) necessary to describe inelastic two-body processes are easily derived by steps similar to the above.

The Transition Amplitude

Our knowledge of interactions, garnered from the decays and scatterings which they produce, is thus concentrated in the transition amplitude $\mathcal{T}$, which we obtain by means of perturbation theory. Let us consider a system which, in the absence of interaction, is described by an unperturbed state $|\psi_0(\alpha)\rangle$, an eigenstate of some Hamiltonian H_0 with quantum numbers labeled by α. Adding an interaction V to the Hamiltonian alters the time development of the system; it will now be described by state vectors $|\psi(\alpha)\rangle$ which are eigenstates of $H = H_0 + V$. We picture this time development as shown in Fig.(1-1), with the states initially and finally non-interacting. Let $|\psi_i(\alpha)\rangle$ denote the initial state which coincides with the "free" state $|\psi_0(\alpha)\rangle$, and $|\psi_f(\alpha)\rangle$ the final state into which $|\psi_i(\alpha)\rangle$ develops. Because of the interaction, $|\psi_f(\alpha)\rangle$ will not coincide with $|\psi_0(\alpha)\rangle$, but will be a superposition of the free states into which the system can be scattered. Time-dependent perturbation theory then tells us that the transition amplitude is given by

$$\mathcal{T}_{\beta\alpha} = \langle\psi_0(\beta)|V|\psi_f(\alpha)\rangle. \tag{1.9}$$

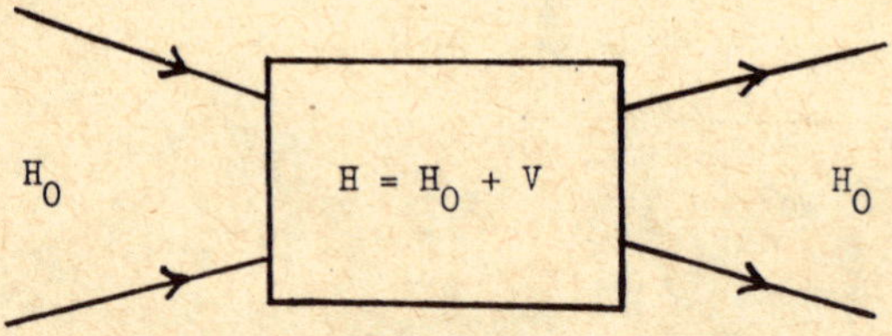

Fig.(1-1) Schematic illustration of the interaction between two particles.

Taking as a first approximation $|\psi_f(a)\rangle = |\psi_0(a)\rangle$ yields the *Born approximation*

$$\mathscr{T}^{(0)}_{\beta a} = \langle\psi_0(\beta)|V|\psi_0(a)\rangle. \tag{1.10}$$

For example, suppose we wish to describe single-particle free states, given in the usual way by plane waves, i.e.

$$\langle \mathbf{r}|\psi_0(\beta)\rangle = e^{i\mathbf{k}_\beta\cdot\mathbf{r}}$$

interacting with a potential $V(\mathbf{r})$. The Born approximation is simply the Fourier transform of the potential,

$$\begin{aligned}\mathscr{T}^{(0)}_{\beta a} &= \int d^3r\, e^{-i\mathbf{k}_\beta\cdot\mathbf{r}}V(\mathbf{r})e^{i\mathbf{k}_a\cdot\mathbf{r}}.\\ &= \widetilde{V}(\mathbf{k}_a - \mathbf{k}_\beta).\end{aligned} \tag{1.11}$$

In particular, if $V(\mathbf{r}) = -e^2/4\pi r$, we obtain

$$\mathscr{T}^{(0)}_{\beta a} = -\frac{e^2}{|\mathbf{k}_\beta - \mathbf{k}_a|}. \tag{1.12}$$

It is instructive, however, to look at an alternate way of obtaining the Coulomb scattering amplitude based on concepts arising in quantum field theory. Let us picture the interaction, not as a potential, but as something which causes a charged particle to emit or absorb a photon, as shown in Fig.(1-2). We may assume that momentum is conserved at the vertex, but it is then easily seen that the total energy must change; therefore this reaction cannot take place in nature. Such

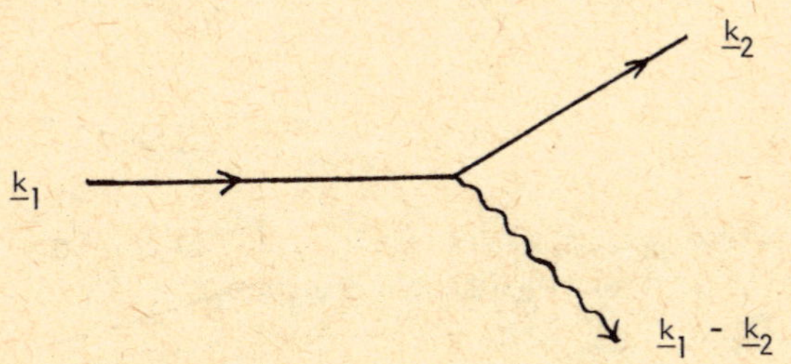

Fig.(1-2) Emission of a photon.

processes are called *virtual* processes. (Equivalently, we may say that if energy were conserved, the photon could not satisfy its mass–energy relation $k = \omega$, and therefore it is called a virtual photon.) The uncertainty relation $\Delta E \Delta t \geqslant \frac{1}{2}$ hints that energy conservation can be abrogated for short periods of time. Consequently a possible mode of interaction for two charged particles is that shown in Fig.(1-3), a virtual photon being exchanged between them.

Such a picture can be formulated quantitatively in perturbation theory. The first order term (1.10) does not contribute, since the states a and β do not contain a photon. The leading term in $\mathscr{T}_{a\beta}$ comes from the next order of perturbation theory,

$$\mathscr{T}^{(1)}_{\beta a} = \langle \psi_0(\beta) | V | \psi_1(a) \rangle, \tag{1.13}$$

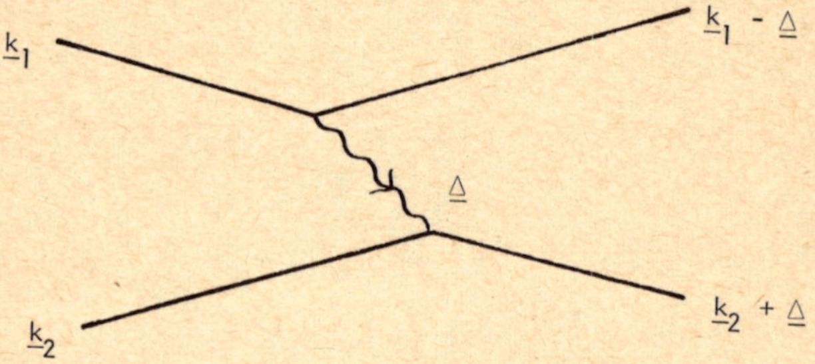

Fig.(1-3) Interaction by virtual photon exchange.

with

$$|\psi_1(a)\rangle = \sum_\gamma{}' \frac{\langle \psi_0(\gamma) | V | \psi_0(a) \rangle}{E_\gamma - E_a} |\psi_0(\gamma)\rangle, \tag{1.14}$$

the summation involving only those states for which $E_\gamma \neq E_a$. Thus

$$\mathscr{T}^{(1)}_{\beta a} = \sum_\gamma{}' V_{\beta\gamma} \frac{1}{E_\gamma - E_a} V_{\gamma a} \tag{1.15}$$

pictures the transition exactly as described above – a transition $V_{\beta\gamma}$, induced by V, to an intermediate state with different energy, followed by another transition $V_{\gamma a}$ to the final state a.

The transition amplitude is then given by summing over all possible processes of this type. Let us assume, for simplicity, that the two particles are identical. Then there are two possible ways of scattering from the initial state β to the final state a, shown in Fig.(1-3) and Fig.(1-4), differing in which particle emits the photon and which absorbs it. For the first of these,

$$E_\gamma - E_a = (E_2 + E_3 + \omega) - (E_3 + E_4) = \omega - (E_4 - E_2), \tag{1.16a}$$

where ω is the energy of the virtual photon and E_i that of the particle labelled i.

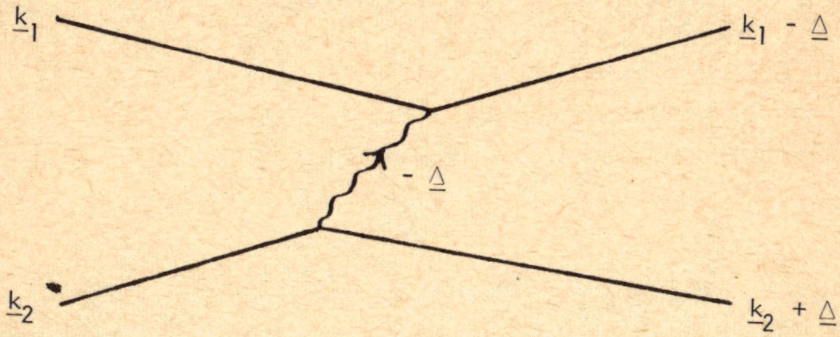

Fig.(1-4) A second process leading to the same final state as that in Fig.(1-3).

Correspondingly, the second diagram yields

$$E_{\gamma'} - E_a = (E_1 + E_4 + \omega) - (E_3 + E_4) = \omega - (E_3 - E_1)$$

$$= \omega + (E_4 - E_2) \tag{1.16b}$$

because $E_1 + E_2 = E_3 + E$. The matrix elements of V which are required all describe either emission or absorption of a photon carrying momentum of magnitude $\Delta = |\mathbf{k}_\beta - \mathbf{k}_a|$, processes which are related by the principle of detailed balance. They measure the strength of the interaction of the photon with the charged particle. Thus we expect $V_{\beta\gamma}$ to be proportional to the charge e. A detailed calculation yields

$$\mathcal{T}^{(1)}_{\beta a} = -\frac{e^2}{2\omega}\left(\frac{1}{\omega - (E_4 - E_2)} + \frac{1}{\omega + (E_4 - E_2)}\right)$$

$$= -e^2 \frac{1}{\omega^2 - (E_4 - E_2)^2}. \tag{1.17}$$

The denominator (2ω) arises from normalizing to a single photon, and taking account of the vector properties of the electromagnetic field produces the negative sign. But in the center of momentum frame, $E_4 = E_2$, and therefore

$$\mathscr{T}_{\beta a}^{(1)} = -\frac{e^2}{\omega^2} = -\frac{e^2}{|\mathbf{k}_a - \mathbf{k}_\beta|^2} \tag{1.18}$$

in agreement with (1.12).

Thus the Coulomb potential may be thought of as representing the exchange of a virtual photon between two charged particles. It is not difficult to see what the higher-order non-relativistic Born approximations describe in this picture; the second Born term is a two-photon exchange, and so on, as shown in Fig.(1-5). The nth Born term is proportional to e^{2n}, and the smallness of $e^2 = 1/137$ produces rapid convergence of the series.

A number of infinities arose in attempts to treat electromagnetic interactions

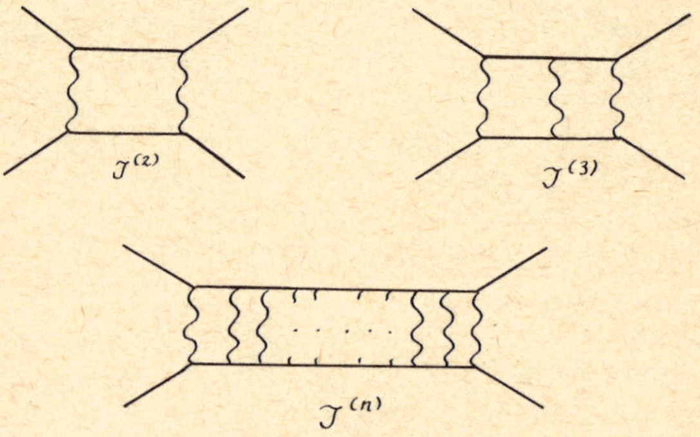

Fig.(1-5) Higher order Born approximations.

relativistically. These difficulties were removed by Feynman, who proved that they could be traced to terms describing the interaction of a charged particle with its *own* field. He showed that the transition amplitude could be obtained by drawing all possible processes, with a certain prescription for writing an integral term corresponding to each; graphs such as Fig.(1-4) are thus called *Feynman diagrams*. His prescription for "renormalizing" the troublesome terms led to the theory of quantum electrodynamics, which seems to provide a complete and quantitatively correct description of the electromagnetic interactions.

The application of such a model to the strong interactions was first attempted in 1935 by Yukawa. The Feynman diagram corresponding to his conjecture, shown in Fig.(1-6), pictures the interaction between two protons being propagated by a virtual π meson, a particle unknown at that time. Yukawa observed that the short range of nuclear forces could be explained if the pion had a non-

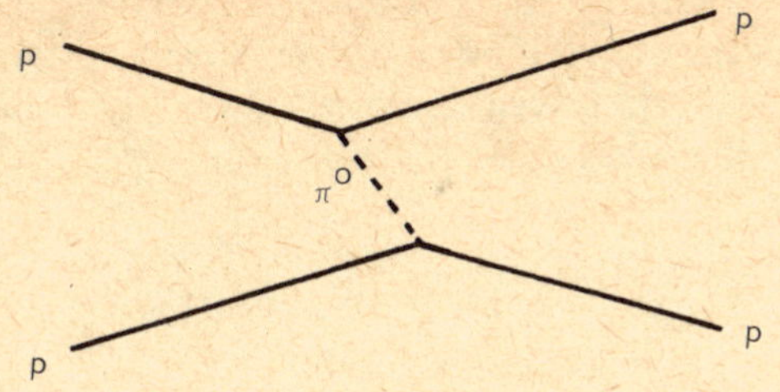

Fig.(1-6) Yukawa picture of strong interactions.

zero mass μ. In that case, $\omega^2 = |\mathbf{k}_\beta - \mathbf{k}_a|^2$ in (1.18) would be replaced by

$$\omega^2 = |\mathbf{k}_\beta - \mathbf{k}_a|^2 + \mu^2,$$

and the corresponding potential, obtained by inverting the Fourier transform, is proportional to

$$\frac{1}{(2\pi)^3} \int d^3k \, . \frac{e^{i\mathbf{k}\cdot\mathbf{r}}}{k^2 + \mu^2} = \frac{1}{4\pi r} e^{-\mu r}. \tag{1.19}$$

Furthermore, if the meson has spin zero, representing a scalar field rather than the vector electromagnetic field, the negative sign in (1.18) does not arise.

That is, if we define the coupling constant g as the generalized "charge" giving the strength of the strong interaction, we find

$$\mathscr{T}^{(1)}_{\beta a} = \frac{g^2}{|\mathbf{k}_\beta - \mathbf{k}_a|^2 + \mu^2} \tag{1.20a}$$

$$V(r) = \frac{g^2}{4\pi r} e^{-\mu r} \tag{1.20b}$$

and an attractive force is produced. The Yukawa potential falls off exponentially; the range of the force may be measured by the rate of this decrease,

$$a = 1/\mu. \tag{1.21}$$

Knowing experimentally that a $\sim 1.5 \times 10^{-15}$ meter, Yukawa predicted the mass of the pion to be about 150 MeV. Although first confused with the muon, the pion was eventually discovered with $\mu \approx 140$ MeV.

Because of the short-range nature of the force, exact measurement of the "charge" g is very difficult. But it was clear from the beginning that the coupling constant should be large, since nuclei are very tightly bound together. The presently accepted value is generally quoted as

$$\frac{g^2}{4\pi} \approx 14.7. \tag{1.22}$$

If we attempted to treat the strong interactions in perturbation theory by means of Feynman diagrams, the result would be a series expansion in $g^2/4\pi$. The large value of this number will clearly not produce a convergent series; instead it will lead to rather drastic divergence. For this reason, our understanding of the strong interactions and our ability to treat them quantitatively remain practically nil. With increasing research into particle physics and scattering theory, however, we have become convinced that this picture is qualitatively correct, despite our inability to solve it quantitatively.

Quantum Numbers

The first step in studying the inner workings of the strong interactions must be to describe the particles which participate. Thus we shall conclude this introduction by reviewing some of the basic quantum numbers which serve as indelible labels for the hadrons. In general, we shall consider these quantum numbers as conserved quantities; in other words, they are eigenvalues of operators which commute with the Hamiltonian. The fact that charge, for example, is strictly conserved means that $[Q, H] = 0$, where Q is the charge operator and H is the total Hamiltonian of the universe. In a similar way, from the observation that other quantities seem experimentally to be conserved we may abstract the existence of corresponding operators. Assuming that these quantum numbers are conserved enables us to specify their values even for short-lived unstable particles, by simply looking at the corresponding values for their decay products. In this way we may consider even resonances with lifetimes of 10^{-23} second to be particles with well-defined quantum numbers.

Such an observation is the motivation for the classification of particles as baryons, mesons, and leptons. It is well established by experiment that, for example, a neutron never changes into a neutral pion π^0, or a proton into a π^+. More generally, let us speak of protons and neutrons jointly as *nucleons* and restrict our consideration to states containing only nucleons and pions. (We shall denote nucleons by capital N, protons and neutrons by lower case p and n.) Then the baryon number B, defined as the number of nucleons minus the number of antinucleons, will be conserved in all known processes. Thus reactions such as

$$\pi^+ p \to pp, \quad \pi^+ p \to pn\pi^+\pi^0,$$

in which B increases from 1 to 2, are not observed, while

$$\pi^+ p \to \pi^+ p \pi^0$$

$$\pi^- p \to \bar{p} n p$$

$$\bar{p} p \to \pi^+ \pi^- \pi^0 \pi^0$$

etc., in which B is constant, are well known. Note, however, that the number of pions may change arbitrarily; therefore we say that they have B = 0. Other baryons and mesons may be assigned values of B by observing the states to which they decay. The Λ^0 hyperon, for instance, decays via $\Lambda^0 \to p\pi^-$, and therefore must have B = 1; while the decay $K^0 \to \pi^+\pi^-$ identifies K^0 as a meson, i.e. B = 0. Defined in this way, the baryon number appears to be conserved absolutely in all interactions. A similar conservation law also seems to hold for the leptons. Curiously, all baryons and all leptons have half-integral spin, while all mesons have integral spin; the reason for this correspondence is not understood.

A second quantum number deduced from observed conservation laws is the strangeness S. Unlike B, however, S is not conserved absolutely. That is, we say that the strangeness operator commutes with that part of the Hamiltonian referring to the strong interactions, but not with the total Hamiltonian – specifically, not with the weak interactions. Thus strangeness is conserved in strong processes, but not in weak ones. It was invented by Gell-Mann to explain the fact that the decays $\Lambda^0 \to p\pi^-$ and $K^0 \to \pi^+\pi^0$ take place much more slowly than one would expect. Typical lifetimes for particles unstable against decay via the strong interactions are about 10^{-23} second, but Λ^0 and K^+ have much longer lifetimes, about 10^{-8} second – a time more characteristic of decays induced by weak interactions. Gell-Mann therefore postulated that both of these reactions are weak, rather than strong. To forbid them in the strong interactions, a non-conserved quantum number may be introduced. Thus we assume that the strong interactions conserve strangeness, and assign S = 0 to nucleons and pions but S = +1 for K^+; then the decay $K^+ \to \pi^+\pi^0$ cannot be strong. Particle and antiparticle must clearly have opposite strangeness in order to annihilate, so the K^- has S = –1. Similarly, the Λ^0 should have non-zero strangeness to prevent the decay $\Lambda^0 \to p\pi^-$ in strong interactions. Because the reaction $\pi^+ n \to K^+\Lambda^0$ occurs with a rate consistent with the strong interactions, we must have S = –1 for the Λ^0. Conservation of strangeness is an important participle in strong interactions. In many cases, it turns out that a more "natural" quantum number is the *hypercharge* Y = B + S; since B is conserved by all interactions, hypercharge and strangeness are equivalent commodities. Studying the symmetries implied by their conservation has led to most of our present ideas about how to classify particles.

Two other important quantum numbers, C and P, originated from theoretical considerations rather than experimental observation. The fact that every particle has an antiparticle suggests defining the charge conjugation operator C transforming between them. For example, if $|p\rangle$ is a proton state, then

$$C|p\rangle = |\bar{p}\rangle \tag{1.23}$$

is the corresponding antiproton state. Similarly, $C|K^+\rangle = |K^-\rangle$, $C|\pi^+\rangle = |\pi^-\rangle$, $C|\Lambda^\circ\rangle = |\bar{\Lambda}^\circ\rangle$, etc. The antiparticle of the antiparticle is the particle, so $C^2 = \mathbb{1}$ is the identity operator. For certain states, however, the antiparticle state and the particle state are identical. The π^0, in particular, is its own antiparticle, and therefore is an eigenstate of C,

$$C|\pi^0\rangle = c|\pi^0\rangle. \tag{1.24}$$

But

$$C^2|\pi^0\rangle = c^2|\pi^0\rangle = |\pi^0\rangle,$$

since $C^2 = \mathbb{1}$, and hence $c = \pm 1$. The principal decay mode of π^0 is $\pi^0 \to 2\gamma$; since the final state is symmetric, it must be even under charge conjugation. Thus it turns out that for π^0, $c = +1$. A second example is provided by a two-particle state containing a proton and an antiproton,

$$|\psi\rangle = |p\rangle|\bar{p}\rangle, \tag{1.25}$$

with

$$|\bar{\psi}\rangle = C|\psi\rangle = |\bar{p}\rangle|p\rangle,$$

distinct from $|\psi\rangle$. Eigenstates of C may be defined, however, as

$$|\psi_\pm\rangle = 1/\sqrt{2}\ [|\psi\rangle \pm |\bar{\psi}\rangle]$$

for which

$$C|\psi_\pm\rangle = \pm|\psi_\pm\rangle. \tag{1.26}$$

Since C reverses the baryon number, the charge, and the strangeness, it is clear that only neutral non-strange mesons can be single-particle eigenstates of C. The eigenvalue of C is known as the charge parity, or C-parity, of the state. If the strong interactions are symmetric under charge conjugation – that is, if antiparticle interacts with antiparticle exactly as particle does with particle – then the C-parity, when it can be defined, will be conserved.

Closely allied with the concept of charge conjugation is that of the parity operation P, which reverses the coordinate axes. Let a complete set of state vectors in configuration space be denoted by $|\mathbf{r}\rangle$, so that any state $|\psi\rangle$ can be written

$$|\psi\rangle = \int d^3r\,|\mathbf{r}\rangle\,\langle\mathbf{r}|\psi\rangle.$$

the wave function being thus identified as

$$\psi(\mathbf{r}) = \langle\mathbf{r}|\psi\rangle.$$

The parity operator P is defined by

$$P|\mathbf{r}\rangle = |-\mathbf{r}\rangle, \tag{1.27}$$

so its effect on an arbitrary state $|\psi\rangle$ is

$$\begin{aligned} P|\psi\rangle &= \int d^3r\,P|\mathbf{r}\rangle\,\langle\mathbf{r}|\psi\rangle \\ &= \int d^3r\,|-\mathbf{r}\rangle\,\langle\mathbf{r}|\psi\rangle \\ &= \int d^3r\,|\mathbf{r}\rangle\psi(-\mathbf{r}). \end{aligned} \tag{1.28}$$

Therefore P replaces $\psi(\mathbf{r})$ by $\psi(-\mathbf{r})$. Clearly $P^2 = \mathbb{1}$, so the eigenvalues of parity, like those of C-parity, are $p = \pm 1$.

A single isolated particle should be in an eigenstate of P, and the corresponding eigenvalue is called its parity. This eigenvalue describes the particle, and must be determined experimentally. If a state with integral angular momentum ℓ is represented by a simple spatial wave function

$$\psi(\mathbf{r}) = R_\ell(r)Y_{\ell m}(\theta, \phi)$$

then the well-known parity of the spherical harmonics

$$Y_{\ell m}(\pi - \theta, \phi + \pi) = (-1)^{\ell} Y_{\ell m}(\theta, \phi)$$

shows that the orbital angular momentum contributes a "natural parity" $(-1)^{\ell}$. In addition, however, the particle may possess an "intrinsic parity" ϵ_p, so that

$$p = \epsilon_p(-1)^{\ell}. \tag{1.29}$$

The possibility that $\epsilon_p = -1$ for a given particle is usually indicated by the prefix "pseudo-". Thus a spin-zero particle with $p = +1$ is called a scalar particle, while if $p = -1$ it is a pseudoscalar; with $\ell = 1$, a vector particle has $p = -1$, and pseudovector (or axial vector) has $p = +1$.

For example, the parity of the π^- was originally determined from the observation that when it is bound by a deuterium nucleus in an s-wave state, the reaction $\pi^- D \to nn$ takes place. The deuteron has two nucleons bound with $\ell = 0$ and a symmetric spin wave function. Therefore the initial state has $\ell = 0$, spin $S = 1$, and total angular momentum $J = \ell$, and its parity is $(-1)^{\ell} p_{\pi} = p_{\pi}$, where p_{π} denotes the intrinsic parity of the pion. The final state consists of two neutrons which, being fermions, must be in an overall antisymmetric state. The only antisymmetric two-neutron state yielding $J = 1$ is p^3, so the two neutrons are in an $\ell = 1$ state, and the parity of the final state is $(-1)^{\ell} = -1$. If parity is conserved, therefore, we must have $p_{\pi} = -1$, i.e. the pion is a pseudoscalar meson.

The parity of particles with half-integral spin cannot be defined unambiguously, for reasons we shall not discuss. It is possible, however, to define the *relative* parity of two fermions; therefore we may simply choose the parity of the nucleons to be positive, and determine all other parities by this convention.

It was assumed without question for many years that parity and charge parity should be conserved in all interactions. Difficulties in understanding the weak interactions led Lee and Yang to suggest in 1957 that parity was not conserved there; when experiment verified this suggestion, they were rewarded with the Nobel Prize. Charge parity is also violated whenever parity fails; but it appeared for a decade that their product CP was conserved. More recently, observations of the decays of K^0 mesons have revealed a violation of CP by one part in one thousand. It has been shown in quantum field theory that for any "well-behaved" theory, the product CPT, where T represents time reversal, must be conserved. If that is so, then the observed CP violation implies the existence of an asymmetry in time reversal also. Theoretical understanding of these problems is minimal. For our purposes, however, none of this matters; we wish to study the strong interactions, for which C, P, and T are all conserved quantum numbers.

Part I

Group theory

Chapter 2

THE ROTATION GROUP

From the days when fire, earth, air and water were considered to be the basic elements, the course of scientific progress in learning the nature of matter has repeated a pattern of reasoning dominated by the identification of similarities and symmetries among the interactions of various materials. At each step, the correct classification of objects by means of these symmetries has eventually prompted the formulation of a theory of the structure of that material. In chemistry, for example, the observation that all of the halides seem to interact similarly with any of the alkalis helped lead to the concept of a quantum number, the valence, which we now recognize as a manifestation of the electron structure of every atom.

In recent years, the physics of strong interactions has reached a level of understanding comparable to that of chemistry after the development of the periodic table but before the discovery of the electron. We have found the quantum numbers which seem to provide a means of classifying the scores of different particles and resonances now known, and a periodic table of sorts which brings order into the various categories. The quantum numbers, which have already been mentioned briefly, arise naturally from the similarities observed in hadronic interactions. The classification of the particles by means of these quantum numbers requires the use of Lie group theory, which predicts that certain multiplets of particles, all having similar properties and specified quantum numbers, should occur. In the first part of this book our purpose is to introduce the theory of unitary symmetry and the particle classification schemes which result from it.

The Rotation Group

The intrinsic angular momentum, or spin, of a particle is the first of the quantum numbers we shall consider. It is the most easily visualized one, because it results from a well-understood symmetry in physical space – the conservation of angular momentum, which is equivalent to the rotational invariance of a physical system. We assume that the reader is already familiar with the operator theory of angular momentum. Because this operator formalism provides the most elementary example of a Lie group and the associated symmetry principles, however, it is appropriate to review it here; the techniques we shall follow, and the conclusions we shall draw, are identical to those which will arise in later cases, where the symmetry principle responsible is not physically apparent.

To begin with, let us show the connection between angular momentum and rotations. Suppose we describe a physical system (an electric potential or a wave function, for example) by assigning a numerical value $\psi(x, y, z)$ to each point (x, y, z) in space. Now we rotate that physical system through an infinitesimal angle $\delta\phi_3$ about the z-axis. Then at the point (x, y, z) we will observe not the value $\psi(x, y, z)$, but a new value $\psi'(x, y, z)$. This new value is just the one which, before the rotation, had been observed at the point $(x + \delta\phi_3 y, y - \delta\phi_3 x, z)$; the rotation brought this value to the point (x, y, z). Thus

$$\psi'(x, y, z) = \psi(x + \delta\phi_3 y, y - \delta\phi_3 x, z), \tag{2.1}$$

and since $\delta\phi_3$ is infinitesimal we can expand this function in Taylor's series about the point (x, y, z) as

$$\psi'(x, y, z) = \psi(x, y, z) + \delta\phi_3 y \frac{\partial}{\partial x} \psi(x, y, z) - \delta\phi_3 x \frac{\partial}{\partial y} \psi(x, y, z). \tag{2.2}$$

The change in the function ψ corresponding to the rotation is therefore

$$\begin{aligned}\delta\psi(x, y, z) &= \psi'(x, y, z) - \psi(x, y, z)\\ &= -\delta\phi_3 \left(x \frac{\partial}{\partial y} - y \frac{\partial}{\partial x} \right) \psi(x, y, z).\end{aligned} \tag{2.3}$$

Now since $\mathbf{p} = (1/i)\boldsymbol{\nabla}$, we recognize the operator

$$\left(x \frac{\partial}{\partial y} - y \frac{\partial}{\partial x} \right)$$

on the right-hand side of (2.3) as $i(xp_y - yp_x) = iJ_3$, a multiple of the z-component of the angular momentum operator. Thus

$$\psi'(x, y, z) = (1 - i\delta\phi_3 J_3)\psi(x, y, z) \tag{2.4}$$

A rotation through a finite angle ϕ_3 can always be considered the product of the N rotations through ϕ_3/N, and if we let N be large and $\delta\phi_3 = \phi_3/N$ the value $\psi_{\phi_3}(x, y, z)$ observed after rotation through ϕ_3 is found through (2.4) to be

$$\psi_{\phi_3}(x, y, z) = \left(1 - \frac{i\phi_3}{N} J_3\right)^N \psi(x, y, z). \tag{2.5}$$

As $N \to \infty$, the limit of this operator equation is

$$\psi_{\phi_3}(x, y, z) = R(\phi_3)\psi(x, y, z),$$

with the operator $R(\phi_3)$ given by

$$R(\phi_3) = e^{-i\phi_3 J_3}, \tag{2.6}$$

which we could equally well have found by "integrating" (2.3).

The effect of rotation about a given direction, such as the z-axis, is therefore obtained by exponentiation of the component of the angular momentum operator in that direction; we can immediately generalize (2.6) to read

$$R(\boldsymbol{\phi}) = e^{-i\boldsymbol{\phi}\cdot\mathbf{J}} \tag{2.7}$$

where the direction of the vector $\boldsymbol{\phi}$ indicates the axis of rotation and its length gives the angle through which the system is rotated. Because of (2.7) we say that the angular momentum operator is the *generator* of rotations.

It is well known that the product of any two rotations is a third rotation; in other words, the set of all possible rotations forms a group. It is an infinite group, since rotations can be made through any angle about any axis. Such groups are called Lie groups, after the Norwegian mathematician Sophus Lie, who first studied them. Every member of the rotation group corresponds to an orthogonal 3×3 matrix which maps (x, y, z) onto (x', y', z'). For that reason the rotation group is usually referred to as O(3), meaning Orthogonal transformations in 3 dimensions.

For Lie groups it is not necessary to study all of the (infinitely many!) members of the group, but only the generators from which they are formed in the manner of (2.7). In addition to its physical significance, then, the angular momentum $\mathbf{J} = \mathbf{r} \times \mathbf{p}$ is important because it generates the rotation group. From either viewpoint, however, we study it by the same methods, i.e. by looking at its representations and its eigenvalue problem. A complete characterization of these can be obtained directly from the commutation relations of the three components.

The angular momentum commutation relations

$$[J_i, J_j] = i\epsilon_{ijk}J_k \tag{2.8}$$

result either directly from the definition $\mathbf{J} = \mathbf{r} \times \mathbf{p}$ or from considering infinitesimal rotations. Since operators cannot be simultaneously diagonalized unless they commute, (2.8) tells us immediately that these three operators generally cannot have simultaneous eigenstates; we can diagonalize only one of them, which is conventionally chosen to be J_3. If, however, we form the bilinear combination

$$\mathbf{J}^2 = J_1^2 + J_2^2 + J_3^2 = J_i J_i, \tag{2.9}$$

then by the usual rules of commutator algebra it is easily verified that

$$\begin{aligned}[J^2, J_i] &= [J_j J_j, J_i] = J_j i\epsilon_{jik} J_k + i\epsilon_{jik} J_k J_j \\ &= i\epsilon_{jik}(J_j J_k + J_k J_j) \\ &= 0 \end{aligned} \tag{2.10}$$

because ϵ_{jik} is antisymmetric while $(J_j J_k + J_k J_j)$ is symmetric. Both $\mathbf{J}^2$ and J_3 can therefore be diagonalized.

Let us denote the normalized simultaneous eigenstates of $\mathbf{J}^2$ and J_3 by $|j, m\rangle$ with

$$\begin{aligned} \mathbf{J}^2 |j, m\rangle &= j(j+1)|j, m\rangle \\ J_3 |j, m\rangle &= m|j, m\rangle. \end{aligned} \tag{2.11}$$

The hermiticity of $\mathbf{J}$, and therefore of $\mathbf{J}^2$, guarantees that j and m are real and that the eigenstates corresponding to different j,m values are orthogonal. The commutation rules alone are sufficient now to determine also that they are both integers or half-integers. To prove this point, we define the raising and lowering operators,

$$J_\pm = J_1 \pm iJ_2, \tag{2.12}$$

for which it follows from (2.1) that

$$[J_3, J_\pm] = \pm J_\pm. \tag{2.13}$$

If we apply the operator equation (2.13) to a state $|j, m\rangle$,

$$(J_3 J_\pm - J_\pm J_3)|j, m\rangle = \pm J_\pm |j, m\rangle, \tag{2.14}$$

we obtain

$$J_3 J_\pm |j, m\rangle = (m \pm 1) J_\pm |j, m\rangle, \tag{2.15}$$

which shows that $J_\pm|j, m\rangle$ is an eigenstate of J_3 with eigenvalue $(m \pm 1)$. It must therefore be a multiple of the state $|j, m \pm 1\rangle$:

$$J_\pm|j, m\rangle = c_\pm |j, m \pm 1\rangle. \tag{2.16}$$

Now (2.12) implies also that

$$\begin{aligned} J_+J_- &= (J_1 + iJ_2)(J_1 - iJ_2) \\ &= J_1^2 + J_2^2 - i[J_1, J_2] \\ &= J_1^2 + J_2^2 + J_3, \end{aligned} \tag{2.17a}$$

and similarly that

$$J_-J_+ = J_1^2 + J_2^2 - J_3. \tag{2.17b}$$

Since $J_- = J_+^\dagger$, $\mathbf{J}^2$ can therefore be written

$$\mathbf{J}^2 = J_\pm^\dagger J_\pm + J_3^2 \pm J_3. \tag{2.18}$$

Rearranging (2.18) and taking its diagonal element in the state $|j, m\rangle$, we find that

$$\begin{aligned} \langle j, m|J_\pm^\dagger J_\pm|j, m\rangle &= |c_\pm|^2 \langle j, m \pm 1|j, m \pm 1\rangle \\ &= \langle j, m|\mathbf{J}^2 - J_3(J_3 \pm 1)|j, m\rangle, \end{aligned}$$

that is,

$$|c_\pm|^2 = j(j + 1) - m(m \pm 1). \tag{2.19}$$

It follows (on choosing the phases) that

$$J_\pm|j, m\rangle = \sqrt{j(j + 1) - m(m \pm 1)}\,|j, m \pm 1\rangle, \tag{2.20}$$

and that $j(j + 1) - m(m \pm 1) \geqslant 0$, which implies that

$$-j \leqslant m \leqslant j. \tag{2.21}$$

If we start with any values of j and m satisfying (2.21), however, we can generate by repeated applications of J_+ (or J_-) states $|j, m'\rangle$ with $m' > j$ (or $m' < -j$). Such states would contradict (2.21), so they must be avoided; this can only happen if there is some state $|j, m_+\rangle$ for which $J_+|j, m_+\rangle = 0$, and likewise a state $|j, m_-\rangle$ for which $J_-|j, m_-\rangle = 0$. Clearly, then, from (2.20) we must have

$$\sqrt{j(j+1) - m_\pm(m_\pm \pm 1)} = 0 \tag{2.22}$$

i.e. $m_\pm = \pm j$. Also, m_+ and m_- must be separated by an integral number of steps, since either must be reached from the state $|j, m\rangle$ if $J_\pm$ is applied a sufficient number of times. Consequently $m_+ - m_- = 2j$ must be an integer, i.e., j is an integer or half-integer, and m takes only the values $-j, -j+1, \ldots j-1, j$.

Thus we have been able to derive the eigenvalue spectrum of the angular momentum operator directly from the commutation relations (2.1). Furthermore, we know the effect of J_3 and $J_\pm$, and hence of J_1 and J_2, on each eigenstate $|j, m\rangle$; therefore we can give a complete specification of the properties of the angular momentum operator **J** for a given value of j. These properties are most conveniently represented by the matrix elements of **J**, i.e. by defining matrices

$$[J_i^{(j)}]_{mm'} = \langle j, m|J_i|j, m'\rangle \tag{2.23}$$

These matrices will have all of the properties ascribed to the operators J_i.

Equation (2.23) defines a particularly simple example of a *matrix representation* of the generators of the rotation group. More generally, we define a matrix representation of dimensionality n as any set of three n-by-n matrices J_i satisfying the commutation relations (2.8). If we know one matrix representation, we can immediately generate an infinite number of others by similarity transformation; if $J_i' = UJ_iU^\dagger$, with U any unitary matrix, then

$$[J_i', J_j'] = U[J_i, J_j]U^\dagger = i\epsilon_{ijk}J_k'. \tag{2.24}$$

Representations related in this way are said to be *equivalent.* Since all equivalent representations contain the same basic information, we need only study a single one of the set for any given purpose.

In particular, an appropriate U can always be chosen to yield a representation with J_3 diagonal, since J_3 is Hermitian. Furthermore, because $\mathbf{J}^2$ commutes with J_3, J^2 can be diagonalized simultaneously. The basis vectors of this diagonal matrix representation will be eigenvectors of J_3 and J^2, and can be labelled by their eigenvalues m and j. For each j, of course, there must be a full set of eigenvalues m; but the representation still need not be as simple as (2.23), since in principle it may contain any mixture of *different* eigenvalues of $\mathbf{J}^2$, while (2.23) corresponds to a single j value.

A representation in which $\mathbf{J}^2$ possesses more than a single eigenvalue is said to be *reducible.* It will leave invariant the subspaces corresponding to different j; that is, no matrix can be formed from **J** which will connect eigenstates with differing j-values. To prove this fact, we let f(**J**) be an arbitrary function of **J**. Since $\mathbf{J}^2$ commutes with **J**, it commutes with f(**J**), so

$$\langle j', m' | [\mathbf{J}^2, f(\mathbf{J})] | j, m\rangle = [j'(j'+1) - j(j+1)] \langle j'm' | f(\mathbf{J}) | jm\rangle = 0$$

and if $j \neq j'$, the matrix element necessarily vanishes. By judicious transformation and labeling of the matrix, therefore, the J_i can always be case in "block diagonal" form

$$J_i = \begin{pmatrix} (j_1) & & 0 \\ & (j_2) & \\ 0 & & \ddots \end{pmatrix} \tag{2.25}$$

in which the submatrices corresponding to each j-value lie along the diagonal, and zeros fill all other positions. The matrix is then said to have been *reduced.* The only situation in which a reduction to block diagonal form will not be possible is when the representation corresponds to a *single* j-value. Thus a representation of the form (2.23) is said to be *irreducible*, and the values of j can be used to label the *irreducible representations of the rotation group.*

The reduction of a reducible representation to block diagonal form is precisely the precedure involved in the addition of angular momenta. Suppose we combine two sets of states $|j_1, m_1\rangle$ and $|j_2, m_2\rangle$ to form a direct product $|j_1m_1\rangle|j_2m_2\rangle$. Operators in the direct product space may be formed from those in the spaces of $|j_1m_1\rangle$ and $|j_2m_2\rangle$ by writing

$$(A^{(1)} \times B^{(2)})|j_1m_1\rangle|j_2m_2\rangle = (A|j_1m_1\rangle)(B|j_2m_2\rangle). \tag{2.26}$$

If $\mathbf{J}^{(1)}$ and $\mathbf{J}^{(2)}$ represent the angular momentum for $|j_1m_1\rangle$ and $|j_2m_2\rangle$, the direct product operator

$$\mathbf{J} = \mathbf{J}^{(1)} \times \mathbb{1}^{(2)} + \mathbb{1}^{(1)} \times \mathbf{J}^{(2)} \tag{2.27}$$

satisfies the angular momentum commutation relations. Therefore the direct product states $|j_1m_1\rangle|j_2m_2\rangle$ furnish the basis of a $(2j_1+1)(2j_2+1)$-dimensional representation. It is easily shown that J_3 is diagonal in this representation, but that $\mathbf{J}^2$ is not.

To reduce the representation we inspect the eigenvalues m of J_3. The maximum value of m is $(j_1 + j_2)$, which is obtained only for the state $|j_1 j_1\rangle|j_2 j_2\rangle$. This state must therefore be an eigenstate $|j_1 + j_2, j_1 + j_2\rangle$ of $\mathbf{J}^2$ and J_3, so the irreducible representation $(j_1 + j_2)$ is present. There are two eigenstates with $m = j_1 + j_2 - 1$, $|j_1, j_1 - 1\rangle|j_2 j_2\rangle$ and $|j_1 j_1\rangle|j_2, j_2 - 1\rangle$; one combination of these corresponds to the state $|j_1 + j_2, j_1 + j_2 - 1\rangle$. The second linearly independent combination must correspond to $|j_1 + j_2 - 1, j_1 + j_2 - 1\rangle$, so the irreducible representation $(j_1 + j_2 - 1)$ is also present. Continuing this process leads to the familiar conclusion that each representation j with $|j_1 - j_2| \leqslant j \leqslant j_1 + j_2$ occurs exactly once in $j_1 \times j_2$. The transformation which effects the reduction provides the definition of the *Clebsch–Gordan coefficients* $\langle j_1 m_1 j_2 m_2 | jm\rangle$:

$$|jm\rangle = \sum_{m_1 m_2} \langle j_1 m_1 j_2 m_2 | jm\rangle |j_1 m_1\rangle |j_2 m_2\rangle. \tag{2.28}$$

Tensor Operators

We can now classify operators by the way in which they transform under rotations. Consider the matrix element $\langle\psi|A|\psi\rangle$ of some operator A in an arbitrary state $|\psi\rangle$. If we rotate the system, $|\psi\rangle$ goes to $|\psi'\rangle = e^{-i\boldsymbol{\phi}\cdot\mathbf{J}}|\psi\rangle$. Defining the transformed operator A' by

$$\langle\psi'|A'|\psi'\rangle = \langle\psi|A|\psi\rangle,$$

we find that

$$A' = e^{-i\boldsymbol{\phi}\cdot\mathbf{J}} A e^{i\boldsymbol{\phi}\cdot\mathbf{J}}. \tag{2.29a}$$

Since a finite rotation can always be obtained from a succession of infinitesimal ones, it is convenient to study (2.29a) for infinitesimal rotations $\delta\boldsymbol{\phi}$. Then

$$A'_{\delta\boldsymbol{\phi}} = (1 - i\delta\boldsymbol{\phi}\cdot\mathbf{J})A(1 + i\delta\boldsymbol{\phi}\cdot\mathbf{J})$$

$$= A - i\delta\boldsymbol{\phi}\cdot[\mathbf{J}, A] \tag{2.29b}$$

to first order in $\delta\boldsymbol{\phi}$. The commutator $[\mathbf{J}, A]$ therefore characterizes the transformation of A under rotations. A *scalar* operator S, for example, is one which is unchanged by rotation, i.e. $[\mathbf{J}, S] = 0$. A *vector* operator $\mathbf{V}$ is a set of three operators V_1, V_2, V_3 which transform under rotations in the same way that the components of the position vector **r** do; it is easily verified that this means

$$[J_i, V_k] = i\epsilon_{ijk}V_k. \tag{2.30}$$

Scalar and vector operators are two particularly simple types. More generally we may define an *irreducible spherical tensor operator* of rank j as a set of 2j + 1 operators T_{jm}, m = –j, . . . j , which transform according to

$$[J_i, T_{jm}] = \sum_{m'} [J^{(j)}]_{mm'}T_{jm'}. \tag{2.31}$$

Notice the formal equivalence of (2.31) to the definition of the matrix elements

$$J_i|j, m\rangle = \sum_{m'} [J^{(j)}]_{mm'}|jm'\rangle.$$

In other words, the T_{jm} "transform like an object with spin j"; the components of an irreducible tensor operator may be put in one-to-one correspondence with the basis states $|jm\rangle$ of an irreducible representation. Every operator can be decomposed into a sum of irreducible tensor operators. Whether the sum is finite or infinite depends, of course, on the nature of the operator.

Let us consider the application of irreducible spherical tensor operator T_{jm} to an angular momentum eigenstate $|j_1m_1\rangle$. If $|\psi\rangle = T_{jm}|j_1m_1\rangle$, then

$$\begin{aligned} J_i|\psi\rangle &= J_iT_{jm}|j_1m_1\rangle \\ &= ([J_i, T_{jm}] + T_{jm}J_i)|j_1m_1\rangle \\ &= [J_i^{(j)}]_{mm'}T_{jm'}|j_1m_1\rangle \\ &\quad + [J_i^{(j)}]_{m_1m_1'}T_{jm}|j_1m_1'\rangle. \end{aligned} \tag{2.32}$$

On the other hand, the direct product state $|jm\rangle|j_1m_1\rangle$ is defined in (2.26) and (2.27) to satisfy

$$\begin{aligned} J_i|jm\rangle|j_1m_1\rangle &= (J_i|jm\rangle)|j_1m_1\rangle + |jm\rangle(J_i|j_1m_1\rangle) \\ &= J^{(j)}_{mm'}|jm'\rangle|j_1m_1\rangle + J^{(j_1)}_{mm1'}|jm\rangle|j_1m_1'\rangle. \end{aligned} \tag{2.33}$$

Comparing (2.32) and (2.33), we see that $T_{jm}|j_1m_1\rangle$ transforms like a direct product state. Therefore it must be a multiple of such a state,

$$T_{jm}|j_1m_1\rangle \propto |jm\rangle|j_1m_1\rangle, \tag{2.34}$$

where the constant of proportionality is independent of the m-values.

From (2.34) there follows immediately the *Wigner–Eckart theorem,* which in extended form will be of basic importance in our study of group theory. Inverting (2.28), we have

$$|jm\rangle|j_1m_1\rangle = \sum_{j_2m_2} \langle j_2m_2|jmj_1m_1\rangle|j_2m_2\rangle, \tag{2.35}$$

so (2.34) is equivalent to

$$T_{jm}|j_1m_1\rangle = \sum_{j_2m_2} T(j_2, j, j_1)\langle j_2m_2|jmj_1m_1\rangle|j_2m_2\rangle, \tag{2.36}$$

i.e.,

$$\langle j_2m_2|T_{jm}|j_1m_1\rangle = T(j_2, j, j_1)\langle j_2m_2|jmj_1m_1\rangle. \tag{2.37}$$

Thus we have shown that:

A matrix element of an irreducible spherical tensor operator can always be separated into the product of a Clebsch–Gordan coefficient and a "reduced matrix element" depending only on the irreducible representations involved.

The Wigner–Eckart theorem implies, in other words, that all of the matrix elements $\langle j_2m_2|T_{jm}|j_1m_1\rangle$ corresponding to the various values of m, m_1, and m_2 are proportional to each other, the constant of proportionality depending only on the order of the tensor operator and the total angular momentum of the two states. These are all quantities which do not depend on the spatial axes we choose to describe **J**. The m-values, on the other hand, clearly depend on how we draw the z-axis. Thus the Wigner–Eckart theorem allows us to separate the physical content of a given operator from the purely geometrical properties describing how it transforms under rotation.

Lie Groups

To the student already well acquainted with the quantum mechanical theory of angular momentum, all of the material presented so far should be familiar. Our purpose in including it has been to show that all of this well-known formalism results entirely from the commutation relations (2.8), which in turn directly reflect the basic operator structure of the rotation group O(3). In the study of particle physics other Lie groups have arisen naturally from the symmetries observed in strong interactions – the special unitary groups SU(n), to be precise – and the investigation of these symmetry groups has led to fundamental insights into the nature of the hadrons. The formal theory of these groups, however, is developed along exactly the same lines as those we have just followed, and the physical conclusions are drawn in exactly the same way.

Thus to close this introduction we shall now give a brief resumé of the theory of Lie groups, which in essence will be simply a more generalized casting of those ideas which were emphasized in our treatment of the rotation group. As in (2.7), all members of the group are obtained by exponentiation of a finite set of n generators X_λ: the structure of the group is completely determined by the commutation relations of these generators

$$[X_\lambda, X_\mu] = i f_{\lambda\mu\nu} X_\nu. \tag{2.38}$$

The constants $f_{\lambda\mu\nu}$, which are the generalizations of ϵ_{ijk} in (2.8), are known as the *structure constants* of the group. The generators are assumed to be hermitian, and the $f_{\lambda\mu\nu}$ are therefore real; they are clearly antisymmetric in λ and μ. In addition, it is easily verified that

$$[X_\lambda[X_\mu, X_\nu]] + [X_\mu[X_\nu, X_\lambda]] + [X_\nu[X_\lambda, X_\mu]] = 0 \tag{2.39a}$$

which implies the Jacobi identity

$$f_{\lambda\mu\sigma}f_{\sigma\nu\tau} + f_{\mu\nu\sigma}f_{\sigma\lambda\tau} + f_{\nu\lambda\sigma}f_{\sigma\mu\tau} = 0. \tag{2.39b}$$

These conditions are sufficiently stringent to limit the number of possible Lie groups. All of them have been studied in great mathematical detail since the pioneering work of E. Cartan, who around the turn of the century succeeded in establishing the formalism we are presenting.

Whereas none of the generators of O(3) commute with each other, in larger groups it is possible to find a subset of the generators which are mutually commutative. Let us denote these generators by H_i, i = 1, . . . , r; the maximum number r of mutually commuting generators is called the *rank* of the group. (The rotation group is therefore of rank 1, with $H_1 = J_3$.) There will also be a number of functions of the generators, analogous to J^2, which commute with all of the generators. These are called the *Casimir operators* of the group, and we shall denote them by C_μ. It turns out that the number of Casimir operators is also equal to the rank of the group. Finally, there may be other functions of the generators, call them S_k, which commute with all of the H_i, but not with the full set of X_λ. The number k of these depends on the group.

Since the H_i, the C_μ, and the S_k are all mutually commutative, they can all be simultaneously diagonalized. That is, we can choose a set of basis vectors which are simultaneous eigenstates of all of these operators. Let us abbreviate them $|chs\rangle$, using c, h, and s to denote the full sets of quantum numbers $c = \{c_1, \ldots, c_r\}$, $h = \{h_1, \ldots, h_r\}$, and $s = \{s_1, \ldots, s_k\}$. Thus

$$C_\mu |chs\rangle = c_\mu |chs\rangle$$

$$H_i |chs\rangle = h_i |chs\rangle$$

$$S_k |chs\rangle = s_k |chs\rangle. \tag{2.40}$$

These eigenstates may be conveniently plotted if we define an r-dimensional *representation space* with coordinates h_i. The eigenstates of the group then correspond to the points $h = \{h_1, \ldots, h_r\}$. The rotation group, for example, has a one-dimensional representation space, and the eigenstates correspond to the integers and half-integers along that single axis. In O(3), furthermore, we have only $C_1 = J^2$, and $H_1 = J_3$; there are no S_k, and an eigenstate is uniquely labeled by giving the two eigenvalues j and m. But in larger groups, there may be several eigenstates having the same sets c and h, distinguished only by s.

There remain n – r generators which do not commute with the H_i's. Cartan was able to show, however, that these can always be put into the form of *step operators* analogous to $J_\pm$; that is, the remaining generators can be linearly transformed into a set of operators E_α, $\alpha = 1, \ldots, (n - r)$, having commutation relations with the H_i's of the form

$$[H_i, E_\alpha] = v_{\alpha i} E_\alpha. \tag{2.41}$$

Applying (2.41) to an eigenstate $|chs\rangle$ leads to

$$H_i E_\alpha |chs\rangle = (h_i + v_{\alpha i}) E_\alpha |chs\rangle, \tag{2.42}$$

implying that $E_\alpha |chs\rangle$ is an eigenstate of H_i with eigenvalue $(h_i + v_{\alpha i})$. Thus $E_\alpha |chs\rangle$ must be a multiple of $|c(h + v_\alpha)s\rangle$; the set of n – r vectors $v_\alpha = \{v_{\alpha i}, \ldots, v_{\alpha r}\}$ are known as the *root vectors* of the group. They illustrate how the generators transform the states of an irreducible representation. For example, since O(3) is of rank r = 1, its root vectors are in a one-dimensional vector space; they are simply $v_\pm = \pm 1$, corresponding to $J_\pm$. The root vector diagram, shown in Fig.(2-1), corresponds exactly with the raising of lowering effects of $J_\pm$ on the value of m.

Now by the manipulations similar to those involved in obtaining (2.21), we can derive relations and inequalities which require that on certain states a given

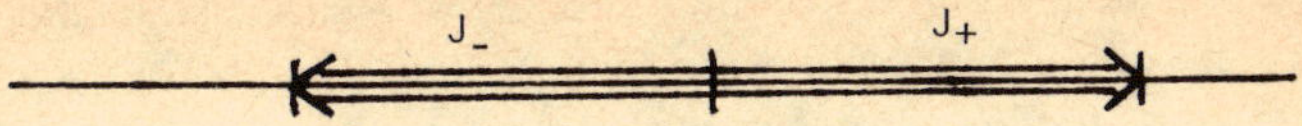

Fig.(2-1) Root vectors of O(3).

step operator E_a must yield zero. The size and shape of the irreducible representations are thus again determined by the eigenvalues of the Casimir operators. These eigenvalues will be the same for all states of an irreducible representation, and may be used to identify the different representations.

Any set of n matrices x_λ which satisfy the commutation relations (2.38) form a matrix representation of the group. As in (2.24), equivalent representations are related by unitary transformations. The irreducible ones are identified by the eigenvalues $c = \{c_1 \ldots c_r\}$ of the Casimir operators, and may be given by writing

$$[x_\lambda(c)]_{hs,h's'} = \langle chs|X_\lambda|ch's'\rangle, \tag{2.43}$$

with the matrix elements subscripted corresponding to the different eigenstates. One irreducible representation of particular interest is provided by the structure constants themselves, with

$$[F_\lambda]_{\mu\nu} \equiv if_{\lambda\mu\nu},$$

for which (2.39b) becomes

$$[F_\lambda, F_\sigma] = if_{\lambda\sigma\tau}F_\tau.$$

Since these matrices are antisymmetric, a unitary transformation must be made in order to obtain an equivalent representation for which the H_i and S_k are diagonal. The resulting representation, which can be shown to be irreducible, is known as the *regular* representation.

It should also be mentioned here that for every representation x_λ there exists a *conjugate* representation x_λ^c defined by

$$[x_\lambda^c]_{hs,h's'} = -[x_\lambda]^*_{hs,h's'} \tag{2.44}$$

as can be shown by merely taking the complex conjugate of (2.38). A representation and its conjugate clearly have the same dimensionality, but whether they are actually equivalent depends on the group; in general, a Lie group is not limited to a single irreducible representation for each dimensionality. (For the rotation group there was only one, but other details which we shall not discuss complicate the question of equivalence there.)

The reduction to block diagonal form of the direct product of two irreducible representations provides a definition of the Clebsch–Gordan coefficients $\langle c_1 h_1 s_1 c_2 h_2 s_2 | chs \rangle$ of the higher group,

$$|chs\rangle = \sum_{h_1 s_1 h_2 s_2} \langle c_1 h_1 s_1 c_2 h_2 s_2 | chs \rangle |c_1 h_1 s_1\rangle |c_2 h_2 s_2\rangle.$$

The usual inversion formula

$$|c_1 h_1 s_1\rangle |c_2 h_2 s_2\rangle = \sum_{chs} \langle chs | c_1 h_1 s_1 c_2 h_2 s_2 \rangle |chs\rangle$$

also holds, with

$$\langle chs | c_1 h_1 s_1 c_2 h_2 s_2 \rangle = \langle c_1 h_1 s_1 c_2 h_2 s_2 | chs \rangle^*.$$

We can also define the irreducible tensor operators T_{chs} of the group, in analogy with (2.31), as a set of operators satisfying

$$[X_\lambda, T_{chs}] = \sum_{h's'} [x_\lambda(c)]_{hs,h's'} T_{ch's'}. \tag{2.45}$$

These tensor operators are in one-to-one correspondence with the eigenstates of an irreducible representation c. A generalized form of the Wigner–Eckhart theorem follows immediately, stating that the matrix element of a tensor operator T_{chs} can be separated into the product form

$$\langle c_1 h_1 s_1 | T_{chs} | c_2 h_2 s_2 \rangle = T(c_1, c, c_2) \langle c_1 h_1 s_1 | chs c_2 h_2 s_2 \rangle, \tag{2.46}$$

in which the reduced matrix element $T(c_1, c, c_2)$ depends only on the irreducible representations c_1, c_2, and c, while the dependence on h_1, h_2, h, s_1, s_2, and s is contained entirely in the Clebsch–Gordan coefficient.

The application of these higher symmetry groups to a physical system proceeds in exactly the same way as does the use of O(3). If the operator A represents a quantity of physical interest – the Hamiltonian, the transition matrix, or any other – we express it in terms of irreducible tensor operators of the group and make use of the Wigner–Eckart theorem. In the simplest case, we might assume A to be a scalar operator; then its matrix elements $\langle chs | A | chs \rangle$ would be independent of h and s. Alternatively, A could be separable into two parts, $A = A_1 + A_2$, with A_1 a scalar under the group, while A_2 is an irreducible tensor of higher order which can be treated as a perturbation. The splittings induced by this perturbation are then related through (2.46).

Application of the theory of Lie groups thus tells us a great deal about the properties of a physical system, both for operators which are invariant under the group and for those that are only approximately so. In the following chapters, we shall encounter numerous examples of both types.

Chapter 3

THE ISOSPIN GROUP SU(2)

With respect to the nuclear forces, the proton and neutron seem to be identical particles. They differ in mass by only 1.3 MeV; this is enough to allow the weak interaction to cause beta-decay, but it is only a 0.14% effect. Although quantitative explanation of this difference remains an unsolved problem, it is clearly of electromagnetic order. Other differences, such as charge or magnetic moment, are also entirely of electromagnetic origin. Insofar as the strong interactions are concerned, it appears that we have here two different states of a single particle, the *nucleon*. Just as a magnetic field is needed to distinguish an electron with spin up from one with spin down, so the difference between "charge up" (proton) and "charge down" (neutron) is only discerned by electromagnetic means.

The charge can thus be considered a multiple of an internal quantum number taking two possible values, 0 and 1, for the nucleon. The state vector of a nucleon must therefore contain, in addition to the spatial description of the particle, its charge eigenvalue. Maintaining the "up–down" analogy, we shall write the proton state vector simply as $|+\rangle$ and the neutron as $|-\rangle$.

Now let us imagine a universe in which all but the strong interactions are "turned off". Then we cannot distinguish whether a given nucleon is proton or neutron; indeed, it may even be a superposition of the two – for example, a neutron which was preparing to beta-decay when we flipped the switch. Thus the state vector $|\psi\rangle$ of the nucleon will, in general, be of the form

$$|\psi\rangle = \psi_p|+\rangle + \psi_n|-\rangle, \qquad (3.1)$$

with $|\psi_p|^2(|\psi_n|^2)$ giving the probability that $|\psi\rangle$ represents a proton (neutron), and $|\psi_p|^2 + |\psi_n|^2 = 1$. Using $|+\rangle$ and $|-\rangle$ as basis states in this way implies that a given system has, in addition to the labeling resulting from spatial variables, a matrix representation in this charge space; that is, every operator M will be described by a 2×2 matrix M_{ij} having elements defined by $\langle +|M|+\rangle$, $\langle +|M|-\rangle$, $\langle -|M|+\rangle$, and $\langle -|M|-\rangle$, and the state $|\psi\rangle$ will correspond to a vector in this space,

$$|\psi\rangle = \psi_p \begin{pmatrix} 1 \\ 0 \end{pmatrix} + \psi_n \begin{pmatrix} 0 \\ 1 \end{pmatrix} = \begin{pmatrix} \psi_p \\ \psi_n \end{pmatrix}. \tag{3.2}$$

The fact that the strong interactions do not distinguish proton from neutron implies that any operator M referring only to the strong interactions must be independent of the charge state of the nucleon. Its matrix elements must be unchanged if a proton is changed into a neutron; or equivalently, it must be insensitive to any transformation altering the relative contributions ψ_p and ψ_n. In the matrix basis (3.2), therefore, M_{ij} must be invariant under any unitary transformation on the two-dimensional state vector $\binom{\psi_p}{\psi_n}$. If U denotes such a transformation, the transformed vector $|\psi'\rangle$ is given by

$$|\psi'\rangle = \begin{pmatrix} \psi'_p \\ \psi'_n \end{pmatrix} = \begin{pmatrix} U_{11} & U_{12} \\ U_{21} & U_{22} \end{pmatrix} \begin{pmatrix} \psi_p \\ \psi_n \end{pmatrix}. \tag{3.3}$$

The transformation of the operator M is given similarly to (2.27a) as

$$M' = UMU^{\dagger};$$

if M is invariant, we must have $M' = M$, which is equivalent to

$$[M, U] = 0. \tag{3.4}$$

Operators involving only the strong interactions therefore commute with all unitary transformations in the two-dimensional nucleon charge space.

These transformations, like rotations, form a group. (Indeed, we shall soon see that these two groups are directly and intimately connected.) Since it refers to *U*nitary transformations in two dimensions, this group is called U(2). Like O(3), it can be investigated by considering an infinitesimal transformation

$$U_\epsilon = e^{i\epsilon H} = [1 + i\epsilon H + \sigma(\epsilon^2)],$$

with ϵ real and infinitesimal. The unitarity of U implies that H is hermitian. A hermitian 2×2 matrix is necessarily of the form

$$H = \begin{pmatrix} a & c + id \\ c - id & b \end{pmatrix} \tag{3.5}$$

which depends on only four real parameters a, b, c, and d; there are therefore only four linearly independent hermitian matrices. It is easily shown that these four can be taken to be the unit matrix plus the three Pauli matrices

$$\tau_1 = \begin{pmatrix} 0 & 1 \\ 1 & 0 \end{pmatrix}, \quad \tau_2 = \begin{pmatrix} 0 & -i \\ i & 0 \end{pmatrix}, \quad \tau_3 = \begin{pmatrix} 1 & 0 \\ 0 & -1 \end{pmatrix}.$$

Therefore H can always be written as

$$H = n_0 \mathbb{1} + n_1\tau_1 + n_2\tau_2 + n_3\tau_3 = n_0 \mathbb{1} + \mathbf{n} \cdot \boldsymbol{\tau}, \tag{3.6}$$

defining the real three-vector $\mathbf{n} = (n_1, n_2, n_3)$.

The unit matrix and the Pauli matrices are thus generators of U(2), and any unitary transformation in two dimensions can be represented by an exponentiation of them,

$$U = \exp\,[i(n_0 \mathbb{1} + i\mathbf{n} \cdot \boldsymbol{\tau})] \tag{3.7}$$

in the same way that any rotation is represented by (2.7). The part of the transformation generated by the unit matrix, however, produces only an overall phase factor, since clearly (3.7) is equivalent to

$$U = e^{in_0} e^{i\mathbf{n}\cdot\boldsymbol{\tau}}. \tag{3.8}$$

Because such phase transformation have a trivial effect on the wave function, and none at all on any physically measurable quantity, it is appropriate to separate them from the ones we are studying by fixing $n_0 = 0$ and concentrate on the subgroup generated by

$$U(\bar{n}) = e^{i\mathbf{n}\cdot\boldsymbol{\tau}} \tag{3.9}$$

only. Matrices of this form are guaranteed, by the general matrix relation $\det e^A = e^{\mathrm{tr} A}$, to have $\det U = 1$. This subgroup of *S*pecial *U*nitary transformations in two dimensions is called SU(2).

The generators of SU(2) are thus proportional to the Pauli matrices $\boldsymbol{\tau}$, which are already familiar in the spin ½ representation of the rotation group, for which $J_i^{(1/2)} = \frac{1}{2}\boldsymbol{\tau}_i$. Thus the analogy between spin states and the nucleon charge states arises naturally, and we define the *isospin* (a shortening of the original term, *isotopic spin*) by writing

$$I_i^{(1/2)} = \tfrac{1}{2}\boldsymbol{\tau}_i. \tag{3.10}$$

The commutation relations of the isospin operators I_i must be the same as those of the representation $I_i^{(1/2)}$,

$$[I_i, I_j] = i\epsilon_{ijk}I_k, \tag{3.11}$$

and identical with those of O(3). Since SU(2) has a Lie group structure determined by those commutation relations, we expect that SU(2) and O(3) are very closely related to each other.

The connection can be made clear by considering a traceless hermitian matrix Λ. By (3.6), we know that Λ can be written

$$\Lambda = \boldsymbol{\lambda} \cdot \boldsymbol{\tau} \tag{3.12a}$$

with some appropriate real three-vector $\boldsymbol{\lambda}$. Now let Λ be transformed as in (3.5) by some U belonging to SU(2), yielding

$$\Lambda' = U^\dagger \Lambda U. \tag{3.13}$$

Then Λ' is also hermitian and traceless, so we can write

$$\Lambda' = \boldsymbol{\lambda}' \cdot \boldsymbol{\tau} \tag{3.12b}$$

Consequently, to each member U of SU(2) there corresponds a rotation R in three dimensions taking $\boldsymbol{\lambda}$ into $\boldsymbol{\lambda}'$,

$$\boldsymbol{\lambda}' = R\boldsymbol{\lambda}.$$

The correspondence thus established is not one-to-one, however, since both U and (−U) will lead to the same Λ' in (3.13) and thus will correspond to the same R.

As a Lie group, SU(2) is therefore "twice as large" as O(3), although their commutation relations are identical. Our entire analysis of the irreducible representations of O(3) was based entirely on these commutation relations, however, so the same results must also hold for SU(2). In particular, a Casimir operator

$$I^2 = \sum_i I_i I_i$$

can be defined and will have eigenvalues $I(I + 1)$, with I an integer or half-integer; and the irreducible representations of SU(2), labeled by these eigenvalues, will be in one-to-one correspondence with those of the rotation group. Thus, in addition to the $I = ½$ nucleon isodoublet, we expect to see sets of particles corresponding to $I = 0, 1, 3/2$, etc., containing $(2I + 1)$ different charge states but otherwise essentially identical. Such isomultiplets do exist; a number of the most important examples are given in Table 3-A, while a more thorough and detailed listing of the hadrons can be found in Appendix A2.

TABLE 3-A Isomultiplets

Particle	I	Charge states	Q	I_3	B	S
N	$1/2$	p	1	$1/2$	1	0
		n	0	$-1/2$	1	0
π	1	π^+	1	1	0	0
		π°	0	0	0	0
		π^-	−1	−1	0	0
Δ	$3/2$	Δ^{++}	2	$3/2$	1	0
		Δ^+	1	$1/2$	1	0
		Δ°	0	$-1/2$	1	0
		Δ^-	−1	$-3/2$	1	0
Λ	0	Λ°	0	0	1	−1
Σ	1	Σ^+	1	1	1	−1
		Σ°	0	0	1	−1
		Σ^-	−1	−1	1	−1
K	$1/2$	K^+	1	$1/2$	0	1
		K°	0	$-1/2$	0	1
$\bar{K}$	$1/2$	$\bar{K}^\circ$	0	$1/2$	0	−1
		$\bar{K}^-$	−1	$-1/2$	0	−1
η	0	η°	0	0	0	0
Ξ	$1/2$	Ξ°	0	$1/2$	1	−2
		Ξ^-	−1	$-1/2$	1	−2
Ω	0	Ω^-	−1	0	1	−3

For each of these charge multiplets the strong interaction properties of the particles are very similar; only the charge distinguishes them. The values of the charge do not coincide directly with the value of I_3, however, but seem to depend also on other quantum numbers of the system. Specifically, the formula

$$Q = I_3 + \tfrac{1}{2}(B + S) = I_3 + \tfrac{1}{2}Y, \tag{3.14}$$

first noted by Gell-Mann and Nishijima, correlates correctly the quantum numbers of all known particles. In Chapter 1 we discussed the charge conjugation operator C, which in changing particle to antiparticle reversed the sign of the charge, the baryon number, and the strangeness. From (3.14) it is clear that the effect of C must also be to reverse the sign of I_3.

Since I_3 is real and diagonal, changing the sign of I_3 is tantamount to transforming to the complex conjugate representation, for which

$$[I_3^c]_{ij} = -[I_3]^*_{ij} = -[I_3]_{ij}.$$

Thus particle and antiparticle may be said to transform according to complex conjugate representations.

In SU(2), however, every representation is equivalent to its conjugate. To prove this, we note that the matrix representations of I_3 and I_1 are necessarily real, while those of I_2 are purely imaginary. Therefore

$$I_1^c = -I_1^* = -I_1$$

$$I_2^c = -I_2^* = I_2$$

$$I_3^c = -I_3^* = -I_3$$

The effect of conjugation is thus to reverse the directions of the I_1 and I_3 axes. But a rotation through π about the I_2 axis has precisely the same effect. Therefore

$$I_i^c = e^{i\pi I_2} I_i e^{-i\pi I_2}, \tag{3.15}$$

and consequently I_i and I_i^c are equivalent.

Let us apply this idea first to the pions. For $I = 1$, it is not difficult to show that

$$e^{i\pi I_2}\begin{pmatrix}\pi^+\\ \pi^\circ\\ \pi^-\end{pmatrix} = \begin{pmatrix}0 & 0 & 1\\ 0 & -1 & 0\\ 1 & 0 & 0\end{pmatrix}\begin{pmatrix}\pi^+\\ \pi^\circ\\ \pi^-\end{pmatrix} = \begin{pmatrix}\pi^-\\ -\pi^\circ\\ \pi^+\end{pmatrix}.$$

Thus the action of $e^{i\pi I_2}$ here is to replace each pion by its antiparticle, with an appropriate phase. We have already seen that $C|\pi^\circ\rangle = |\pi^\circ\rangle$; the effect of C on the charged pions, however, may be consistently defined by either $C|\pi^+\rangle = |\pi^-\rangle$ or $C|\pi^+\rangle = -|\pi^-\rangle$. Let us choose the latter convention for the moment. Then

$$Ce^{i\pi I_2}\begin{pmatrix}\pi^+\\ \pi^\circ\\ \pi^-\end{pmatrix}=C\begin{pmatrix}\pi^-\\ -\pi^\circ\\ \pi^+\end{pmatrix}=-\begin{pmatrix}\pi^+\\ \pi^\circ\\ \pi^-\end{pmatrix}$$

shows that the pions are eigenstates of the operator

$$G = Ce^{i\pi I_2} \tag{3.16}$$

with eigenvalue $g_\pi = -1$. The same conclusion is reached if we use the "normal" charge conjugation convention $C|\pi^+\rangle = |\pi^-\rangle$, provided we assign appropriate phases *within* the pion isomultiplet, namely $(-\pi^+, \pi^\circ, \pi^-)$ instead of $(\pi^+, \pi^\circ, \pi^-)$. Such phases do not, of course, affect any physical result.

The operation $G = Ce^{i\pi I_2}$, known as "G-conjugation" is thus an important extension of the concept of charge conjugation. It changes an isomultiplet into the charge-conjugate isomultiplet, properly arranged. Non-strange mesons, and multiparticle states with $B = S = 0$, can be eigenstates of G. Since $G^2 = \mathbb{1}$, the eigenvalues of G are ± 1, and are correspondingly identified as the "G-parity" of the state. Thus the G-parity of the pion is $g_\pi = -1$, and consequently an n-pion state must have $g_{n\pi} = (-1)^n$. Since both isosymmetry and charge conjugation symmetry are conserved in strong interactions, however, G-parity must also be. Therefore the G-parity of a state can be determined from the number of pions into which it decays via G-conserving strong processes. The ρ meson, for example, is observed to decay by $\rho \to 2\pi$, so it must have $g_\rho = +1$; the decay $\rho \to 3\pi$ is therefore forbidden by G-parity, and indeed is not observed.

The effect of G on other states also produces certain phase effects. On the nucleons, for example, with $C|p\rangle = |\bar{p}\rangle$ and $C|n\rangle = |\bar{n}\rangle$, we have

$$C\begin{pmatrix}p\\ n\end{pmatrix}=\begin{pmatrix}\bar{p}\\ \bar{n}\end{pmatrix}$$

reversing the eigenvalues of I_3. They are rearranged appropriately by

$$e^{i\pi I_2}=\begin{pmatrix}0 & 1\\ -1 & 0\end{pmatrix},$$

yielding

$$G\begin{pmatrix}p\\ n\end{pmatrix}=e^{i\pi I_2}\begin{pmatrix}\bar{p}\\ \bar{n}\end{pmatrix}=\begin{pmatrix}\bar{n}\\ -\bar{p}\end{pmatrix}$$

A phase appears in the antinucleon isodoublet, implying that in any calculation involving both isosymmetry and charge conjugation we must define its members as $(\bar{n}, -\bar{p})$ in order to use G-parity. The same will be true of any other isodoublet; for example, identifying (K^+, K°) as the kaon isodoublet states requires the antikaon doublet to be $(\bar{K}^\circ, -K^-)$.

Applications of Isosymmetry

If the strong interactions are isosymmetric, as suggested by experiment, the strong interactions Hamiltonian commutes with all transformations of the isospin group SU(2); it is a scalar operator under SU(2), and so is the transition operator corresponding to it. The implications of isosymmetry follow from this statement through the Wigner–Eckart theorem, exactly as those of rotational invariance do.

The simplest example of the ideas and techniques involved is the three-particle vertex shown in Fig.(3-1), which may be thought of either as a coupling constant, as in Chapter 1, or as describing the decay of a resonance A into products B and C. In either case, it will be described by the matrix element $\langle BC|\mathscr{T}|A\rangle$ of the transition operator $\mathscr{T}$. Let us assume that A, B, and C represent isomultiplets, denoting the states therefore by

$$|A\rangle = |A; I^A, I_3^A\rangle$$

$$|B\rangle = |B; I^B, I_3^B\rangle$$

$$|C\rangle = |C; I^C, I_3^C\rangle$$

Thus the initial state is an eigenstate of isospin, but the final state is a direct product state containing all values of total isospin from $|I^B - I^C|$ to $I^B + I^C$. Using the Clebsch–Gordan coefficients, we may write

$$\langle B; I^B, I_3^B|\langle C; I^C, I_3^C| = \sum_I \langle I^B I_3^B I^C I_3^C | I I_3\rangle \langle BC; I, I_3|.$$

If $\mathscr{T}$ is an isoscalar operator, then it must conserve total isospin, and

$$\langle BC|\mathscr{T}|A\rangle = \langle I^B I_3^B I^C I_3^C | I^A I_3^A\rangle\, \mathscr{T}_{A\to BC} \tag{3.17}$$

where $\mathscr{T}_{A\to BC}$ is a reduced matrix element. Consequently the amplitudes for all isospin-related $A \to BC$ decays are proportional, the constant of proportionality being simply the appropriate Clebsch–Gordan coefficient.

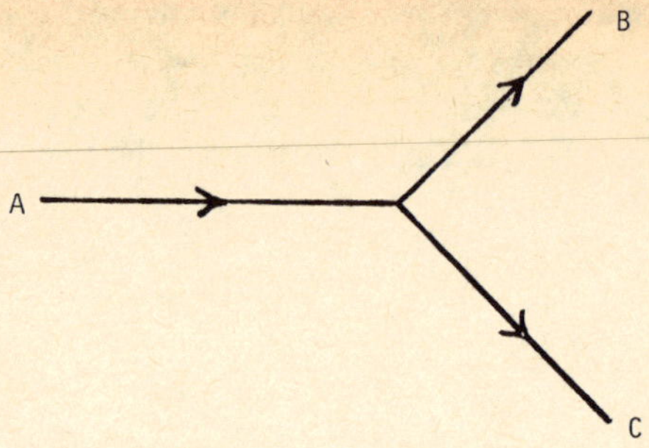

Fig.(3-1) Coupling of A to BC.

For example, let us consider the decays of the resonance leading to final stages consisting of a nucleon plus a pion. There are six possible decays: $\Delta^{++} \to p\pi^+$, $\Delta^+ \to n\pi^+$, $\Delta^+ \to p\pi^\circ$, $\Delta^\circ \to n\pi^\circ$, $\Delta^\circ \to p\pi^-$, and $\Delta^- \to n\pi^-$. These decays will be described by the matrix elements $\langle N\pi | \mathscr{T} | \Delta \rangle$ of the transition operator $\mathscr{T}$ between the initial and final states. Since $\mathscr{T}$ is an isoscalar, these matrix elements will vanish unless total isospin is conserved. We must therefore express the $N\pi$ combinations in terms of eigenstates of total isospin, which we shall denote by $|N\pi; I, I_3\rangle$. By the usual rules of addition of angular momenta, combining the nucleon (isospin $\frac{1}{2}$) with the pion (isospin 1) leads to both $I = \frac{1}{2}$ and $I = \frac{3}{2}$ in the final state; specifically,

$$|N\pi; \tfrac{3}{2}, \tfrac{3}{2}\rangle = |p\pi^+\rangle$$

$$|N\pi; \tfrac{3}{2}, \tfrac{1}{2}\rangle = \frac{1}{\sqrt{3}} \left[|n\pi^+\rangle + \sqrt{2}\,|p\pi^\circ\rangle\right]$$

$$|N\pi; \tfrac{3}{2}, -\tfrac{1}{2}\rangle = \frac{1}{\sqrt{3}} \left[|p\pi^-\rangle + \sqrt{2}\,|n\pi^\circ\rangle\right]$$

$$|N\pi; \tfrac{3}{2}, -\tfrac{3}{2}\rangle = |n\pi^-\rangle$$

$$|N\pi; \tfrac{1}{2}, \tfrac{1}{2}\rangle = \frac{1}{\sqrt{3}} \left[\sqrt{2}\,|n\pi^+\rangle - |p\pi^\circ\rangle\right]$$

$$|N\pi; \tfrac{1}{2}, -\tfrac{1}{2}\rangle = \frac{1}{\sqrt{3}} \left[|n\pi^\circ\rangle - \sqrt{2}\,|p\pi^-\rangle\right]. \tag{3.18}$$

The initial state, which we shall label $|\Delta; I, I_3\rangle$, is a pure $I = \frac{3}{2}$ state. Therefore the Wigner–Eckart theorem implies that if the $N\pi$ state has $I = \frac{1}{2}$, the matrix element vanishes; if the $N\pi$ state has $I = \frac{3}{2}$, the matrix element is given by a single reduced matrix element multiplied by an appropriate Clebsch–Gordan coefficient. It follows, for example, that

$$\frac{1}{\sqrt{3}}[\langle p\pi^\circ| - \sqrt{2}\langle n\pi^+|]\,\mathscr{T}\,|\Delta^+\rangle = \langle N\pi; \tfrac{1}{2}, \tfrac{1}{2}|\,\mathscr{T}\,|\Delta; \tfrac{3}{2}, \tfrac{1}{2}\rangle$$

$$= 0,$$

that is,

$$\langle p\pi^\circ|\,\mathscr{T}\,|\Delta^+\rangle = \sqrt{2}\langle n\pi^+|\,\mathscr{T}\,|\Delta^+\rangle. \tag{3.19}$$

Since the decay probability is proportional to $|\langle\mathscr{T}\rangle|^2$, this means that twice as many Δ^+'s should decay to $p\pi^\circ$ as to $n\pi^+$. This ratio must be corrected slightly to allow for the electromagnetic mass differences in the two final states, but is very well confirmed by experimental data.

Similarly, we find

$$\langle n\pi^+|\,\mathscr{T}\,|\Delta^+\rangle = \langle p\pi^-|\,\mathscr{T}\,|\Delta^\circ\rangle = \frac{1}{\sqrt{3}}\langle p\pi^+|\,\mathscr{T}\,|\Delta^{++}\rangle,$$

$$\langle p\pi^\circ|\,\mathscr{T}\,|\Delta^+\rangle = \langle n\pi^\circ|\,\mathscr{T}\,|\Delta^\circ\rangle = \frac{2}{\sqrt{3}}\langle p\pi^+|\,\mathscr{T}\,|\Delta^{++}\rangle,$$

$$\langle n\pi^-|\,\mathscr{T}\,|\Delta^-\rangle = \langle p\pi^+|\,\mathscr{T}\,|\Delta^{++}\rangle, \tag{3.20}$$

which make corresponding predictions for the relative strengths of the other decays. The same technique can be applied to any strong decay or resonance production process.

A two-body scattering reaction can also be treated using isosymmetry. For example, we consider first the scattering of pions by deuterons. The deuteron is an isosinglet, so the πD state is purely $I = 1$:

$$|\pi D; 1, I_3\rangle = |\pi; 1, I_3\rangle|D; 0, 0\rangle.$$

Elastic πD scattering is described by the matrix element $\langle\pi D|\,\mathscr{T}\,|\pi D\rangle$, and if $\mathscr{T}$ is isoscalar it follows immediately that

$$\langle \pi^+ D | \mathcal{T} | \pi^+ D \rangle = \langle \pi^- D | \mathcal{T} | \pi^- D \rangle = \langle \pi^\circ D | \mathcal{T} | \pi^\circ D \rangle .$$

Thus the scattering amplitude for πD scattering does not depend on the charge of the pion. This prediction must also be corrected for the electromagnetic interactions between charged pions and the deuteron; after these corrections are made, the prediction of isosymmetry is verified to within 1%.

In general, however, scattering situations are more complicated; when the particles involved are not isosinglets there will be several different reactions possible, and neither the initial nor the final state need be a pure I-state. For example, there are eight possible pion–nucleon scattering processes. Let us consider the charge exchange $\pi^- p \to \pi^\circ n$. Both initial and final states are combinations of $I = ½$ and $I = 3/2$; inverting (3.18), we find that

$$|\pi^- p\rangle = \frac{1}{\sqrt{3}} [\, |N\pi; \tfrac{3}{2}, -\tfrac{1}{2}\rangle - \sqrt{2}\, |N\pi; \tfrac{1}{2}, -\tfrac{1}{2}\rangle]$$

$$|\pi^\circ n\rangle = \frac{1}{\sqrt{3}} [\sqrt{2}\, |N\pi; \tfrac{3}{2}, -\tfrac{1}{2}\rangle + |N\pi; \tfrac{1}{2}, -\tfrac{1}{2}\rangle]$$

Because $\mathcal{T}$ is an isoscalar, matrix elements such as $\langle N\pi; \tfrac{3}{2}, -\tfrac{1}{2} | \mathcal{T} | N\pi; \tfrac{1}{2}, -\tfrac{1}{2}\rangle$ will vanish, and we obtain

$$\langle \pi^\circ n | \mathcal{T} | \pi^- p\rangle = \frac{\sqrt{2}}{3} [\mathcal{T}^{3/2} - \mathcal{T}^{1/2}], \tag{3.21}$$

where $\mathcal{T}^I = \langle N\pi; I | \mathcal{T} | N\pi; I\rangle$ is the reduced matrix element for the process. Similar calculations lead to

$$\langle \pi^+ p | \mathcal{T} | \pi^+\rangle = \langle \pi^- n | \mathcal{T} | \pi^- n\rangle = \mathcal{T}^{3/2}$$

$$\langle \pi^\circ p | \mathcal{T} | \pi^\circ p\rangle = \langle \pi^\circ n | \mathcal{T} | \pi^\circ n\rangle = \frac{1}{3} [2\mathcal{T}^{3/2} + \mathcal{T}^{1/2}]$$

$$\langle \pi^- p | \mathcal{T} | \pi^- p\rangle = \langle \pi^+ n | \mathcal{T} | \pi^+ n\rangle = \frac{1}{3} [\mathcal{T}^{3/2} + 2\mathcal{T}^{1/2}], \tag{3.22}$$

implying that

$$\langle \pi^\circ n | \mathcal{T} | \pi^- p\rangle = \frac{1}{\sqrt{2}} [\langle \pi^+ p | \mathcal{T} | \pi^+ p\rangle - \langle \pi^- p | \mathcal{T} | \pi^- p\rangle]. \tag{3.23}$$

Comparison of (3.23) with experiment is less straightforward than the preceding examples, but isosymmetry is again verified.

Reactions in which the final state particles belong to isomultiplets different from those in the initial state can also be treated in this way. Let us consider the strangeness exchange processes $\pi N \to K\Sigma$, and in particular $\pi^- p \to K^\circ\Sigma^\circ$. The final state $K^\circ\Sigma^\circ$ must also be decomposed into total isospin eigenstates; once again we are combining isospin 1 with isospin ½, so the algebra is identical to (3.18). The result is

$$|K^\circ\Sigma^\circ\rangle = \frac{1}{\sqrt{3}}\,[\sqrt{2}\,|K\Sigma;\tfrac{3}{2},-\tfrac{1}{2}\rangle + |K\Sigma;\tfrac{1}{2},-\tfrac{1}{2}\rangle]\,. \tag{3.24}$$

Once again the matrix elements between $I = \frac{1}{2}$ and $I = \frac{3}{2}$ states will vanish, and we find

$$\langle K^\circ\Sigma^\circ|\mathscr{T}|\pi^- p\rangle = \frac{\sqrt{2}}{3}\,[\mathscr{Y}^{3/2} - \mathscr{Y}^{1/2}]\,, \tag{3.25}$$

where $\mathscr{Y}^I = \langle K\Sigma; I|\mathscr{T}|\pi N; I\rangle$ denotes the reduced matrix element for total isospin I for the strangeness exchange reaction.

The matrix elements for other $\pi N \to K\Sigma$ processes can be similarly calculated; the results are

$$\langle K^+\Sigma^+|\mathscr{T}|\pi^+ p\rangle = \langle K^\circ\Sigma^-|\mathscr{T}|\pi^- n\rangle = \mathscr{Y}^{3/2}$$

$$\langle K^\circ\Sigma^+|\mathscr{T}|\pi^\circ p\rangle = \langle K^+\Sigma^-|\mathscr{T}|\pi^\circ n\rangle = \frac{\sqrt{2}}{3}\,[\mathscr{Y}^{3/2} - \mathscr{Y}^{1/2}]$$

$$\langle K^+\Sigma^\circ|\mathscr{T}|\pi^\circ p\rangle = \langle K^\circ\Sigma^\circ|\mathscr{T}|\pi^\circ n\rangle = \frac{1}{3}\,[2\,\mathscr{Y}^{3/2} + \mathscr{Y}^{1/2}]$$

$$\langle K^+\Sigma^\circ|\mathscr{T}|\pi^+ n\rangle = \langle K^\circ\Sigma^\circ|\mathscr{T}|\pi^- p\rangle = \frac{\sqrt{2}}{3}\,[\mathscr{Y}^{3/2} - \mathscr{Y}^{1/2}]$$

$$\langle K^+\Sigma^-|\mathscr{T}|\pi^- p\rangle = \langle K^\circ\Sigma^+|\mathscr{T}|\pi^+ n\rangle = \frac{1}{3}\,[\mathscr{Y}^{3/2} + 2\,\mathscr{Y}^{1/2}]\,. \tag{3.26}$$

In a later chapter we shall see that the process $\pi^- p \to K^+\Sigma^-$, which would require the exchange of a doubly charged meson, is effectively forbidden at high energy.

Thus its matrix element should vanish, which implies that $\mathscr{Y}^{3/2} = -2\,\mathscr{Y}^{1/2}$ and that all of these amplitudes are proportional to each other. In particular, it follows that $\langle K^+\Sigma^+|\mathscr{T}|\pi^+p\rangle = \sqrt{2}\langle K^\circ\Sigma^\circ|\mathscr{T}|\pi^-p\rangle$, implying that the cross section for $\pi^+p \to K^+\Sigma^+$ should be twice that for $\pi^-p \to K^\circ\Sigma^\circ$, in agreement with experiment.

The assumption that the strong interactions are invariant under the isospin group SU(2) thus leads to a great many relations between various two-body processes. It can also be used in considering more complicated reactions leading to more than two particles in the final state, to other resonances, etc. We shall not give any further comparisons of isosymmetry predictions with experiment; in general, they hold to at least the accuracy expected, allowing for symmetry breaking due to electromagnetic interactions. Thus it appears that the strong interactions are indeed isosymmetric.

Chapter 4

SU(3)

In the last chapter, we introduced SU(2) by studying unitary transformations on the two-dimensional nucleon charge space. Although initially only the $I = ½$ proton–neutron representation was considered, we found that the group obtained also had larger irreducible representations corresponding exactly with those of the rotation group O(3). Furthermore, isomultiplets with $I = 0$, 1 and 3/2 do seem to exist as well as those with $I = ½$.

A simple way to visualize the connection between the nucleon isodoublet and these other isomultiplets is to think again of the spin ½ representation of O(3). This is the smallest representation of O(3), but from it we can construct, by combination and reduction, all of the larger irreducible representations. That is, we can produce a state with any desired value of angular momentum by combining enough spin ½ states. Representations having this "building block" property are called elementary representations. They exist in all Lie groups; the nucleons provide the elementary representation of SU(2).

States of higher isospin can therefore be built up from the nucleons. For example, the two-nucleon state will be a combination of two $I = ½$ states, and so will have $I = 0$ or $I = 1$. The deuteron is the isosinglet; the $I = 1$ system, which would include the diproton and dineutron, does not have a bound state. But our interest is in particles, not nuclei; how can we combine nucleons to produce elementary particle states? We merely use, in addition to the nucleons, the antinucleons, which also form an isodoublet. They transform according to the complex conjugate representation, which in SU(2) is equivalent to the elementary (nucleon) representation, as shown in Chapter 3. By combining nucleon with antinucleon, we produce states with $I = 0$ or $I = 1$ and baryon number B = 0.

For example, the two-particle state containing a proton and an antineutron will have $I = 1$, $I_3 = 1$, and B = S = 0. These are precisely the quantum numbers of the π^+ meson. We may therefore say that the π^+ is a bound state of p and $\bar{n}$, just as the deuteron is a bound state of p and n. It is a very tightly bound state, to be sure, since the mass of proton plus antineutron is nearly 2 GeV; but it can indeed be so described. While conservation of energy forbids the process $\pi^+ \to p\bar{n}$ from occurring directly, we can see $\pi^+A \to (p\bar{n})A$ on an arbitary target A if the initial state has sufficient kinetic energy.

The ρ^+ meson also has these same quantum numbers; consequently it also is a bound state of p and $\overline{n}$. It differs from the pion only in that it is a bound state with different angular momentum, having j = 1. Similarly, the π°, π^-, ρ°, ρ^-, η°, ϕ°, and ω° mesons may be pictured as bound states of the $N\overline{N}$ system with appropriate isospin and angular momentum. In fact, all known non-strange mesons may be accounted for in this way. Since every nucleon must be accompanied by an antinucleon to produce a bound state with B = 0, we expect to see only integral values of I among the mesons. In particular, the $N\overline{N}$ system can have only $I = 0$ or 1; states with higher isospin will require larger combinations such as $NN\overline{N}\overline{N}$. Significantly, however, no mesons with $I > 1$ have yet been found.

The non-strange baryons can be similarly conceived as combinations of nucleons and mesons. We will therefore expect to see only half-integral values of I for the baryons, and only $I \leqslant 3/2$ unless complicated combinations (such as $NNN\overline{N}\overline{N}$, for example) are involved. As with the mesons, simplicity seems to prevail; no baryons with $I > 3/2$ are yet firmly established. Let us consider the nucleon–pion system, which can have isospin 3/2 or 1/2. The $I = 3/2$ system has quantum numbers identical with those of the Δ resonance, which can thus be thought of as an $N\pi$ state (or even as an $(NN\overline{N})$ state). Since the Δ decays strongly through $\Delta \rightarrow N\pi$, this description seems completely natural. The $I = ½$ system, on the other hand, has exactly the same quantum numbers as those of the nucleon. There are higher nucleon resonances with these quantum numbers which can be so identified; but it is also completely consistent, if somewhat confusing, to think of the nucleon itself as a "bound state of pion and nucleon"! (This curious notion is basic to the "bootstrap theory", which we shall discuss in some detail in a later chapter.)

Generating the Strange Particles

Can the strange particles be similarly conceived as combinations of a small basic set of particles? They can, provided we introduce a single strange particle into that basic set. For example, the K^- meson has the quantum numbers B = 0, $S = -1$, $I = ½$, $I_3 = -½$, which are obtained from the combination of an antiproton with a Λ° or a Σ°, or of an antineutron with a Σ^-. Using the isosinglet Λ° is clearly the most economical choice. Thus we picture the isodoublet $(\overline{K}^-, \overline{K}^{\circ})$ as a $(\Lambda^{\circ}\overline{N})$ bound state. Likewise the strange baryons are considered bound states of (p, n, Λ°) with strange or non-strange mesons; for example, the Σ can be obtained either from $(\Lambda^{\circ}\pi)$ or from $(N\overline{K})$ combinations, and the Ξ from $(\Lambda^{\circ}\overline{K})$.

Therefore, by enlarging the nucleon isodoublet – the basis of the elementary representation of SU(2) – to form a basic triplet (p, n, Λ°), we can generate, in the above manner, the quantum numbers of both non-strange and strange hadrons. It is then natural to ask whether this triplet forms the basis of the elementary representation in some higher group. If so, we might hope to classify the remaining hadrons neatly according to the irreducible representations of that group.

Just as considering transformations on the two nucleon states led us to SU(2), however, a consideration of transformations on the (p, n, Λ°) triplet will lead us to SU(3), the special unitary group in three dimensions. Instead of (3.2), we define a three-dimensional complex basis on these states,

$$|\psi\rangle = \psi_p \begin{pmatrix} 1 \\ 0 \\ 0 \end{pmatrix} + \psi_n \begin{pmatrix} 0 \\ 1 \\ 0 \end{pmatrix} + \psi_\Lambda \begin{pmatrix} 0 \\ 0 \\ 1 \end{pmatrix} = \begin{pmatrix} \psi_p \\ \psi_n \\ \psi_\Lambda \end{pmatrix}. \tag{4.1}$$

By studying transformations on (4.1), we can investigate the structure of SU(3). Because of the relatively large mass difference between the Λ° and the nucleons, we should not expect that the strong interactions are invariant under all transformations on (4.1); but it may be that SU(3) nonetheless provides an *approximate* symmetry, with the symmetry breaking effects small enough to be treated by perturbative methods.

The SU(3) Algebra

We wish therefore to study the group SU(3). Its structure, like that of SU(2), can be investigated by considering the traceless hermitian generators from which all unitary transformations U with det U = 1 can be generated. Thus we write $U = e^{iH}$; in three dimensions, the most general traceless hermitian matrix H is of the form

$$H = \begin{pmatrix} a_3 + a_8 & a_1 - ia_2 & a_4 - ia_5 \\ a_1 + ia_2 & -a_3 + a_8 & a_6 - ia_7 \\ a_4 + ia_5 & a_6 + ia_7 & -2a_8 \end{pmatrix} \tag{4.2}$$

depending on eight real parameters $a_1, \ldots, a_8$. Consequently there are eight linearly independent traceless hermitian 3×3 matrices, which will provide the generators of SU(3). By studying these matrices we can obtain the commutation relations of SU(3); from the commutation relations, the structure of irreducible representations of other dimensionalities can be determined.

Since (4.2) contains two independent diagonal elements, a_3 and a_8, there will be two diagonal generators. Thus SU(3) is a group of rank 2; its root-vectors will be two-dimensional, and its eigenstates will be labeled in representation space by the corresponding two eigenvalues. According to the results of Cartan discussed in Chapter 2, we expect that the six non-diagonal generators can be combined to form step operators on these eigenstates.

The structure of these generators is most easily seen if we note that the three-dimensional vector space defined by (4.1) has three equivalent two-dimensional subspaces corresponding to (p, n), (n, Λ°), and (Λ°, p). That is, the operators of SU(3) must include as a subgroup all transformations of the form

$$U_I = \begin{pmatrix} I_{11} & I_{12} & 0 \\ I_{21} & I_{22} & 0 \\ 0 & 0 & 1 \end{pmatrix} \tag{4.3}$$

which affects only the nucleon states. Since the Λ° is an isoscalar, this subgroup is nothing more than the isospin SU(2) group, and will be generated by operators analogous to those of (3.11). Therefore we expect that three of the generators are given by

$$I_1 = \tfrac{1}{2}\begin{pmatrix} 0 & 1 & 0 \\ 1 & 0 & 0 \\ 0 & 0 & 0 \end{pmatrix}, \quad I_2 = \tfrac{1}{2}\begin{pmatrix} 0 & -i & 0 \\ i & 0 & 0 \\ 0 & 0 & 0 \end{pmatrix}, \quad I_3 = \tfrac{1}{2}\begin{pmatrix} 1 & 0 & 0 \\ 0 & -1 & 0 \\ 0 & 0 & 0 \end{pmatrix}. \tag{4.4}$$

Transformations on the (n, Λ°) and (Λ°, p) subspaces will lead similarly to SU(2) subgroups; by analogy to isospin, we call these subgroups U-spin and V-spin, respectively. They will be generated by matrices equivalent to (4.4) in the appropriate subspaces, namely

$$U_1 = \tfrac{1}{2}\begin{pmatrix} 0 & 0 & 0 \\ 0 & 0 & 1 \\ 0 & 1 & 0 \end{pmatrix}, \quad U_2 = \tfrac{1}{2}\begin{pmatrix} 0 & 0 & 0 \\ 0 & 0 & -i \\ 0 & i & 0 \end{pmatrix}, \quad U_3 = \tfrac{1}{2}\begin{pmatrix} 0 & 0 & 0 \\ 0 & 1 & 0 \\ 0 & 0 & -1 \end{pmatrix} \tag{4.5}$$

and

$$V_1 = \tfrac{1}{2}\begin{pmatrix} 0 & 0 & 1 \\ 0 & 0 & 0 \\ 1 & 0 & 0 \end{pmatrix}, \quad V_2 = \tfrac{1}{2}\begin{pmatrix} 0 & 0 & i \\ 0 & 0 & 0 \\ -i & 0 & 0 \end{pmatrix}, \quad V_3 = \tfrac{1}{2}\begin{pmatrix} -1 & 0 & 0 \\ 0 & 0 & 0 \\ 0 & 0 & 1 \end{pmatrix}. \tag{4.6}$$

(The choice of signs in (4.6) should be noted; we have defined them so that cyclic permutations of (p, n, Λ°) correspond to cyclic permutations of (I, U, V).)

Equations (4.4), (4.5), and (4.6) define nine matrices acting as generators of the SU(2) subgroups, whereas we expected only eight generators for SU(3). It is easily seen, however, that

$$I_3 + U_3 + V_3 = 0, \tag{4.7}$$

so only eight of these generators are independent. As expected, (4.7) shows that there are only two diagonal generators, for which any two linearly independent combinations of I_3, U_3, and V_3 can be chosen. The non-diagonal generators are I_1, I_2, U_1, U_2, V_1, and V_2, and the step operators of SU(3) are identical with those of the SU(2) subgroups. Analogously to $I_\pm = I_1 \pm iI_2$, we define $U_\pm = U_1 \pm iU_2$ and $V_\pm = V_1 \pm iV_2$; the effect of U_+, for example, is to convert a Λ° into an n.

Thus SU(3) is made up of three interwoven SU(2) subgroups. Its commutation relations therefore include those of the SU(2) subgroups, which we write using the step operators as

$$\begin{aligned} &[I_3, I_\pm] = \pm I_\pm && [I_+, I_-] = 2I_3 \\ &[U_3, U_\pm] = \pm U_\pm && [U_+, U_-] = 2U_3 \\ &[V_3, V_\pm] = \pm V_\pm && [V_+, V_-] = 2V_3 \end{aligned} \tag{4.8a}$$

plus other relations giving the commutators of isospin with U-spin and V-spin and of U-spin with V-spin. These can be evaluated directly from the matrix representations (4.4)–(4.6). The results are:

$$[I_3, U_3] = [U_3, V_3] = [V_3, I_3] = 0 \tag{4.8b}$$

$$[I_3, U_\pm] = \mp\tfrac{1}{2}U_\pm \qquad [I_3, V_\pm] = \mp\tfrac{1}{2}V_\pm$$

$$[U_3, V_\pm] = \mp\tfrac{1}{2}V_\pm \qquad [U_3, I_\pm] = \mp\tfrac{1}{2}I_\pm$$

$$[V_3, I_\pm] = \mp\tfrac{1}{2}I_\pm \qquad [V_3, U_\pm] = \mp\tfrac{1}{2}U_\pm \tag{4.8c}$$

$$[I_\pm, U_\pm] = \pm\tfrac{1}{2}V_\mp \qquad [U_\pm, V_\pm] = \pm\tfrac{1}{2}I_\mp \qquad [V_\pm, I_\pm] = \pm\tfrac{1}{2}U_\mp \tag{4.8d}$$

$$[I_\pm, U_\mp] = [U_\pm, V_\mp] = [V_\pm, I_\mp] = 0 \tag{4.8e}$$

Just as the angular momentum commutation relations determine the structure of the rotation group, so the twenty-eight commutation relations (4.8) determine the structure of SU(3).

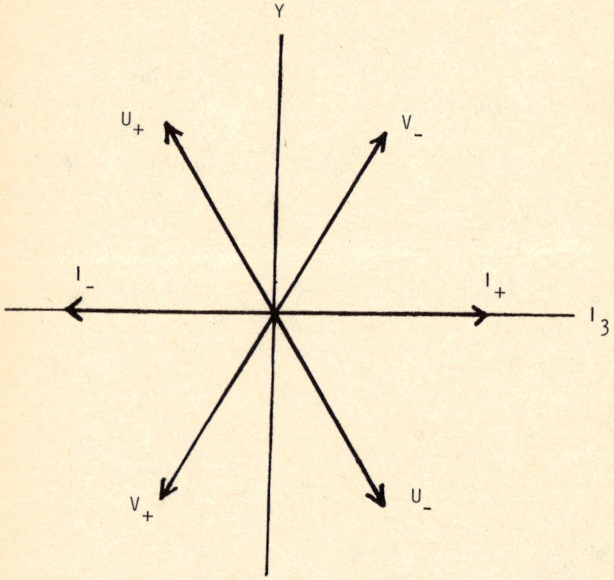

Fig.(4-1) Root vectors of SU(3).

In order to define a set of basis vectors for the group, however, we must choose the two diagonal generators. Since isosymmetry seems to be valid in the strong interactions, let us choose them in such a way that the SU(2) subgroup generated by **I** is maintained. Then I_3 is one diagonal generator. The other must be chosen to commute with **I**; that is, isosymmetry requires that its eigenvalue be the same for all states within an isomultiplet. Thus U_3, for example, is unacceptable because the commutation relation $[U_3, I_\pm] = \mp\frac{1}{2}I_\pm$ implies that $I_\pm$

changes the eigenvalue of U_3 as well as that of I_3. The only satisfactory choice is

$$I_8 \equiv \tfrac{2}{3}(U_3 - V_3) \tag{4.9}$$

(or any multiple thereof). From (4.8b) we have $[I_3, I_8] = 0$, and since

$$[(U_3 - V_3), I_\pm] = (\pm I_\pm \mp I_\pm) = 0 \tag{4.10}$$

it follows that $[I_8, \mathbf{I}] = 0$ and I_8 commutes with the entire isospin subgroup. The commutation relations of I_8 with $U_\pm$ and $V_\pm$ are obtained using (4.9):

$$[I_8, U_\pm] = \pm U_\pm$$

$$[I_8, V_\pm] = \mp V_\pm. \tag{4.8f}$$

The matrix representation of I_8 is

$$I_8 \doteq \tfrac{1}{3}\begin{pmatrix} 1 & 0 & 0 \\ 0 & 1 & 0 \\ 0 & 0 & -2 \end{pmatrix}; \tag{4.11}$$

the utility of the factor $\tfrac{1}{3}$ is that it allows us to write a simple relation between I_8 and the hypercharge, namely

$$Y = I_8 + \tfrac{2}{3}B \tag{4.12}$$

where B(=1) is the baryon number.

With I_3 and I_8 as the diagonalized generators, the basis states will occur naturally in isomultiplets. It follows from (4.10), in fact, that if we define $I^2 = I_i I_i$, exactly as in SU(2), then $[I^2, I_8] = [I^2, I_3] = 0$. Thus there are actually three mutually commuting operators, I_3, I_8, and I^2, and the basis states can be taken as simultaneous eigenstates of all three. This does *not* imply that I^2 is a Casimir operator of SU(3), of course; it does not commute with U-spin or V-spin. It is a Casimir operator of the conserved SU(2) subgroup, however, and that is why it can be diagonalized. The existence of additional diagonal operators such as I^2 reflects the fact that, as we pointed out in Chapter 2, in groups of higher rank there may be more than one state having the same eigenvalues for the diagonal generators.

The choice of isospin as the conserved subgroup was prompted by physics, not be group theory; we could equally well choose U-spin or V-spin, and in certain applications they are more appropriate. To maintain U-spin, for example, U_3 and

$$U_8 \equiv \tfrac{2}{3}(V_3 - I_3) \tag{4.13}$$

must be chosen as the diagonal generators. Then $[\mathbf{U}, U_8] = 0$, and $\mathbf{U}^2 = U_i U_i$ can also be diagonalized. The basis states will occur naturally in U-multiplets rather than isomultiplets. By using (4.7), (4.9), and (4.12) it can be shown that

$$U_8 = \tfrac{1}{3}B - (I_3 + \tfrac{1}{2}Y); \tag{4.14}$$

The Gell-Mann–Nishijima formula $Q = e(I_3 + \tfrac{1}{2}Y)$ is therefore equivalent to

$$Q = e(\tfrac{1}{3}B - U_8). \tag{4.15}$$

Constant U_8 thus implies constant charge; consequently all members of a U-multiplet have the same electromagnetic interactions.

V-spin can be treated similarly, using as diagonal operators V_3 and $V_8 = \tfrac{2}{3}(I_3 - U_3) = -(I_8 + U_8)$. No simple physical connection comparable to (4.12) and (4.15) seems to exist in this case, however, and we shall not explore it further.

In order to identify a state in SU(3) we must specify the eigenvalues of five operators – two Casimir invariants identifying the representation, two diagonal generators, and an SU(2) invariant. Since the first two are the same for every member in a given irreducible representation, let us omit them for the present and assume that the isospin SU(2) group is chosen. Then the basis eigenstates may be denoted by $|I, I_3, I_8\rangle$, with

$$I^2|I, I_3, I_8\rangle = I(I+1)|I, I_3, I_8\rangle$$

$$I_3|I, I_3, I_8\rangle = I_3|I, I_3, I_8\rangle$$

$$I_8|I, I_3, I_8\rangle = I_8|I, I_3, I_8\rangle \tag{4.16a}$$

Furthermore, the usual results for $I_\pm$ are immediate:

$$I_\pm|I, I_3, I_8\rangle = \sqrt{I(I+1) - I_3(I_3 \pm 1)}\,|I, I_3 \pm 1, I_8\rangle. \tag{4.16b}$$

Equations defining the effect of the other operators on $|I, I_3, I_8\rangle$ may be obtained from the definitions and commutations relations. For example, since $U_3 = \tfrac{1}{4}(3I_8 - 2I_3)$ and $U_8 = -(I_3 + \tfrac{1}{2}I_8)$, we have

$$U_3|I, I_3, I_8\rangle = \tfrac{1}{4}(3I_8 - 2I_3)|I, I_3, I_8\rangle$$

$$U_8|I, I_3, I_8\rangle = -(I_3 + \tfrac{1}{2}I_8)|I, I_3, I_8\rangle. \tag{4.17a}$$

The action of $U_\pm$ is obtained from the commutators $[I_3, U_\pm] = \mp\tfrac{1}{2}U_\pm$ and. $[I_8, U_\pm] = \pm U_\pm$, which imply that

$$I_3U_\pm|I, I_3, I_8\rangle = (I_3 \mp \tfrac{1}{2})U_\pm|I, I_3, I_8\rangle$$

$$I_8U_\pm|I, I_3, I_8\rangle = (I_8 \pm 1)U_\pm|I, I_3, I_8\rangle$$

Therefore $U_\pm|I, I_3, I_8\rangle$ is an eigenstate of I_3 and I_8 with eigenvalues $(I_3 \mp \tfrac{1}{2})$ and $(I_8 \pm 1)$. It is easily shown, however, that it is *not* generally an eigenstate of $\mathbf{I}^2$. Consequently we can only say that

$$U_\pm|I, I_3, I_8\rangle = \sum_{I'} u_\pm(I, I', I_3, I_8)|I', I_3 \mp \tfrac{1}{2}, I_8 \pm 1\rangle, \tag{4.17b}$$

where the coefficient $u_\pm(I, I', I_3, I_8)$ and the allowed values of I' depend on the irreducible representation to which $|I, I_3, I_8\rangle$ belongs. Similarly it can be shown that

$$V_3|I, I_3, I_8\rangle = (-\tfrac{1}{2}I_3 - \tfrac{3}{4}I_8)|I, I_3, I_8\rangle$$

$$V_8|I, I_3, I_8\rangle = (I_3 - \tfrac{1}{2}I_8)|I, I_3, I_8\rangle$$

$$V_\pm|I, I_3, I_8\rangle = \sum_{I'} v_\pm(I, I', I_3, I_8)|I', I_3 \mp \tfrac{1}{2}, I_8 \mp 1\rangle \tag{4.18}$$

The determination of $u_\pm$ and $v_\pm$ is a difficult problem in general, and we shall not enter into it here.† It should be noticed, however that (4.17) and (4.18) imply that integral and half-integral values of I must alternate as I_8 varies by integer steps.

Generators and Casimir Invariants

The eight hermitian generators of SU(3) are thus given essentially by the eight operators I_1, I_2, U_1, U_2, V_1, V_2, I_3, and I_8. In fact, the infinitesimal generators

† General formulae for $u_\pm$ and $v_\pm$ can be obtained from the article by L. C. Biedenharn in reference 3.

as defined in (2.38) are usually taken to be simple multiples of these operators, namely

$$
\begin{aligned}
X_1 &= I_1 & X_5 &= -V_2 \\
X_2 &= I_2 & X_6 &= U_1 \\
X_3 &= I_3 & X_7 &= U_2 \\
X_4 &= V_1 & X_8 &= \sqrt{3/2}\, I_8 .
\end{aligned}
\tag{4.19}
$$

With this definition of the operators X_α, the commutation relations (4.8) can be put in the form (2.38), yielding

$$[X_\alpha, X_\beta] = i f_{\alpha\beta\gamma} X_\gamma ; \tag{4.20}$$

the $f_{\alpha\beta\gamma}$ are real, totally antisymmetric, and vanish except for the following and their permutations:

$$
\begin{aligned}
f_{123} &= 1 & f_{246} &= \tfrac{1}{2} & f_{367} &= -\tfrac{1}{2} \\
f_{147} &= \tfrac{1}{2} & f_{257} &= \tfrac{1}{2} & f_{458} &= \sqrt{3/2} \\
f_{156} &= -\tfrac{1}{2} & f_{345} &= \tfrac{1}{2} & f_{678} &= \sqrt{3/2}.
\end{aligned}
\tag{4.21}
$$

As we have stressed, the commutation relations (4.20) characterize the SU(3) algebra generally. Although they were derived using the three-dimensional elementary representation, we expect them to hold for *every* representation of SU(3), with the same structure constants $f_{\alpha\beta\gamma}$ given in (4.21).

It is also possible to consider the anticommutators $\{X_\alpha, X_\beta\} \equiv X_\alpha X_\beta + X_\beta X_\alpha$ of the X's. Representing them appropriately by matrices obtained from equations (4.4)–(4.6), we find that for the elementary representation

$$\{X_\alpha, X_\beta\} = \tfrac{1}{3}\delta_{\alpha\beta} + d_{\alpha\beta\gamma} X_\gamma \tag{4.22}$$

with $d_{\alpha\beta\gamma}$ real and totally symmetric; the non-vanishing $d_{\alpha\beta\gamma}$ are the permutations of

$$
\begin{aligned}
d_{118} &= 1/\sqrt{3} & d_{297} &= -\tfrac{1}{2} & d_{355} &= \tfrac{1}{2} & d_{558} &= -1/(2\sqrt{3}) \\
d_{146} &= \tfrac{1}{2} & d_{256} &= \tfrac{1}{2} & d_{366} &= -\tfrac{1}{2} & d_{668} &= -1/(2\sqrt{3})
\end{aligned}
$$

$d_{157} = ½$ $\quad d_{338} = 1/\sqrt{3}$ $\quad d_{377} = -½$ $\quad d_{778} = -1/(2\sqrt{3})$

$$d_{228} = 1/\sqrt{3} \quad d_{344} = ½ \quad d_{448} = -1/(2\sqrt{3}) \quad d_{888} = -1/\sqrt{3}. \tag{4.23}$$

The anticommutation relations (4.22) cannot be generalized to hold for all representations, however; this result is valid only for the elementary representation. Despite its lack of generality, however, (4.22) is nonetheless important because it defines the $d_{\alpha\beta\gamma}$, which will be used in connection with the Casimir invariants.

From (4.20) and (4.22), or directly from (4.21) and (4.23), some interesting relations among the f's and d's can be obtained. For example, it can be shown that $f_{\alpha\beta\gamma}$ is totally antisymmetric by noting that, since the X_α are traceless, (4.22) implies that $\mathrm{Tr}(X_\alpha X_\beta) = ½\delta_{\alpha\beta}$. Furthermore $f_{\alpha\beta\gamma}$ is clearly antisymmetric in α and β, from (4.20), and

$$\begin{aligned}\mathrm{Tr}(X_\alpha, [X_\beta, X_\gamma]) &= \mathrm{Tr}(X_\alpha X_\beta X_\gamma - X_\alpha X_\gamma X_\beta)\\ &= \mathrm{Tr}([X_\alpha, X_\beta] X_\gamma)\end{aligned} \tag{4.24}$$

since Tr(AB) = Tr(BA). But (4.24) implies immediately that

$$\mathrm{Tr}(f_{\beta\gamma\delta} X_\alpha X_\delta) = \mathrm{Tr}(f_{\alpha\beta\delta} X_\delta X_\gamma)$$

i.e.,

$$f_{\beta\gamma\alpha} = f_{\alpha\beta\gamma} = -f_{\beta\alpha\gamma}. \tag{4.25}$$

In the same way, it is easily shown that $d_{\alpha\beta\gamma}$ is totally symmetric. A number of other relations can be obtained similarly by manipulating commutators, anticommutators and traces, e.g.

$$f_{\alpha\beta\gamma} f_{\alpha\beta\delta} = 3\delta_{\gamma\delta} \tag{4.26a}$$

$$d_{\alpha\beta\gamma} d_{\alpha\beta\delta} = \tfrac{5}{3}\delta_{\gamma\delta} \tag{4.26b}$$

$$d_{\beta\gamma\lambda} f_{\alpha\lambda\delta} + f_{\alpha\lambda\gamma} d_{\beta\lambda\delta} + f_{\beta\alpha\lambda} d_{\lambda\gamma\delta} = 0 \tag{4.26c}$$

$$f_{\alpha\beta\lambda} f_{\lambda\gamma\delta} + f_{\lambda\gamma\lambda} f_{\lambda\alpha\delta} + f_{\gamma\alpha\lambda} f_{\lambda\beta\delta} = 0. \tag{4.26d}$$

The last of these is the Jacobi identity (2.39).

We note that the two Casimir invariants of SU(3) can be constructed using the $f_{\alpha\beta\gamma}$ and $d_{\alpha\beta\gamma}$. By analogy with $J^2 = J_i J_i$, we might expect that one of the two will be given by

$$X^2 \equiv X_\alpha X_\alpha. \tag{4.27}$$

This is indeed the case, since

$$[X_\alpha X_\alpha, X_\beta] = if_{\alpha\beta\gamma}(X_\alpha X_\gamma + X_\gamma X_\alpha) = 0$$

because of the antisymmetry for $f_{\alpha\beta\gamma}$. In terms of **I**, **U**, and **V**, it can be shown that

$$X^2 = \mathbf{I}^2 + \mathbf{U}^2 + \mathbf{V}^2 - \tfrac{1}{3}(I_3^2 + U_3^2 + V_3^2). \tag{4.28}$$

By using (4.26a), X^2 can also be written

$$X^2 = 2/(3i)\, f_{\alpha\beta\gamma} X_\alpha X_\beta X_\gamma, \tag{4.29}$$

which suggests the form of the second invariant,

$$X^3 \equiv d_{\alpha\beta\gamma} X_\alpha X_\beta X_\gamma. \tag{4.30}$$

That $[X^3, X_\lambda] = 0$ follows from (4.26c).

Chapter 5

PARTICLE CLASSIFICATION IN SU(3)

If the Hamiltonian describing the strong interactions is approximately invariant under the transformations of SU(3), then the hadrons should occur naturally in multiplets corresponding to its irreducible representations. The structure of these representations is determined by the two Casimir invariants in essentially the same way as in O(3). The rotation group, being of rank 1, has a single Casimir invariant $\mathbf{J}^2$ and a one-dimensional representation space, and the determination of its irreducible representations is correspondingly simple. SU(3) is only of rank 2, but the manipulations involved in obtaining its irreducible representations are much more than twice as cumbersome as those in O(3).

Four representations are already known to us or easily constructed, however, and we shall discuss these briefly before describing the general structure. One is the trivial singlet representation in which all of the generators are represented by zero and all SU(3) operators by unity. This representation is one-dimensional; its single eigenstate has $I = I_3 = I_8 = 0$. A second irreducible representation is the three-dimensional representation used in the last chapter. Its three eigenstates correspond to (p, n, Λ°), and are plotted in representation space in Fig.(5-1a).

A third irreducible representation may always, as we pointed out in Chapter 2, be constructed by simply using the structure constants $f_{\alpha\beta\gamma}$. Suppose we define a set of eight matrices R_α by $(R_\alpha)_{\beta\gamma} = -if_{\alpha\beta\gamma}$. Then, since Jacobi's identity requires that

$$f_{\alpha\beta\lambda}f_{\lambda\gamma\delta} + f_{\gamma\alpha\lambda}f_{\lambda\beta\delta} + f_{\beta\gamma\lambda}f_{\lambda\alpha\delta} = 0, \qquad (4.26d)$$

we have, on using the antisymmetry of $f_{\alpha\beta\gamma}$,

$$(R_\alpha R_\delta)_{\beta\gamma} - (R_\delta R_\alpha)_{\beta\gamma} = if_{\alpha\delta\lambda}(R_\lambda)_{\beta\gamma},$$

i.e.

$$[R_\alpha, R_\delta] = if_{\alpha\delta\lambda}R_\lambda \, . \qquad (5.1)$$

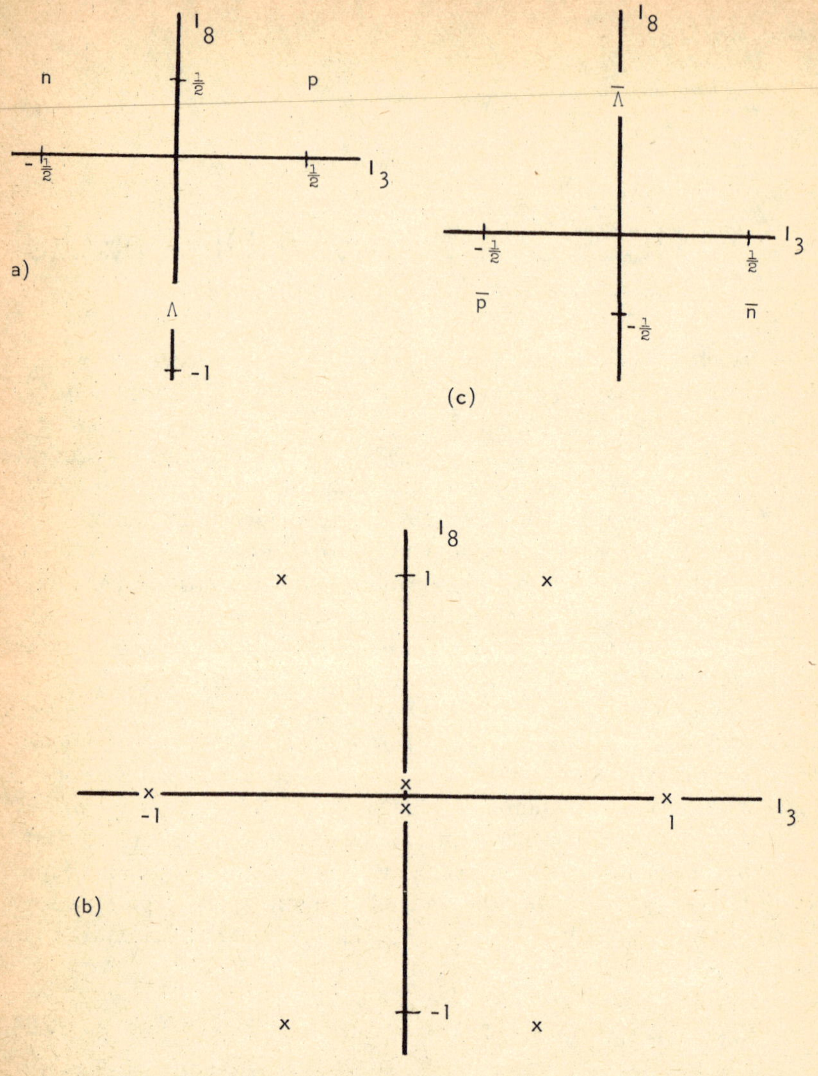

Fig.(5-1) Eigenstates of (a) the elementary representation 3, (b) the regular representation 8, and (c) the conjugate representation $\bar{3}$.

Therefore the matrices R_a provide a representation, which in fact is irreducible. All of the R_a are antisymmetric, so none of them are diagonal; but by a unitary transformation $UR_aU^\dagger \equiv F_a$ they can be transformed into an equivalent representation in which F_3 and F_8 are diagonal, namely

$$F_3 = \begin{pmatrix} \tfrac{1}{2} & & & & & & & \\ & -\tfrac{1}{2} & & & & & 0 & \\ & & 1 & & & & & \\ & & & 0 & & & & \\ & & & & -1 & & & \\ & & & & & 0 & & \\ & 0 & & & & & \tfrac{1}{2} & \\ & & & & & & & -\tfrac{1}{2} \end{pmatrix} \quad F_8 = \begin{pmatrix} 1 & & & & & & & \\ & 1 & & & & & 0 & \\ & & 0 & & & & & \\ & & & 0 & & & & \\ & & & & 0 & & & \\ & & & & & 0 & & \\ & 0 & & & & & -1 & \\ & & & & & & & -1 \end{pmatrix} \tag{5.2}$$

This octet representation is called the regular representation; its eigenstates are shown in representation space in Fig.(5-1b). It should also be noted that the $d_{\alpha\beta\gamma}$ can be used to construct a set of eight matrices T_α by $(T_\alpha)_{\beta\gamma} = -id_{\alpha\beta\gamma}$. It follows then from (4.26d) that

$$[R_\alpha, T_\beta] = if_{\alpha\beta\gamma}T_\gamma,$$

or defining $D_\alpha \equiv UT_\alpha U^\dagger$ as above, that

$$[F_\alpha, D_\beta] = if_{\alpha\beta\gamma}D_\gamma. \tag{5.3}$$

Since the $f_{\alpha\beta\gamma}$ provide the regular representation, it follows from (5.3) and the definition (2.45) that the D_α form an irreducible tensor operator transforming according to this representation.

Finally, we know that every representation has a complex conjugate representation; if a set of matrices X_α satisfy

$$[X_\alpha, X_\beta] = if_{\alpha\beta\gamma}X_\gamma,$$

then if we define X^c_α by $(X^c_\alpha)_{\beta\gamma} = -(X_\alpha)^*_{\beta\gamma}$, it follows that also

$$[X^c_\alpha, X^c_\beta] = if_{\alpha\beta\gamma}X^c_\gamma, \tag{5.4}$$

In SU(2) we saw that every representation was equivalent to its complex conjugate, the transformation between them being essentially the G-parity. In SU(3), however, this is not necessarily so. Clearly the singlet representation is self-conjugate, and the reality of $f_{\alpha\beta\gamma}$ can be used to show that the regular representation is equivalent to its conjugate. The elementary representation is distinct from its conjugate, however. This can be seen by noting that I_8 is real and $\det(I_8) = -2/27$ in this representation; but $I^c_8 = -I^*_8 = -I_8$, so $\det(I^c_8) = +2/27$. An equivalence transformation cannot change the determinant, since det $(UI_8U^\dagger)$ = det $(I_8U^\dagger U$ = det (I_8). Therefore I_8^c cannot be equivalent to I_8. It follows that there are two inequivalent three-dimensional representations.

The complex conjugate representation here has

$$I_3^c = -I_3 = \tfrac{1}{2}\begin{pmatrix} -1 & & \\ & 1 & \\ & & 0 \end{pmatrix}, \quad I_8^c = -I_8 = \tfrac{1}{3}\begin{pmatrix} -1 & & \\ & -1 & \\ & & 2 \end{pmatrix} \tag{5.5}$$

so its eigenstates, plotted in Fig.(5-1c), have exactly the quantum, numbers of the antibaryons $(\bar{p}, \bar{n}, \bar{\Lambda}^\circ)$. In order to make (5.5) resemble the (p, n, Λ°) representation it has become conventional to make a similarity trnasformation changing the basis states $(\bar{p}, \bar{n}, \bar{\Lambda}^\circ)$ to $(\bar{n}, -\bar{p}, \bar{\Lambda}^\circ)$. The first two states are then the antiparticle isodoublet as defined in (3.16). The matrices for $I_\pm$, $U_\pm$, and $V_\pm$ in this representation are given by

$$I_+ = \begin{pmatrix} 0 & 1 & 0 \\ 0 & 0 & 0 \\ 0 & 0 & 0 \end{pmatrix} \qquad I_- = \begin{pmatrix} 0 & 0 & 0 \\ 1 & 0 & 0 \\ 0 & 0 & 0 \end{pmatrix}$$

$$U_+ = -\begin{pmatrix} 0 & 0 & 0 \\ 0 & 0 & 1 \\ 0 & 0 & 0 \end{pmatrix} \qquad U_- = -\begin{pmatrix} 0 & 0 & 0 \\ 0 & 0 & 0 \\ 0 & 1 & 0 \end{pmatrix}$$

$$V_+ = \begin{pmatrix} 0 & 0 & 0 \\ 0 & 0 & 0 \\ 1 & 0 & 0 \end{pmatrix} \qquad V_- = \begin{pmatrix} 0 & 0 & 1 \\ 0 & 0 & 0 \\ 0 & 0 & 0 \end{pmatrix} \tag{5.6}$$

The minus signs in $U_\pm$ should be noted; they are the opposite of the usual Condon–Shortley sign convention (2.20) for raising and lowering operators. No equivalence transformation can be made which will avoid the appearance of these minus signs in one of the SU(2) subgroups. Changing to $(\bar{n}, -\bar{p}, -\bar{\Lambda}^\circ)$, for example, only transfers them to $V_\pm$. Phase difficulties are therefore to be expected in SU(3), and great care is required to ensure that a consistent set of phases is used throughout any calculation.

Other irreducible representations can be constructed using techniques based on permutation symmetry. The details are given in Appendix A; here we shall only describe, and justify where possible, some of the important conclusions. Much can be said about the *shape* of an irreducible representation (that is, of the plot of its eigenstates in representation space) as a consequence of the three SU(2) subgroups of SU(3). For example, since all states occur as isomultiplets, the representations must be symmetric about the I_8 axis. Furthermore, the commutation relations (4.8) are not affected by cyclic permutations of (I, U, V), so representations must be symmetric under these transformations. It follows that they must also be symmetric about the U_8 and V_8 axes. In Fig.(5-2) we show the representation space with the three sets of axes (I_3, I_8), (U_3, U_8), and (V_3, V_8). All representations must have threefold reflection symmetry in this space.

In fact, it can be shown that every irreducible representation must have pre-

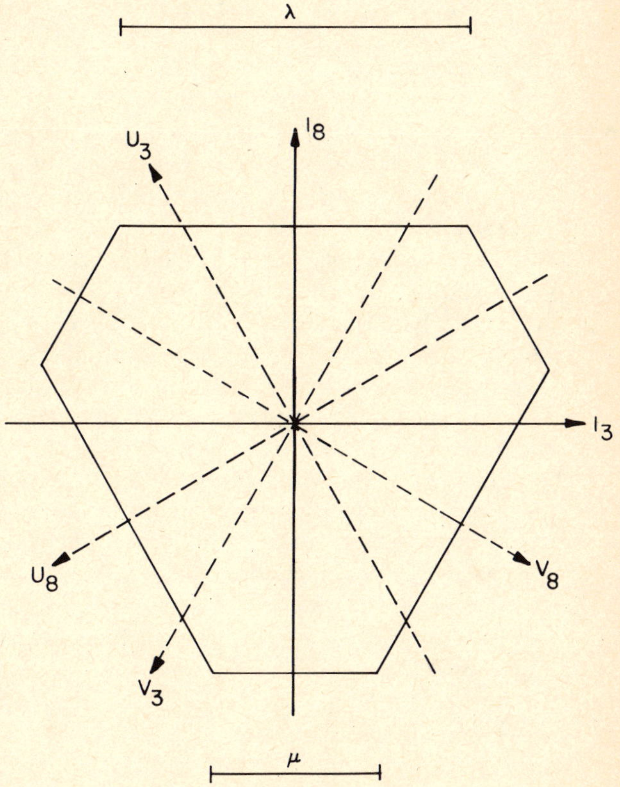

Fig.(5-2) Possible sets of axes in SU(3) configuration space, showing the symmetric shape of an irreducible representation.

cisely the hexagonal shape shown in Fig.(5-2). The lengths of the sides of the hexagon may be specified by λ and μ, as shown; specifically, $\lambda = 2I_{max}$ and $\mu = 2I_{min}$, where I_{max} and I_{min} are the total isospins corresponding to the maximum and minimum values of I_8 within the representation. Giving these two parameters specifies an irreducible representation uniquely; for every pair of integers (λ, μ) there is exactly one irreducible representation. We have said that the Casimir invariants label the irreducible representations, so it should be expected that λ and μ are related to X^2 and X^3. This is indeed the case. Suppose we apply X^2 to the state with $I_3 = I = \frac{1}{2}\lambda$ and $I_8 = I_{max}$. Using (4.28), we find that the eigenvalue of X^2 is

$$c_1 = \tfrac{1}{3}(\lambda^2 + \mu^2 + \lambda\mu + 3(\lambda + \mu)). \tag{5.7}$$

Evaluating X^3 is more difficult; the result is an eigenvalue

$$c_2 = \tfrac{1}{18}(\lambda - \mu)(2\lambda + \mu + 3)(\lambda + 2\mu + 3). \tag{5.8}$$

Clearly it is more economical to label an irreducible representation by (λ, μ) than by the eigenvalues c_1 and c_2. (Recall that in O(3) we use j rather than j(j + 1).)

In addition to the shape of the representation it is necessary to know the number of independent states occurring at each value of I_3 and I_8. The answer is that for a given value of I_8, the total isospin ranges by integer steps from $I_< = \frac{1}{2}|I_8 - \frac{2}{3}(\mu - \lambda)|$ up to $I_> = \frac{1}{2}(\lambda + \mu) - \frac{1}{2}|I_8 - \frac{1}{3}(\lambda - \mu)|$, a result which is most easily pictured by the following algorithm: The outside states are unique. At each step toward the origin, the number of independent states increases by one, until the hexagon shrinks to a triangle; thereafter the number of states remains constant. For example, let us consider the irreducible representation (6,3), shown in Fig.(5-3). At the outside edge there is a single state. One step toward the center, there is a hexagon with sides $\lambda' = 5$, $\mu' = 2$, on which there are two states. Another step in, there are three states; but after a third step, the hexagon has become a triangle, so all of the remaining points have four independent states. The representation (6,3) therefore has a total of 120 eigenstates, i.e. it is a 120-dimensional representation.

A similar construction can be made for an arbitrary representation (λ, μ), with the result that its dimensionality is given by

$$N(\lambda, \mu) = \tfrac{1}{2}(\lambda + 1)(\mu + 1)(\lambda + \mu + 2) \tag{5.9}$$

It is apparent in (5.9) that $N(\lambda, \mu) = N(\mu, \lambda)$. In fact (μ, λ) is the complex conjugate representation of (λ, μ), so we would expect them to have the same dimensionality. With this exception, however, (5.9) will ordinarily (although not always)

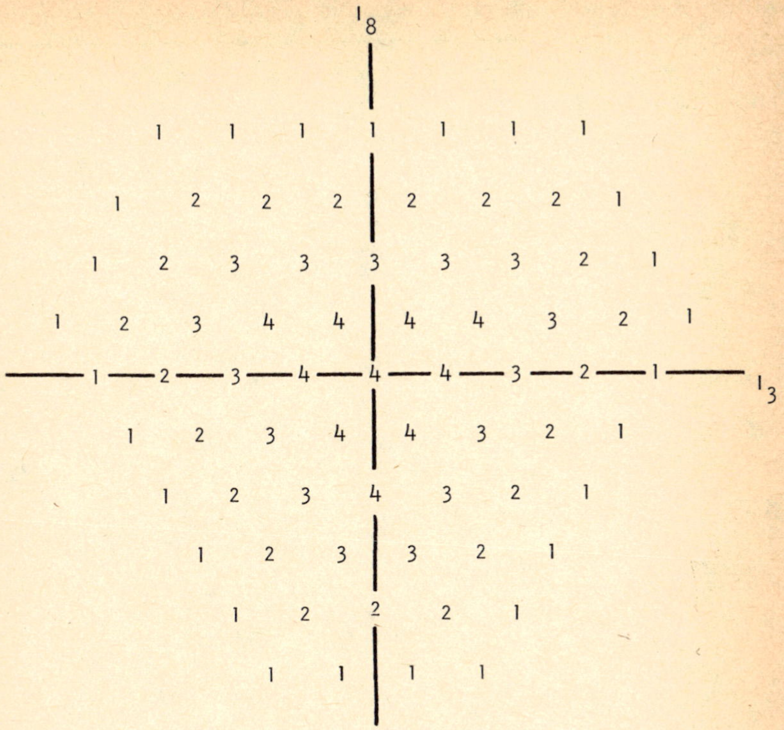

Fig.(5-3) Multiplicities of the eigenstates of the (6,3) representation. Each state is labeled by the appropriate value.

lead to different dimensionalities for different irreducible representations. For this reason it has become common practice to label an irreducible representation (λ, μ) with $\lambda \geqslant \mu$ by simply giving its dimensionality $n = N(\lambda, \mu)$; the complex conjugate representation (μ, λ) is identified by a bar, i.e. $\overline{n}$. Thus the four representations discussed above are called the 1, 3, $\overline{3}$, and 8. The eigenstates are therefore appropriately labeled by giving n, the eigenvalue I of total isospin, I_3, and I_8, so we shall hereafter write them in the form $|n, I, I_3, I_8\rangle$.

Table 5-A lists a number of the smaller irreducible representations, including the ones which will be of interest in the following chapters.

The Sakata Model

The model on which we have based our derivation of SU(3), using the baryon triplet (p, n, Λ°) as the elementary 3 representation, was first proposed by

TABLE 5-A Some irreducible representations of SU(3)

Dimensionality n	Young Diagram (App. A1)	(λ, μ)	Triality t
3		(1, 0)	1/3
$\overline{3}$		(0, 1)	2/3
6		(2, 0)	2/3
8		(1, 1)	0
10	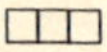	(3, 0)	0
15	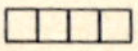	(4, 0)	1/3
15′	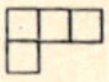	(2, 1)	1/3
21	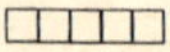	(5, 0)	2/3
24	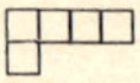	(3, 1)	2/3
27	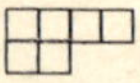	(2, 2)	0
28	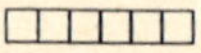	(6, 0)	0
35	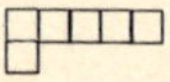	(4, 1)	0

Sakata: it is thus generally called the Sakata model, and (p, n, Λ°) are often referred to as "sakatons" in this context. In SU(2) we pictured the non-strange mesons as nucleon–antinucleon bound states; now we assume that both strange and non-strange mesons are analogously composed of sakaton–antisakaton pairs. It follows that they should be classified in irreducible representations which occur in the direct product $3 \otimes \overline{3}$. Likewise, the remaining baryons are considered sakaton–meson combinations, and therefore should occur in multiplets resulting from the reduction of $3 \otimes (3 \otimes \overline{3})$.

The general techniques needed to reduce a direct product of SU(3) irreducible representations are quite complicated, and we shall not give them here. For simple cases, however, it is not difficult to find the correct reduction formula. The procedure involved is analogous to that used in Chapter 2 for the rotation group. In combining $j_1 \otimes j_2$, we found a maximal value of $m = j_1 + j_2$, which implied the presence of the irreducible representation $j = j_1 + j_2$. Removing it left one state with $m = j_1 + j_2 - 1$, so $j = j_1 + j_2 - 1$ was also present. Stripping away successively the largest irreducible representation remaining led eventually to the reduction

$$j_1 \otimes j_2 = (j_1 + j_2) \oplus (j_1 + j_2 - 1) \oplus \ldots \oplus |j_1 - j_2|. \tag{5.10}$$

In SU(3) we can apply the same technique.

As an example, the direct product $3 \otimes \overline{3}$ contains nine independent sakaton–antisakaton states. Three of these – $p\overline{p}$, $n\overline{n}$, and $\Lambda^\circ\overline{\Lambda}^\circ$ – appear at the origin in representation space, as shown in Fig.(5-4). The remaining six states correspond exactly with the outside states of the 8. An 8 must therefore be present; anything larger would not be filled. Since the octet includes only two linearly independent

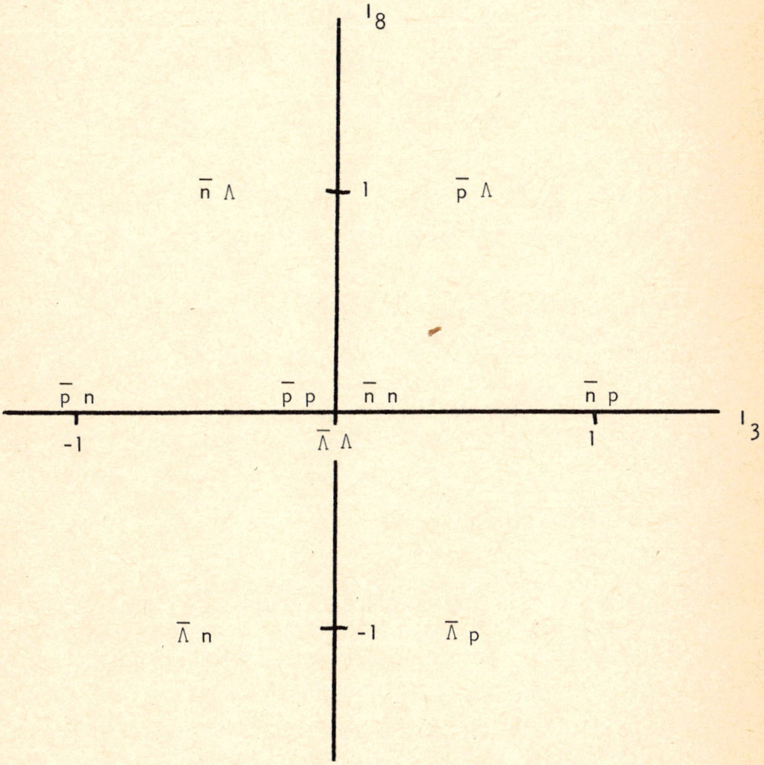

Fig.(5-4) Sakaton–antisakaton states in configuration space.

states at the origin, the third state is clearly a singlet. Thus we obtain the reduction formula

$$3 \otimes \overline{3} = 8 \oplus 1. \tag{5.11a}$$

For the states away from the origin, the Clebsch–Gordan coefficients corresponding to (5.11) are trivial except for phases:

$$|8, 1, 1, 0\rangle = |p\rangle|\overline{n}\rangle \qquad |8, 1, -1, 0\rangle = -|n\rangle|\overline{p}\rangle$$

$$|8, \tfrac{1}{2}, \tfrac{1}{2}, 1\rangle = |p\rangle|\overline{\Lambda}^\circ\rangle \qquad |8, \tfrac{1}{2}, -\tfrac{1}{2}, 1\rangle = |n\rangle|\overline{\Lambda}^\circ\rangle$$

$$|8, \tfrac{1}{2}, \tfrac{1}{2}, -1\rangle = |\Lambda^\circ\rangle|\overline{n}\rangle \qquad |8, \tfrac{1}{2}. -\tfrac{1}{2}, -1\rangle = -|\Lambda^\circ\rangle|\overline{p}\rangle. \tag{5.12a}$$

The states with $I_3 = I_8 = 0$ are not so simple. The $I = 1$ state is easily obtained by applying I_+ to $|8, 1, -1, 0\rangle$, yielding the SU(2) result

$$|8, 1, 0, 0\rangle = \frac{1}{\sqrt{2}} [|n\rangle|\overline{n}\rangle - |p\rangle|\overline{p}\rangle]. \tag{5.12b}$$

To obtain the other two states, we note first that the singlet representation must be invariant under permutations of (p, n, Λ°), so it must be the symmetric combination,

$$|1, 0, 0, 0\rangle = \frac{1}{\sqrt{3}} [|p\rangle|\overline{p}\rangle + |n\rangle|\overline{n}\rangle + |\Lambda^\circ\rangle|\overline{\Lambda}^\circ\rangle]. \tag{5.12c}$$

The third state is orthogonal to these two, so it can only be

$$|8, 0, 0, 0\rangle = \frac{1}{\sqrt{6}} [|p\rangle|\overline{p}\rangle + |n\rangle|\overline{n}\rangle - 2|\Lambda^\circ\rangle|\overline{\Lambda}^\circ\rangle]. \tag{5.12d}$$

Equations (5.12) provide the definitions of the Clebsch–Gordan coefficients for the reduction $3 \otimes \overline{3} = 8 \oplus 1$.

This stripping technique can be applied to the reduction of any direct product of SU(3) irreducible representations. In general, however, the results cannot be expressed as neatly as in (5.10). In particular, it is possible for a given representation to appear more than once in a reduction; for example,

$$3 \otimes 3 \otimes 3 = 10 \oplus 8 \oplus 8 \oplus 1 \tag{5.11b}$$

contains the octet twice. More complete discussions of general reduction techniques can be found in the references cited in the Reading List. The remaining reductions which will be of particular importance to us are

$$3 \otimes 8 = 15 \oplus \overline{6} \oplus 3 \tag{5.11c}$$

$$8 \otimes 8 = 27 \oplus 10 \oplus \overline{10} \oplus 8 \oplus 8 \oplus 1 \tag{5.11d}$$

$$8 \otimes 10 = 35 \oplus 27 \oplus 10 \oplus 8 \tag{5.11e}$$

$$10 \otimes 10 = 35 \oplus 28 \oplus 27 \oplus \overline{10}. \tag{5.11f}$$

The Clebsch–Gordan coefficients of SU(3) are defined by writing an abbreviated form

$$|\mathrm{n}_\gamma, \nu\rangle = \sum_{\nu_1\nu_2} \begin{pmatrix} \mathrm{n}' & \mathrm{n}'' & \mathrm{n}_\gamma \\ \nu' & \nu'' & \nu \end{pmatrix} |\mathrm{n}', \nu'\rangle|\mathrm{n}'', \nu''\rangle, \tag{5.13}$$

where n, n$'$, and n$''$ denote the irreducible representations involved, while ν stands for all three of the quantum numbers I, I_3, and I_8, and similarly ν' and ν''. The label γ is included to allow for the possibility that the representation n appears more than once in the reduction, as in (5.11b).

The presence of the isospin subgroup allows us to separate

$$\begin{pmatrix} \mathrm{n}' & \mathrm{n}'' & \mathrm{n}_\gamma \\ \nu' & \nu'' & \nu \end{pmatrix}$$

into the product of an SU(2) Clebsch–Gordán coefficient and an "isoscalar factor" which is independent of I_3, I_3', and I_3'',

$$\begin{pmatrix} \mathrm{n}' & \mathrm{n}'' & \mathrm{n}_\gamma \\ \nu' & \nu'' & \nu \end{pmatrix} = \langle I'I_3'I''I_3''|II_3\rangle \left(\begin{matrix} \mathrm{n}' & \mathrm{n}'' \\ I'I_8' & I''I_8'' \end{matrix} \,\middle|\, \begin{matrix} \mathrm{n}_\gamma \\ II_8 \end{matrix}\right). \tag{5.14}$$

The proof of (5.14) follows directly on considering (5.13) for different states in the same isomultiplet. Tables of the isoscalar factors

$$\left(\begin{matrix} \mathrm{n}' & \mathrm{n}'' \\ I'I_8' & I''I_8'' \end{matrix} \,\middle|\, \begin{matrix} \mathrm{n}_\gamma \\ II_8 \end{matrix}\right)$$

are given by de Swart (see ref.3); by using them along with known SU(2)

Clebsches $\langle I'I_3'I''I_3''|II_3\rangle$, any desired SU(3) Clebsch–Gordan coefficient can be obtained.

The meson classifications in the Sakata model are based on $3 \otimes \overline{3}$, so (5.11a) predicts that only octets and singlets of mesons will be observed. Their structure, in terms of sakaton–antisakaton pairs, will be given by wave functions (5.12). In a similar way, baryons pictured as sakaton–meson bound states will be classified in irreducible representations resulting from $3 \otimes 8$; from (5.11c), therefore, we expect them to occur in multiplets corresponding to 15, $\overline{6}$, and 3. If SU(3) were an exact symmetry of the strong interactions Hamiltonian, we would expect all eigenstates of a given irreducible representation to be degenerate, i.e. to have the same mass, spin, parity, etc. The symmetry is certainly not exact, since the Λ°-nucleon mass difference is far from negligible. The spin and parity, however, are apparently independent of SU(3) (at least for the sakatons), so we may assume that all members of a multiplet will have a common J^P.

Meson Classification in the Sakata Model

We therefore explore the possibility that mesons occur naturally only in singlets and octets. The octets will contain non-strange mesons with $I = 0$ and $I = 1$ and isodoublets with $S = +1$ and $S = -1$. Eight pseudoscalar ($J^P = 0^-$) mesons with exactly these quantum numbers are well known: the pions, the kaons, and the isosinglet $\eta^\circ(549)$. They are now accepted as the members of the P (for pseudoscalar) octet. To each of these particles a sakaton–antisakaton state may be assigned as in (5.12):

$$|K^+\rangle = |p\overline{\Lambda}^\circ\rangle \qquad |K^\circ\rangle = |n\overline{\Lambda}^\circ\rangle$$

$$|\overline{K}^\circ\rangle = |\Lambda^\circ\overline{n}\rangle \qquad |K^-\rangle = |\Lambda^\circ\overline{p}\rangle$$

$$|\pi^+\rangle = |p\overline{n}\rangle \qquad |\pi^-\rangle = |n\overline{p}\rangle$$

$$|\pi^\circ\rangle = \frac{1}{\sqrt{2}}\,[|n\overline{n}\rangle - |p\overline{p}\rangle] \tag{5.15}$$

$$\eta^\circ = \frac{1}{\sqrt{6}}\,[|p\overline{p}\rangle + |n\overline{n}\rangle - 2|\Lambda^\circ\overline{\Lambda}^\circ\rangle]$$

The masses of the kaons and the η° are more than triple the pion mass, so the symmetry is badly broken; but the classification of this octet is nonetheless a considerable success for SU(3). Table 3-A lists two other pseudoscalar mesons, both considerably heavier than the others: the $\eta'(958)$ and the E(1420). Both of these seem to be SU(3) singlets, although it is an open question which of them, if either, represents the singlet occurring in equation (5.11a).

A second octet, containing the vector mesons with $J^P = 1^-$ and therefore called the V octet, may also be identified. It includes the K* resonances as the strange isodoublets and the ρ as the non-strange isotriplet. For the non-strange isosinglet, however, there are two possibilities: the $\omega(783)$ and the $\phi(1020)$. Only experimental information can tell us which one to assign as the octet number and which as an SU(3) singlet. As we shall see in Chapter 6, it turns out that it is more appropriate to take both of these particles as *mixtures* of the octet and singlet states. That is, we assume that there is "ω–ϕ mixing"; the physical states are

$$|\omega\rangle = \cos\theta\,|8, 0, 0, 0\rangle + \sin\theta\,|1, 0, 0, 0\rangle$$

$$|\phi\rangle = -\sin\theta\,|8, 0, 0, 0\rangle + \cos\theta\,|1, 0, 0, 0\rangle, \tag{5.16}$$

with θ a "mixing angle". In the next chapter we shall show that this angle seems experimentally to be about 40°.

The same situation is repeated again for the tensor mesons with $J^P = 2^+$. The particles classified in the T octet are the $K_V(1420)$ resonances, the $A_2(1300)$, and as the two isoscalars the f(1260) and the f′(1514). The latter two, like ω and ϕ, are mixtures of octet and singlet, with a mixing angle of about 30°.

Thus it appears that the Sakata model leads to correct classification in SU(3) multiplets of at least three different octets. A number of other mesons have also been observed, but the spin-parity assignments of most of them are not yet sufficiently well determined to group them into complete multiplets.

Baryon Classification in SU(3)

Next we turn to the classification of baryon states in the Sakata model of SU(3). In SU(2) the nucleon resonances could be pictured as nucleon–pion states; here we assume similarly that the higher baryons are sakaton–meson combinations. The possible baryonic irreducible representations, in addition to the 3, are therefore those resulting from reduction of the direct product $3 \otimes 8 = 15 \oplus \overline{6} + 3$, as given in equation (5.14). We thus expect to classify baryons in the Sakata model according to the irreducible representations 15, $\overline{6}$, and 3.

Prominent in the baryon tables are five $J^P = \frac{1}{2}^+$ baryons with relatively low masses and long lifetimes: the Σ isotriplet, with strangeness -1, and the Ξ isodoublet with strangeness -2. In this model, the Ξ must be classified in a 15; the

Σ could go into a 6, but presumably would accompany the Ξ in a 15. If so, there should be ten other $J^P = \frac{1}{2}^+$ baryons filling the remaining states.

In particular, a baryon with strangeness +1 and isospin 1, corresponding to a $(\overline{K}N)$ state, should exist. The $\overline{6}$ requires likewise an isosinglet S = +1 state. No baryons with positive strangeness have been observed; the kaon–nucleon system seems to be completely free of low-mass resonances.

Furthermore, the other states occurring in 15 or in $\overline{6}$ do not correspond easily with known particles. It seems therefore that the Sakata model, which worked extremely well for mesons, cannot provide a correct classification of the baryons. This conclusion is reinforced by looking at other J^P values; all of them are lacking in positive strangeness states, and none occur in multiplets resembling either the 15 or the $\overline{6}$. A number of more sophisticated arguments lead to the same result: the Sakata model fails for the baryons.

In fact, the basic triplet (p, n, Λ°) of the Sakata model does not differ drastically from the five states listed above; like the Σ and the Ξ, they have $J^P = \frac{1}{2}^+$, low masses, and long lifetimes. What is more, these eight particles are identical with the states expected in an octet!

For this reason, Gell-Mann suggested in 1961 that the $\frac{1}{2}^+$ baryons, like the mesons, are appropriately classified into the octet representation of SU(3). Like the "Eightfold Way" to selflessness taught by the Buddha, there seems to be an "eightfold way" to understanding the hadrons, and Gell-Mann gave this name to the octet classification theory. It is now recognized as correct; indeed, at least three other octets, with $J^P = 3/2^-$, $5/2^+$ and $5/2^-$, seem to be present among the baryon resonances. The members of these octets are listed in Table 5-B.

In the eightfold way the identification of the hypercharge in terms of I_8, given in equation (4.11), is no longer correct. Instead, we have simply

$$Y = I_8, \qquad (5.17a)$$

and correspondingly for the charge,

$$Q = e(I_3 + \tfrac{1}{2} I_8) = -eU_8. \qquad (5.17b)$$

Substituting (5.17a) for the earlier relation (4.11) does not affect the meson classification, since there we had B = 0.

An octet of baryons cannot, however, contain any $I = 3/2$ states such as the familiar $J^P = 3/2^+$ Δ resonance. To account for such states a larger irreducible representation is needed. The smallest satisfactory one is the 10. It contains no states with positive strangeness; if the Δ is classified in it, we expect also to see an S = −1 isotriplet, an S = −2 isodoublet, and an S = −3 isosinglet, all with the same spin and parity as the Δ. Assuming that the first two of these were correctly

TABLE 5-B Mesons and baryons classified in the eightfold way

Meson octets			Singlet–octet mixing		
J^P	I = 1 Y = 0	I = ½ Y = 1	I = 0 Y = 0	I = 0 Y = 0	Mixing angle
0^-	π	K	$\eta(549)$ (alternatively	$\eta(958)$ E(1420)	10.4° 6.2°)
1^-	$\rho(765)$	K*(890)	$\phi(1020)$	$\omega(784)$	39.4°
2^+	$A_2(1300)$	$K_v(1420)$	f(1514)	f(1260)	30.2°

Baryon octets				
J^P	I = ½ Y = 1	I = 0 Y = 0	I = 1 Y = 0	I = ½ Y = −1
½⁺	N	Λ	Σ	Ξ
$3/2^-$	N(1520)	$\Lambda(1690)$	$\Sigma(1670)$	$\Xi(1820)$
$5/2^+$	N(1688)	$\Lambda(1815)$	$\Sigma(1915)$	$\Xi(2030)$
$5/2^-$	N(1670)	$\Lambda(1830)$	$\Sigma(1765)$	$\Xi(1930)$

filled by the $Y_1^*(1385)$ and the $\Xi^*(1530)$, Gell-Mann predicted the existence of the isosinglet, which he christened Ω^-. Observing also that the masses of Δ, Y_1^*, and Ξ^* seem to be equally spaced with a difference of about 145 MeV, he predicted the mass of the Ω^- to be approximately 1675 MeV. The Ω^- was subsequently observed with almost exactly that mass, providing striking confirmation of the decuplet classification. (We shall investigate mass splittings in more detail in Chapter 6.)

Although no other decuplets of particles have yet been firmly established, it does seem that the octet–decuplet classification scheme for the baryons is the correct one. The Sakata model must therefore be discarded. How then can we maintain the successful meson classification that resulted from it? It is true that they can still be pictured as baryon–antibaryon bound states, but the SU(3) classification would then be based on $8 \otimes 8$ rather than $3 \otimes \overline{3}$. The appropriate reduction, which we shall discuss in more detail later, is given in (5.11):

$$8 \otimes 8 = 27 \oplus 10 \oplus \overline{10} \oplus 8 \oplus 8 \oplus 1.$$

While $8 \otimes 8$ does lead to meson octets and singlets, it contains in addition 27, 10, and $\overline{10}$, which do not seem to be needed. Constructing the higher baryon states by combining octet baryons with octet mesons will likewise require this reduction; here also the physically observed multiplets 8 and 10 are accompanied by 27 and $\overline{10}$, which do not occur among the known particles. A model using a basic octet is thus at best a clumsy one, requiring nature to discriminate against some representations while favoring others.

Quarks

In particular, nature has apparently discriminated against the elementary representation; no basic triplet of particles, corresponding to the 3, is observed. The absence of the elementary representation is most uncomfortable. It is as if there were particles of higher spins but none having spin ½. We should therefore like particularly to know the physical properties of the 3, and whether particles belonging to the basic triplet do exist.

The structure of the 3 is, of course, the same as that we associated with the sakatons – an isodoublet plus an isosinglet. Maintaining this analogy, we label the particles of the isodoublet $\mathscr{P}$ and $\mathscr{N}$, and call the isosinglet λ. The physical properties that are assigned to these objects are altered, however, because the definitions of charge and hypercharge as functions of I_3 and I_8 were charged. From (5.7a) and (4.11) we have for the hypercharge

$$Y = \tfrac{1}{3}\begin{pmatrix} 1 & & \\ & 1 & \\ & & -2 \end{pmatrix}. \tag{5.18}$$

Thus the hypercharge of $\mathscr{P}$ and $\mathscr{N}$ is 1/3, and that of λ is −2/3. The charges are given correspondingly by (5.7); the $\mathscr{P}$ has a charge of +2/3e, the $\mathscr{N}$ and λ −1/3e.

Objects belonging to the triplet will thus be unlike any known particles. They will have fractional hypercharge and fractional charge. These strange properties were pointed out by Gell-Mann, who christened the three particles "quarks",† and simultaneously by Zweig, who called them "aces". Despite its rather harsh sound, the former term has come into general usage.

The mesons are therefore pictured as bound states of a quark–antiquark system. Their classification according to $3 \otimes \overline{3}$ is exactly as in the Sakata model, since B = 0, and the octet wave functions are those given in (5.5) provided that the

† The quoted source of this name is a line in Joyce's *Finnegan's Wake*, "Three quarks for Muster Mark"; in German, however, "quark" is either cheese or nonsense. This translation often seems more appropriate.

sakatons (p, n, Λ°) are replaced by the quarks ($\mathscr{P}$, $\mathscr{N}$, λ). To classify the baryons, however, we must obtain octets and decuplets from multiquark states. Combining two quarks provides neither of these representations; but if we take three-quark states, the appropriate reduction is

$$3 \otimes 3 \otimes 3 = 10 \oplus 8 \oplus 8 \oplus 1. \tag{5.11b}$$

Thus we obtain naturally singlets, octets, and decuplets, but no undesirable larger representations, if we assume that the baryons are bound states of three quarks. In addition to fractional charge and hypercharge, therefore, the quarks must have fractional baryon number, B = 1/3.

With these peculiar quantum numbers, free quarks would be quite easy to recognize. Experimentalists therefore began soon after quarks were postulated to search intensively for fractionally charged particles. No conclusive evidence for their existence has yet appeared. Attempts to observe quark–antiquark pair production in accelerator beams have led to negative results, indicating that either the quarks are too heavy to produce with existing accelerators or the pair production cross section is miniscule. In this way a lower limit of ~6 GeV has been estimated for the quark mass. Other experiments have looked for quarks in cosmic rays, in sea water, in moon dust, and elsewhere, but none have been found.

Thus it seems likely that quarks do not exist. Nature must discriminate somehow against the elementary representation of SU(3); a selection rule of some kind must exist which forbids the 3 representation and all others leading to fractional values of charge, hypercharge, and baryon number. Only states in which the number of quarks exceeds the number of antiquarks by 3N can have B = N; these states also are the only ones with integral values of Q and Y. If we label the irreducible representations (λ, μ), this means that only those states for which

$$t \equiv (\lambda - \mu)\,(\mathrm{mod}\ 3) = 0 \tag{5.19}$$

are allowed. The value of t is known as the *triality* of the representation; nature seems to choose only those states having triality zero. It is interesting to note that these are precisely the states for which the eigenvalue of the Casimir operator c_1 is an integer. One possible mechanism for outlawing the fractionally charged states is therefore to require the reality of $e^{i\pi t}$ or of $e^{i\pi c_1}$. No physical reason for such a requirement can be given, however, and the question of why there might be no quarks will probably remain unanswered until one is finally found.

Chapter 6

CONSEQUENCES OF SU(3)

In the preceding chapter we have seen how the hadrons may be classifed as eigenstates corresponding to irreducible representations of SU(3). Even though the symmetry is broken, as reflected in the mass differences among SU(3) multiplets, it seems to provide a natural explanation of the hypercharge and isospin values associated with a given J^P. If the strong interactions were completely invariant with respect to SU(3), all operators referring to them would be SU(3)-scalars; that is, they would commute with the entire SU(3) group. Even though this is not so, it may be a good first approximation. For example, assuming that the transition operator is invariant with respect to SU(3) leads to the prediction of relations between various scattering amplitudes.

In this chapter we shall investigate the consequences of assigning simple SU(3) properties to the pertinent operators describing the strong interactions. The properties of single-particle states – the mass and electromagnetic properties, in particular – will be considered first. Here the symmetry breaking is our primary interest. The properties of three- and four-particle vertices, i.e. coupling constants and scattering amplitudes, will then be considered by assuming the transition matrix essentially scalar under SU(3).

Baryon Mass Relations in SU(3)

We have already remarked that the assignment of $\Delta(1238)$, $Y_1^*(1385)$ and $\Xi^*(1530)$ to the baryon decuplet, combined with the observation that their masses are split by about 145 MeV, led Gell-Mann to predict correctly the existence of the $\Omega^-(1675)$. The masses of the decuplet baryons are thus equally spaced as a function of the hypercharge; we neglect electromagnetic effects and assume them independent of I_3. A first guess as a mass operator for the baryons would therefore be

$$M(n, I, I_3, Y) = m_n + a_n Y \tag{6.1}$$

the constants m_n and a_n being reduced matrix elements depending on n. The ½⁺ baryon octet shows that (6.1) is incorrect, however, since there it would predict $M(\Sigma^\circ) = M(\Lambda^\circ)$, a relation which fails by 7%.

Suppose, however, that we may consider only the negatively charged members $(\Delta^-, Y_1^{*-}, \Xi^{*-}, \Omega^-)$ of the decuplet. Then we are using only the U-spin SU(2) subgroup of SU(3). The mass operator should be diagonal, since we do not expect it to connect different states; therefore it must be combination of the unit operator and U_3, i.e.

$$M = M_U + b_U U_3, \tag{6.2}$$

With M_U and b_U dependent only on the total U-spin. This formulation is equivalent to (6.1) for the decuplet, and yields equal spacing:

$$M(\Delta^-) = M_{3/2} + \tfrac{3}{2} b_{3/2}$$

$$M(Y_1^{*-}) = M_{3/2} + \tfrac{1}{2} b_{3/2}$$

$$M(\Xi^{*-}) = M_{3/2} - \tfrac{1}{2} b_{3/2}$$

$$M(\Omega^-) = M_{3/2} - \tfrac{3}{2} b_{3/2}, \tag{6.3}$$

corresponding to $M_{3/2} \approx 1457$ MeV, $b_{3/2} \approx -145$ MeV.

For the octet, however, (6.1) and (6.2) lead to different results. Clearly (6.2), having two parameters, leads to relations between the masses of different particles only when applied to three or more states, i.e. to a system with $U \geqslant 1$. Meaningful application in the octet can therefore be made only to the U-triplet $(n, \Sigma^\circ_u, \Xi^\circ)$. The $U_3 = 0$ state, which we have labeled Σ°_u, is neither Σ° nor Λ°, but a mixture of the two; so is the corresponding U-singlet, which we call Λ°_u. Thus we write

$$|\Sigma^\circ_u\rangle = \alpha|\Sigma^\circ\rangle - \beta|\Lambda^\circ\rangle$$

$$|\Lambda^\circ_u\rangle = \beta|\Sigma^\circ\rangle + \alpha|\Lambda^\circ\rangle \tag{6.4}$$

with $|\alpha|^2 + |\beta|^2 = 1$. The coefficients α and β can be determined in a number of ways. For example, applying the commutation relation $[U_-, I_+] = 0$ to the state $|n\rangle$ yields

$$U_- I_+ |n\rangle = I_+ U_- |n\rangle,$$

i.e.

$$U_-|p\rangle = I_+\sqrt{2}(a|\Sigma^\circ\rangle - \beta|\Lambda^\circ\rangle).$$

Then since $U_-|p\rangle = |\Sigma^+\rangle$, we have

$$|\Sigma^+\rangle = 2a|\Sigma^+\rangle \tag{6.5}$$

yielding $a = \frac{1}{2}$, $\beta = \pm\sqrt{3}/2$. Choosing the positive β leads to

$$|\Sigma_u^\circ\rangle = \tfrac{1}{2}|\Sigma^\circ\rangle - \sqrt{3}/2\ |\Lambda^\circ\rangle$$

$$|\Lambda_u^\circ\rangle = \sqrt{3}/2\ |\Sigma^\circ\rangle + \tfrac{1}{2}|\Lambda^\circ\rangle. \tag{6.6}$$

The mass formula (6.2) predicts for the masses of the U-triplet

$$M(n) = M_1 + b$$

$$M(\Sigma_u^\circ) = M_1$$

$$M(\Xi^\circ) = M_1 - b, \tag{6.7}$$

from which it follows that

$$\tfrac{1}{2}[M(n) + M(\Xi^\circ)] = M(\Sigma_u^\circ). \tag{6.8}$$

Using (6.6) for $|\Sigma_u^\circ\rangle$, we have

$$M(\Sigma_u^\circ) = \langle\Sigma_u^\circ|M|\Sigma_u^\circ\rangle$$

$$= \tfrac{1}{4}[\langle\Sigma^\circ| - \sqrt{3}\langle\Lambda^\circ|]\ M[|\Sigma^\circ\rangle - \sqrt{3}|\Lambda^\circ\rangle]$$

$$= \tfrac{1}{4}M(\Sigma^\circ) + \tfrac{3}{4}M(\Lambda^\circ) - \tfrac{1}{4}\sqrt{3}\ \mathrm{Re}\ \langle\Sigma^\circ|M|\Lambda^\circ\rangle; \tag{6.9}$$

the cross term $\langle\Sigma^\circ|M|\Lambda^\circ\rangle$ vanishes provided we assume that M commutes with the isospin group. Thus (6.8) becomes

$$\tfrac{1}{2}[M(n) + M(\Xi^\circ)] = \tfrac{1}{4}M(\Sigma^\circ) + \tfrac{3}{4}M(\Lambda^\circ). \tag{6.10}$$

The observed values of the masses yield

$\frac{1}{2}[M(n) + M(\Xi^\circ)] = 1128$ MeV

$\frac{1}{4}M(\Sigma^\circ) + \frac{3}{4}M(\Lambda^\circ) = 1134$ MeV.

The U-spin prediction (6.10) is therefore accurate to a surprisingly good degree; the error is 3% of the splitting term b_1, and less than 1% of the total masses. A result analogous to (6.10) should also hold in other octets, of course. For the $J^P = 3/2^+, 5/2^-$, and $5/2^+$ baryon octets in particular, the agreement is quite good, as shown in Table 6-A; this agreement provides the principal supporting evidence for those classifications.

Despite this apparent success, however, the formula (6.2) cannot be valid for an SU(3) mass operator. The reason is that U_3 does not commute with the isospin group; as we have seen in (4.8c),

TABLE 6-A Comparison of baryon octet masses with the Gell-Mann–Okubo mass formula

J^P	Particles	2(M(N) + M(Ξ)) (MeV)	3M(Λ) + M(Σ) (MeV)
$\frac{1}{2}^+$	N Λ Σ Ξ	4514	4538
$\frac{3}{2}^-$	N(1520) Λ(1690) Σ(1670) Ξ(1820)	6680	6740
$\frac{5}{2}^+$	N(1688) Λ(1815) Σ(1915) Ξ(2030)	7436	7360
$\frac{5}{2}^-$	N(1670) Λ(1830) Σ(1765) Ξ(1930)	7200	7255

$$[U_3, I_\pm] = \mp\tfrac{1}{2}I_\pm.$$

Therefore the cross term cannot be omitted from (6.9); instead,

$$[M, I_\pm] = [M_u + b_u U_3, I_\pm] = \mp\tfrac{1}{2}b_u I_\pm,$$

so

$$MI_\pm = I_\pm(M \mp \tfrac{1}{2}b_u), \tag{6.11}$$

implying that the masses in an isomultiplet are split as well, by half the amount of the U-splitting. Thus (6.3) contradicts the observed equality of masses within an isomultiplet.

If we wish the mass operator to avoid difficulties such as those implied by (6.11), we must construct it correctly using SU(3) tensor operators. That is, it must be expressed in terms of operators $T_{n\nu} = T(n, I, I_3, Y)$, with n labeling an irreducible representation and ν denoting the eigenvalues I, I_3, and Y, which satisfy the commutation relations

$$[X_a, T_{n\nu}] = \sum_{\nu'} (X_a^{(n)})_{\nu\nu'} T_{n\nu'} \tag{6.12}$$

with $(X_a^{(n)})_{\nu\nu'}$ denoting the matrix representation of F_a in n. The mass operator should not connect different states, so it must be diagonal. It is easily seen from (6.12) that $T(n, I, I_3, Y)$ is diagonal only if $I_3 = Y = 0$; using $F_3 = I_3$, for example, we have $(X_3^{(n)})_{\nu\nu'} = I_3(\nu)\delta_{\nu\nu'}$, so that (6.12) becomes

$$X_3 T(n, I, I_3, Y)|n', I_3', Y'\rangle = (I_3 + I_3')T(n, I, I_3, Y)|n', I', I_3', Y'\rangle.$$

Thus $T(n, I, I_3, Y)$ changes the eigenvalue I_3' by I_3. A similar proof shows that it changes Y' by Y. Thus the only diagonal tensor operators are $T(n, I, 0, 0)$.

Since we are neglecting the small mass differences within an isomultiplet, however, we also wish the mass operator to commute with **I**, i.e. to be an isoscalar. Thus the only satisfactory mass operators are of the form $T(n, 0, 0, 0)$. The only irreducible representations having $I = I_3 = Y = 0$ members are n = 1, 8, 27, etc. The scalar term $T(1, 0, 0, 0)$ will not lead to any mass splitting, so the simplest tensor which provides a satisfactory mass operator for SU(3) is of the form

$$M = m_1 \mathbb{1} + m_8 T(8, 0, 0, 0). \tag{6.13}$$

Thus the mass splitting is expected to transform like the isoscalar member of an octet, i.e. like the hypercharge. Equation (6.2) does not follow immediately from (6.13), however, because the hypercharge itself is not necessarily the only operator having the correct transformation properties. That is, higher-order combinations of generators may satisfy the same commutations relations as Y. It was originally shown by Okubo that there is a single operator of this type, given by

$$Y^{(2)} = (I^2 - \tfrac{1}{4}Y^2 - g(n)\mathbb{1}). \qquad (6.14)$$

The constant term $g(n)\mathbb{1}$ is included to guarantee that $Y^{(2)}$, like Y, is traceless, and $g(n)$ can be determined from this requirement.

Thus, in fact, there should be two independent terms in (6.13), corresponding to these two tensor operators,

$$M = m_1\mathbb{1} + aY + bY^{(2)}.$$

We therefore write the mass operator as

$$M = m_1\mathbb{1} + aY + b(I^2 - \tfrac{1}{4}Y^2 - g(n)\mathbb{1})$$

$$= m(n)\mathbb{1} + aY + b(I^2 - \tfrac{1}{4}Y^2),$$

which immediately yields the *Gell-Mann–Okubo mass formula*

$$M(n, I, Y) = m_n + a_n Y + b_n(I(I+1) - \tfrac{1}{4}Y^2). \qquad (6.15)$$

The particles in a decuplet have I and Y values related simply by $Y = 2(I - 1)$, so (6.15) leads to the equal spacing law

$$M(10, I, Y) = (m_{10} - b_{10}) + (a_{10} + 3b_{10})Y.$$

For the $J^P = \tfrac{1}{2}^+$ baryon octet, (6.15) becomes

$$M(n) = m_8 + a_8 + \tfrac{1}{2}b_8$$

$$M(\Sigma) = m_8 + 2b_8$$

$$M(\Lambda^\circ) = m_8$$

$$M(\Xi) = m_8 - a_8 + \tfrac{1}{2}b_8$$

from which the relation

$$\tfrac{1}{2}[M(n) + M(\Xi)] = \tfrac{1}{4}M(\Sigma) + \tfrac{3}{4}M(\Lambda^\circ) \tag{6.10}$$

is again obtained.

The Gell-Mann–Okubo mass formula thus successfully accounts for the equally spaced masses in the 10 and the experimentally verified relation (6.10) in the octets. An alternative method of deriving these same results would be to assume the transformation law implied by (6.13) and use the Wigner–Eckart theorem directly to evaluate the matrix elements. In the decuplet, we have

$$\langle 10, I, I_3, Y|T(8,0,0,0)|10, I, I_3, Y\rangle = \langle 10\|T(8)\|10\rangle \begin{pmatrix} 10 & 8 & 10 \\ \nu & 0 & \nu \end{pmatrix}, \tag{6.16}$$

where $\langle 10\|T(8)\|10\rangle$ is a reduced matrix element, and the mass splittings are proportional to the Clebsch–Gordan coefficients

$$\begin{pmatrix} 10 & 8 & 10 \\ \nu & 0 & \nu \end{pmatrix}.$$

The detailed properties of these coefficients lead to equal spacing, as expected. For the octet, however, we must evaluate $\langle 8, I, I_3, Y|T(8,0,0,0)|8, I, I_3, Y\rangle$. The reduction of $8 \otimes 8$ contains two 8's, so there are two independent sets of Clebsches, and we have

$$\langle 8, I, I_3, Y|T(8,0,0,0)|8, I, I_3, Y\rangle =$$

$$\langle 8\|T(8)\|8\rangle_1 \begin{pmatrix} 8 & 8 & 8_1 \\ \nu & 0 & \nu \end{pmatrix} + \langle 8\|T(8)\|8\rangle_2 \begin{pmatrix} 8 & 8 & 8_2 \\ \nu & 0 & \nu \end{pmatrix}. \tag{6.17}$$

The two reduced matrix elements $\langle 8\|T(8)\|8\rangle_{1,2}$ are analogous to the two constants a_n and b_n occurring in (6.15). Evaluation of the Clebsches leads, as expected, to (6.10).

These two methods of arriving at the results of (6.15) are essentially equivalent. Okubo's technique is more powerful, however, since it does not require us to use a detailed knowledge of Clebschery; the operator $Y^{(2)}$ contains the crucial features of the coefficients in its formulation.

Meson Mass Relations

The derivation of the Gell-Mann–Okubo formula made no assumptions regarding the baryon number, so we would intuitively expect it to apply equally well to the mesons. For the pseudoscalar meson octet, for example, the equality of K and $\overline{K}$ masses leads to

$$\tfrac{1}{2}[M(K) + M(\overline{K})] = M(K) = \tfrac{1}{4}M(\pi) + \tfrac{3}{4}M(\eta) \qquad (6.18)$$

as the Gell-Mann–Okubo result. Using the kaon and pion masses to calculate $M(\eta)$ we find that (6.18) predicts $M(\eta) = 615$ MeV, in serious disagreement with the experimental value 549 MeV.

In order to remedy this failure, it has been suggested that for mesons we should consider the square of the mass, rather than the mass itself, as the tensor operator in (6.13). The reason usually given is that the mesons satisfy the Klein–Gordon equation, which contains m^2, while the baryons satisfy the Dirac equation, which is linear in the mass. Then instead of (6.18), we have

$$M^2(K) = \tfrac{1}{4}m^2(\pi) + \tfrac{3}{4}m^2(\eta). \qquad (6.19)$$

Experimentally, $m^2(K) = 0.246\ \text{GeV}^2$ while $\tfrac{1}{4}m^2(\pi) + \tfrac{3}{4}m^2(\eta) = 0.231\ \text{GeV}^2$, so (6.19) is in fairly good agreement with the data.

A similar result would also hold in the other meson octets. For the vector mesons, for example, the analogue of (6.19) is

$$m^2(K^*) = \tfrac{1}{4}m^2(\rho) + \tfrac{3}{4}m^2(\omega_8) \qquad (6.20)$$

where ω_8 denotes the $I = Y = 0$ of the octet. Using the known masses of K^* and ρ, we find from (6.17) that $m(\omega_8) = 929$ MeV. No vector meson with that mass has been observed; the only candidates having the correct quantum numbers are $\omega(783)$ and $\phi(1020)$.

In order to explain this discrepancy the concept of ω–ϕ mixing, already mentioned in Chapter 5, has been introduced. In the absence of SU(3) symmetry breaking, (5.1) indicates that we might expect an octet plus a singlet of mesons, presumably all with the same mass. Then there would be two degenerate $I = Y = 0$ mesons, $|8, 0, 0, 0\rangle \equiv |\omega_8\rangle$ and $|1, 0, 0, 0\rangle \equiv |\omega_1\rangle$. The symmetry-breaking interaction, treated as a perturbation, will mix these two degenerate states; the resulting physical states $|\omega\rangle$ and $|\phi\rangle$ will therefore be given by

$$|\omega\rangle = \cos\theta_v|\omega_8\rangle + \sin\theta_v|\omega_1\rangle$$

$$|\phi\rangle = -\sin\theta_v|\omega_8\rangle + \cos\theta_v|\omega_1\rangle \qquad (6.21)$$

for some mixing angle θ_V. To determine θ_V, (6.21) can be inverted and $m^2(\omega_8)$ calculated:

$$m^2(\omega_8) = \langle \omega_8 | m^2 | \omega_8 \rangle$$

$$= [\cos\theta_V \langle\omega| - \sin\theta_V \langle\phi|]\, m^2 [\cos\theta_V |\omega\rangle - \sin\theta_V |\phi\rangle]$$

$$= \cos^2\theta_V m^2(\omega) + \sin^2\theta_V m^2(\phi). \qquad (6.22)$$

Using $m^2(\omega_8)$ from (6.20) and the experimental ω and ϕ masses, we find $\theta_V = 40.1°$ (assuming θ_V is in the first quadrant).

A similar situation arises in the T octet. The analogue of (6.19) in this case is

$$m^2(K_A(1320)) = \tfrac{1}{4} m^2(A_2(1300)) + \tfrac{3}{4} m^2(f_8),$$

yielding 1456 MeV for the mass of f_8. Once again singlet–octet mixing is needed, with the f(1264) and f′(1514) being given analogously to (6.21) by

$$|f\rangle = \cos\theta_T |f_8\rangle + \sin\theta_T |f_1\rangle$$

$$|f'\rangle = -\sin\theta_T |f_8\rangle + \cos\theta_T |f_1\rangle.$$

The resulting value of θ_T is 29.9°.

Treating the mass splitting operator as an SU(3)-tensor transforming like T_8 therefore leads to a seemingly correct description of hadronic masses. Two questions may be asked about this approach. The first is why the V and T meson octets need singlet–octet mixing, while the baryon octets and the P meson octet do not. A partial answer to this question will be given when we study SU(6). The second question arises if we consider the mass splitting a perturbation, which we have treated in first order in writing (6.13). The first order effects are then quite large, typically 10–20%. We would therefore expect to see appreciable second order effects as well; why are they apparently absent? No good answer to this question has yet been given.

Electromagnetic Properties

The electromagnetic properties of the hadrons can be treated in a similar manner. The fact that all members of a U-spin multiplet have the same charge suggests that any operator describing electromagnetic effects should be a U-scalar, just as the mass operator was isoscalar. Any electromagnetic quantity E will therefore satisfy relations such as

$$E(\Delta^-) = E(Y_1^{*-}) = E(\Xi^{*-}) = E(\Omega^-)$$

$$E(\Delta^\circ) = E(Y_1^{*\circ}) = E(\Xi^{*\circ})$$

$$E(\Delta^+) = E(Y_1^{*+}) \tag{6.23}$$

for the decuplet and

$$E(p) = E(\Sigma^+)$$

$$E(\Sigma^-) = E(\Xi^-)$$

$$E(n) = E(\Xi^\circ) = E(\Sigma_u^\circ) \tag{6.24}$$

for the baryon octet. The state Σ_u° here must be taken to represent the $U = 1$ combination of Σ° and Λ° given in (6.9); therefore

$$\begin{aligned} E(\Sigma_u^\circ) &= \tfrac{1}{4}[\langle\Sigma^\circ| - \sqrt{3}\langle\Lambda^\circ|]\, E[|\Sigma^\circ\rangle - \sqrt{3}\,|\Lambda^\circ\rangle] \\ &= \tfrac{1}{4}E(\Sigma^\circ) + \tfrac{3}{4}E(\Lambda^\circ) - \sqrt{3}/2\, E(\Sigma\Lambda), \end{aligned} \tag{6.25}$$

where $E(\Sigma\Lambda) = \mathrm{Re}\langle\Sigma^\circ|E|\Lambda^\circ\rangle$. (The comparable cross term in the mass splitting vanished because that operator conserved isospin; here it must be included, since electromagnetic operators do not commute with **I**.)

In addition to U-scalarity, however, we may assume specific SU(3) transformation properties for E. In most cases electromagnetic effects are important only to first order (at least in comparison with strong interactions), so they depend linearly on the charge. It then follows from the identification $Q = -eU_8$ that an electromagnetic operator should be an irreducible SU(3)-tensor transforming like U_8. Since U_8 is the U-spin analogue of Y, the general form of such a tensor can be written down by making the appropriate changes in (6.13):

$$E = E_1 U_8 + E_2(U^2 - \tfrac{1}{4}U_8^2 - g(n)\mathbb{1}). \tag{6.26}$$

For the octet, g(8) = 1, and it follows from (6.26) that

$$E(p) = E(\Sigma^+) = -E_1 - \tfrac{1}{2}E_2$$

$$E(n) = E(\Xi^\circ) = E(\Sigma_u^\circ) = E_2$$

$$E(\Lambda_u^\circ) = -E_2$$

$$E(\Sigma^-) = E(\Xi^-) = E_1 - \tfrac{1}{2}E_2 \qquad (6.27)$$

with $E(\Sigma_u^\circ)$ given by (6.25) and

$$E(\Lambda_u^\circ) = \tfrac{1}{4}[\sqrt{3}\langle\Sigma^\circ| + \langle\Lambda^\circ|]\,E\,[\sqrt{3}\,|\Sigma^\circ\rangle + |\Lambda^\circ\rangle]$$

$$= \tfrac{3}{4}E(\Sigma^\circ) + \tfrac{1}{4}E(\Lambda^\circ) + \sqrt{3}/2\,E(\Sigma\Lambda). \qquad (6.28)$$

From (6.25), (6.27), and (6.28) it is easily shown that

$$E(\Sigma^\circ) + E(\Lambda^\circ) = 0$$

$$\tfrac{1}{2}E(\Sigma^\circ) + \sqrt{3}/2\,E(\Sigma\Lambda) = E(\Sigma^+) + E(\Sigma^-) = -E(n). \qquad (6.29)$$

A simpler relation among the Σ's can be obtained by assuming that under isospin transformations E behaves as a combination of isoscalar and isovector, yielding

$$E(\Sigma^\circ) = \tfrac{1}{2}[E(\Sigma^+) + E(\Sigma^-)]. \qquad (6.30)$$

Combining this result with (6.29) leads to

$$E(\Lambda^\circ) = -E(\Sigma^\circ) = \tfrac{1}{2}E(n) = -1/\sqrt{3}\,E(\Sigma\Lambda). \qquad (6.31)$$

Other than the charge, the primary electromagnetic operator of interest is the magnetic moment μ, which describes the strength of the coupling of the spin to an external magnetic field. The best values, in nuclear magnetons, of those which have been measured are:

$$\mu(p) = 2.79 \qquad \mu(\Sigma^+) = 2.62 \pm 0.41$$

$$\mu(n) = -1.91 \qquad \mu(\Lambda^\circ) = -0.672 \pm 0.061. \qquad (6.32)$$

These experimental values are thus in reasonable, although not perfect, agreement with the above predictions.

Similar results can be obtained, using g(10) = 2, for the decuplet baryons. A particularly simple set of results follow, which can be summarized as

$$E_1 = (E_1 - \tfrac{3}{2}E_2)U_8 = E'Q. \tag{6.33}$$

The magnetic moments of the decuplet baryons, for example, should therefore be directly proportional to the charge. It does not appear, however, that any measurements of these magnetic moments are likely within the foreseeable future.

Three-Particle Vertices

Next we consider the implications of SU(3) for strong interactions coupling constants. As described in Chapter 1, these constants describe the coupling strengths of three-particle vertices. If the masses involved are such that the vertex describes a kinematically allowed decay, the coupling constants determine the decay widths. Otherwise, they will appear in one-particle exchange amplitudes; for example, $\Lambda^\circ \to pK^-$ is forbidden by energy conservation, but the $\Lambda N\bar{K}$ coupling constant enters into the description of $\pi^- p \to \bar{K}^{*\circ}\Lambda^\circ$, as shown in Fig.(6-1).

There are two ways in which SU(3) may affect these vertices. First, the transition operator can have definite tensor properties manifested directly in the coupling constants; second, the transition probability will depend on the masses, which are split by SU(3). The latter effect can be accounted for kinematically, by using the correct masses in calculating phase space. It is therefore appropriate to obtain relations among the coupling constants by assuming that the transition operator has definite tensor properties – namely, that it is a scalar – under SU(3). Symmetry breaking can then be included later by using the observed masses in the dynamical parts of the calculations, while maintaining the SU(3) symmetry of the coupling constants. Thus we assume that the coupling constant $g_{(n'\nu')(n''\nu'')(n\nu)}$ describing the $|n\nu\rangle \to |n'\nu''\rangle\,|n''\nu''\rangle$ vertex strength is an SU(3) scalar. The Wigner–Eckart theorem then implies that different coupling constants may be related by

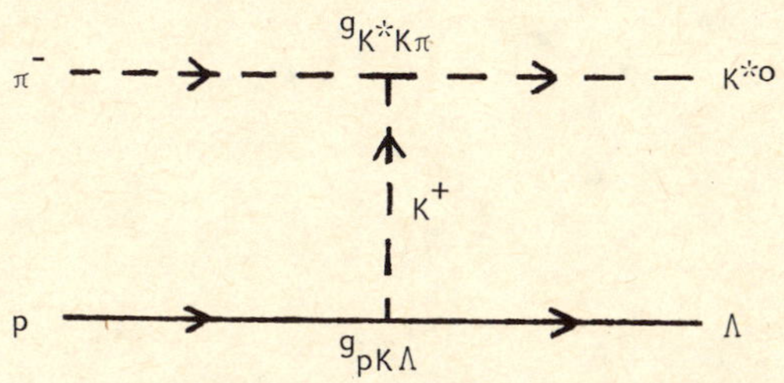

Fig.(6-1) Appearance of SU(3) coupling constants in one-particle exchange diagrams.

$$g_{(n'\nu')(n''\nu'')(n\nu)} = \begin{pmatrix} n' & n'' & n \\ \nu' & \nu'' & \nu \end{pmatrix} g(n', n'', n), \qquad (6.34)$$

where $g(n', n'', n)$ is a reduced matrix element.

The simplest example of the implications of (6.34) is a consideration of the decays of the decuplet baryons. Suppose first that the final state consists of an octet baryon and a *singlet* meson. Then the Clebsch–Gordan coefficient in (6.34) vanishes, since $8 \otimes 1 = 8$ does not connect to the initial 10 state. Recalling that the vector meson singlet is a mixture of the physical ω and ϕ states,

$$|\omega_1\rangle = \sin\theta_V|\omega\rangle + \cos\theta_V|\phi\rangle,$$

we have

$$g_{\omega_1 BD} = 0 = \sin\theta_V g_{\omega BD} + \cos\theta_V g_{\theta BD}, \qquad (6.35)$$

so

$$\frac{g_{\phi BD}}{g_{\omega BD}} = -\tan\theta_V, \qquad (6.36)$$

where D is any decuplet baryon and B the corresponding octet baryon.

If the final state contains an octet meson rather than a singlet, all of the coupling constants are proportional to the Clebsch–Gordan coefficients

$$\begin{pmatrix} 8 & 8 & 10 \\ \nu' & \nu'' & \nu \end{pmatrix}.$$

It is instructive here to note that these relations can be obtained by using isospin and U-spin. For example, relations among the various $\pi N\Delta$ coupling constants following from isospin have already been obtained in equations (3.18) and (3.19), and results for other isospin-related decays are similarly obtained. All of these relations derive from the SU(2) part of

$$\begin{pmatrix} 8 & 8 & 10 \\ \nu' & \nu'' & \nu \end{pmatrix}.$$

The isoscalar factors can be obtained using U-spin analogously. Since $(\Delta^-, Y_1^{*-}, \Xi^{*-}, \Omega^-)$ form a $U = 3/2$ system, they can decay only to baryon–meson combinations with $U = 3/2$. Using the baryon U-triplet

$(n, \Sigma_u^\circ = \tfrac{1}{2}\Sigma^\circ - \sqrt{3/2}\Lambda^\circ, \Xi^\circ)$ with the U-doublet mesons (π^-, K^-), we find U-spin states

$$|\tfrac{3}{2}, \tfrac{3}{2}\rangle = |\pi^- n\rangle$$

$$|\tfrac{3}{2}, \tfrac{1}{2}\rangle = \sqrt{2/3}\,|\pi^-\Sigma_u^\circ\rangle + \sqrt{1/3}\,|K^- n\rangle \qquad |\tfrac{1}{2}, \tfrac{1}{2}\rangle = \sqrt{1/3}\,|\pi^-\Sigma_u^\circ\rangle - \sqrt{2/3}\,|K^- n\rangle$$

$$|\tfrac{3}{2}, -\tfrac{1}{2}\rangle = \sqrt{2/3}\,|K^-\Sigma_u^\circ\rangle + \sqrt{1/3}\,|\pi^-\Xi^\circ\rangle \quad |\tfrac{1}{2}, -\tfrac{1}{2}\rangle = \sqrt{2/3}\,|\pi^-\Xi^\circ\rangle - \sqrt{1/3}\,|K^-\Sigma_u^\circ\rangle$$

$$|\tfrac{3}{2}, -\tfrac{3}{2}\rangle = |K^-\Xi^\circ\rangle, \tag{6.37}$$

from which it follows that

$$g_{\pi^- n\Delta^-} = g_{K^-\Xi^\circ\Omega^-} = g(\tfrac{1}{2}, 1, \tfrac{3}{2})$$

$$g_{\pi^-\Lambda^\circ Y_1^{*-}} = g_{K^-\Lambda^\circ\Xi^{*-}} = 1/\sqrt{2}\,g(\tfrac{1}{2}, 1, \tfrac{3}{2})$$

$$g_{\pi^-\Sigma^\circ Y_1^{*-}} = g_{K^-\Sigma^\circ\Xi^{*-}} = -1/\sqrt{6}\,g(\tfrac{1}{2}, 1, \tfrac{3}{2})$$

$$g_{K^- n Y_1^{*-}} = g_{\pi^-\Xi^\circ\Xi^{*-}} = 1/\sqrt{3}\,g(\tfrac{1}{2}, 1, \tfrac{3}{2}) \tag{6.38a}$$

A similar procedure using the $U = 1$ mesons and the $U = \tfrac{1}{2}$ baryons leads to

$$g_{K^\circ\Sigma^-\Delta^-} = g_{\bar{K}^\circ\Xi^-\Omega^-} = g(1, \tfrac{1}{2}, \tfrac{3}{2})$$

$$g_{\eta^\circ\Sigma^- Y_1^{*-}} = g_{\eta^\circ\Xi^-\Xi^{*-}} = 1/\sqrt{2}\,g(1, \tfrac{1}{2}, \tfrac{3}{2})$$

$$g_{\pi^\circ\Xi^- Y_1^{*-}} = g_{\pi^\circ\Xi^-\Xi^{*-}} = 1/\sqrt{6}\,g(1, \tfrac{1}{2}, \tfrac{3}{2})$$

$$g_{K^\circ\Xi^- Y_1^{*-}} = g_{\bar{K}^\circ\Sigma^-\Xi^{*-}} = 1/\sqrt{3}\,g(1, \tfrac{1}{2}, \tfrac{3}{2}). \tag{6.38b}$$

A relation between (6.38a) and (6.38b) is given by isospin,

$$g_{\bar{K}^\circ\Xi^-\Omega^-} = -g_{K^-\Xi^\circ\Omega^-}$$

$$\therefore\ g(\tfrac{1}{2}, 1, \tfrac{3}{2}) = -g(1, \tfrac{1}{2}, \tfrac{3}{2}). \tag{6.38c}$$

Using the coupling constants in equations (6.38) plus the implications of isosymmetry, we may obtain g_{PBD} for all decuplet–octet-pseudoscalar couplings.

If all three particles at the vertex belong to octets, the situation is more complicated, since there are two linearly independent 8's in $8 \otimes 8$ and thus two independent couplings contributing to (6.34). These two couplings correspond to symmetry and antisymmetry of the two 8's. The f and d matrices defined in (4.21) and (4.23) are antisymmetric and symmetric, respectively, and transform like octets. Therefore let us consider the vectors

$$|f, \nu_a\rangle = \sum_{\beta\gamma} f_{a\beta\gamma} |8, \nu_\beta\rangle |8, \nu_\gamma\rangle$$

$$|d, \nu\rangle = \sum_{\beta\gamma} d_{a\beta\gamma} |8, \nu_\beta\rangle |8, \nu_\gamma\rangle, \tag{6.39}$$

where $|8, \nu_\beta\rangle$ and $|8, \nu_\gamma\rangle$ stand for the basis eigenstates providing the octet representation with $(R_a)_{\beta\gamma} = f_{a\beta\gamma}$. Specifically

$$|8, \nu_1\rangle = 1/\sqrt{2}[|8, 1, 1, 0\rangle + |8, 1, -1, 0\rangle]$$

$$|8, \nu_2\rangle = \tfrac{1}{2}[|8, 1, 1, 0\rangle - |8, 1, -1, 0\rangle]$$

$$|8, \nu_3\rangle = |8, 1, 0, 0\rangle \qquad |8, \nu_4\rangle = 1/\sqrt{2}[|8, \tfrac{1}{2}, \tfrac{1}{2}, 1\rangle + |8, \tfrac{1}{2}, -\tfrac{1}{2}, -1\rangle]$$

$$|8, \nu_5\rangle = [|8, \tfrac{1}{2}, \tfrac{1}{2}, 1\rangle - |8, \tfrac{1}{2}, -\tfrac{1}{2}, -1\rangle]$$

$$|8, \nu_6\rangle = 1/\sqrt{2}[|8, \tfrac{1}{2}, -\tfrac{1}{2}, 1\rangle + |8, \tfrac{1}{2}, \tfrac{1}{2}, -1\rangle]$$

$$|8, \nu_7\rangle = i/\sqrt{2}[|8, \tfrac{1}{2}, -\tfrac{1}{2}, 1\rangle - |8, \tfrac{1}{2}, \tfrac{1}{2}, -1\rangle]$$

$$|8, \nu_8\rangle = |8, 0, 0, 0\rangle. \tag{6.40}$$

These eigenvectors transform under infinitesimal transformations according to

$$(1 + i\epsilon_\lambda X_\lambda)|8, \nu_\mu\rangle = |8, \nu_\mu\rangle + i\epsilon_\lambda f_{\lambda\mu\rho} |8, \nu_\rho\rangle$$

and on using (4.26d) we find that

$$(1 + i\epsilon_\lambda X_\lambda)|f, \nu_a\rangle = |f, \nu_a\rangle + i\epsilon_\lambda f_{\lambda a\rho} |f, \nu_\rho\rangle$$

$$(1 + i\epsilon_\lambda X_\lambda)|d, \nu_a\rangle = |d, \nu_a\rangle + i\epsilon_\lambda f_{\lambda a\rho} |d, \nu_\rho\rangle. \tag{6.41}$$

Therefore both $|f, \nu_a\rangle$ and $|d, \nu_a\rangle$ are octets. In other words, the two independent sets of Clebsch–Gordan coefficients connecting $8 \otimes 8$ to 8 may be chosen as

$$\begin{pmatrix} 8 & 8 & 8_f \\ \nu_\beta & \nu_\gamma & \nu_a \end{pmatrix} = f_{a\beta\gamma}$$

$$\begin{pmatrix} 8 & 8 & 8_d \\ \nu_\beta & \nu_\gamma & \nu_a \end{pmatrix} = d_{\alpha\beta\gamma}. \tag{6.42}$$

Instead of (6.35), then, we will have

$$g_{(8,\nu_\beta)(8,\nu_\gamma)(8,\nu_a)} = \begin{pmatrix} 8 & 8 & 8_f \\ \nu_\beta & \nu_\gamma & \nu_a \end{pmatrix} g^{(f)} + \begin{pmatrix} 8 & 8 & 8_d \\ \nu_\beta & \nu_\gamma & \nu_a \end{pmatrix} g^{(d)}$$

$$\equiv \frac{2}{F+D}\,[Ff_{\alpha\beta\gamma} + Dd_{\alpha\beta\gamma}]\,g. \tag{6.43}$$

Thus the relative contributions of symmetric and antisymmetric coupling are given by "D/F ratio". (Since the states $|8, \nu_a\rangle$ given in (6.40) are not eigenstates of I_3 and Y, of course, appropriate unitary transformations must be made on (6.42) and (6.43) in order to obtain the physically meaningful coupling constants.)

An important example of (6.43) is the pseudoscalar meson–baryon coupling shown in Fig.(6-2). A particular case is the familiar pion–nucleon coupling constant $g_{\pi NN}$, for which $g_{\pi NN}^2/4\pi \approx 14$; inserting the appropriate α, β, γ we find that $g_{\pi NN} = g$, independently of D/F. For other PBB vertices, however, there is explicit dependence on D/F, and experimental information on these couplings can be used to determine D/F. For example, the η°-nucleon vertex yields

$$g_{\eta^\circ NN} = 2/\sqrt{3}\left(\frac{3F - D}{F + D}\right) g; \tag{6.44}$$

since the data indicate that this coupling is very weak, we have $D/F \sim 3$. Other PBB coupling constants can be similarly evaluated and compared with experiment, leading to the conclusion that $D/F \approx 2$.

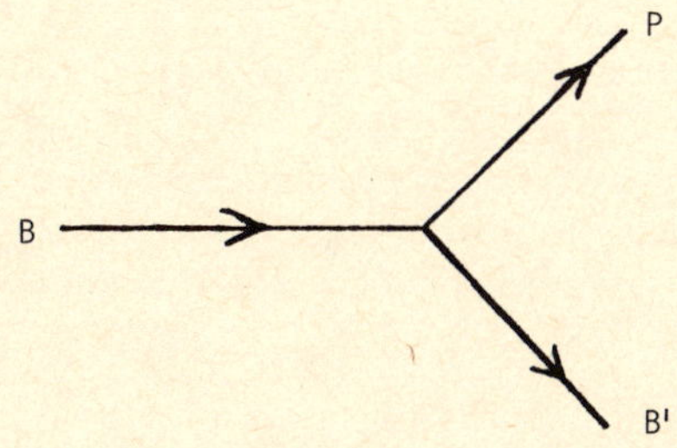

Fig.(6-2) PBB coupling.

Other three-octet vertices will also lead to (6.43), of course; we could consider the couplings between baryons and vector (or tensor) mesons, or any of a variety of three-meson vertices. Each case will have its own values of g and D/F. Determining g will always require experimental data, but in certain instances D/F can be obtained directly from symmetry arguments.

For example, the VPP vertex describes the decay of a vector meson to two pseudoscalar mesons. Since the initial state has $J^P = 1^-$, the pseudoscalar mesons in the final state must have orbital angular momentum $\ell = 1$. A spatial wave function with $\ell = 1$ is antisymmetric under interchange of the two particles' positions. Since the overall wave function must be symmetric, it follows that the SU(3) part of the wave function must also be antisymmetric. Thus the PPV coupling must be purely f-type, i.e. D/F = 0. (By a similar logic, PPT coupling must be entirely d-type.) From

$$g_{P_\alpha P_\beta V_\gamma} = 2f_{\alpha\beta\gamma}g', \tag{6.45}$$

a number of interesting relations between the decay widths of the vector mesons follow, including for example

$$g_{\pi^+\pi^-\omega_8} = 0,$$

$$g_{K^+\pi^-K^{*\circ}} = 1/\sqrt{2}\, g_{\pi^+\pi^-\rho^\circ},$$

$$g_{K^\circ\pi^\circ K^{*\circ}} = -\tfrac{1}{2} g_{\pi^+\pi^-\rho^\circ}, \tag{6.46}$$

as well as a number of others. All of these results can also be derived directly by using isospin and U-spin plus G-parity, since this set of assumptions is fully equivalent to SU(3) with pure f-type coupling. (A more detailed discussion is given by Lipkin in reference 1 of the Reading List.)

Two-Body Scattering

Finally we consider two-body scattering processes in SU(3). Here again we assume that mass splitting effects may be accounted for in phase space and that otherwise the transition operator is a scalar operator with respect to SU(3). Then the scattering amplitude for the process ab → cd is obtained by reducing the initial and final two-particle states

$$\langle n_c\nu_c|\langle n_d\nu_d|\mathscr{T}|n_a\nu_a\rangle|n_b\nu_b\rangle =$$

$$\sum_{nn'} \begin{pmatrix} n_c & n_d & n' \\ \nu_c & \nu_d & \nu' \end{pmatrix}^* \langle n'\nu' | \mathscr{T} | n, \nu\rangle \begin{pmatrix} n_a & n_b & n \\ \nu_a & \nu_b & \nu \end{pmatrix} \tag{6.47}$$

yielding for scalar $\mathscr{T}$

$$= \sum_{n} \begin{pmatrix} n_c & n_d & n \\ \nu_c & \nu_d & \nu \end{pmatrix}^* \begin{pmatrix} n_a & n_b & n \\ \nu_a & \nu_b & \nu \end{pmatrix} \mathscr{T}^{(n)} \tag{6.48}$$

according to the Wigner–Eckart theorem, with $\mathscr{T}^{(n)}$ being the reduced scattering amplitude for the irreducible representation n in the given process.

At present, the only available target particles are protons and neutrons; the beam particles are pions, kaons, nucleons, antiprotons, and lambdas. In every case the scattering processes accessible to experiment involve two octet particles in the initial state, so the values of n in (6.47) follow from $8 \otimes 8 = 27 \oplus 10 \oplus \overline{10} \oplus 8_f \oplus 8_d \oplus 1$. The final states may contain singlet or octet mesons and octet or decuplet baryons (and antibaryons). The values of n′ in (6.47) will be obtained from the appropriate reduction formulas. The summation in (6.48) will clearly contain only those irreducible representations appearing in both initial and final states.

Suppose first the final state contains a singlet meson and a decuplet baryon. Then the final state is entirely 10, so the summation in (6.48) contains a single term, and since

$$\begin{pmatrix} 10 & 1 & 10 \\ \nu & 0 & \nu \end{pmatrix} \equiv 1$$

we have

$$\langle 10, \nu_c | \langle 1, 0 | \mathscr{T} | 8, \nu_a\rangle | 8, \nu_b\rangle = \begin{pmatrix} 8 & 8 & 10 \\ \nu_a & \nu_b & \nu_c \end{pmatrix} \mathscr{T}^{(10)}. \tag{6.49}$$

The relations which follow from (6.49) are thus identical to those obtained for the coupling constants g_{cab} in (6.38). That is, the four-particle vertices in which one particle is an SU(3) singlet are related in the same way as the corresponding three-particle vertices. The same result is clearly obtained if the final state contains octet and singlet; the amplitudes are related analogously to (6.43), with appropriate D/F ratios.

If the final state is made up of two decuplets, the values of n′ in (6.47) are $10 \otimes 10 = 35 \oplus 28 \oplus \oplus 27 \oplus \overline{10}$. Consequently there are two independent reduced matrix elements, $\mathscr{T}^{(27)}$ and $\mathscr{T}^{(10)}$, in (6.48), and

$\langle 10, \nu_c | \langle 10, \nu_d | \mathcal{T} | 8, \nu_a \rangle | 8, \nu_b \rangle =$

$$\begin{pmatrix} 10 & 10 & 27 \\ \nu_c & \nu_d & \nu \end{pmatrix}^* \begin{pmatrix} 8 & 8 & 27 \\ \nu_a & \nu_b & \nu \end{pmatrix} \mathcal{T}^{(27)} + \begin{pmatrix} 10 & 10 & \overline{10} \\ \nu_c & \nu_d & \nu \end{pmatrix}^* \begin{pmatrix} 8 & 8 & \overline{10} \\ \nu_a & \nu_b & \nu \end{pmatrix} \mathcal{T}^{(\overline{10})} . \tag{6.50}$$

All possible processes BB → DD are described in terms of the two amplitudes $\mathcal{T}^{(27)}$ and $\mathcal{T}^{(10)}$ by (6.50), and a large number of relations between them follow, including

$$\mathcal{T}(pp \to \Delta^{++}\Delta^{\circ}) = \sqrt{2}\,\mathcal{T}(p\Sigma^{+} \to \Delta^{++}Y_1^{*\circ})$$

$$\mathcal{T}(p\Lambda^{\circ} \to \Delta^{++}Y_1^{*-}) = -\,\mathcal{T}(p\Sigma^{\circ} \to \Delta^{++}Y_1^{*-}). \tag{6.51}$$

When the final state contains octet and decuplet, then, since $8 \otimes 10 = 35 \oplus 27 \oplus 10 \oplus 8$, we have four terms in (6.50), corresponding to $\mathcal{T}^{(27)}$, $\mathcal{T}^{(10)}$, and the two independent octet couplings $\mathcal{T}^{(8f)}$ and $\mathcal{T}^{(8d)}$. By writing relations analogous to the above in this case a number of results can be obtained; as before, however, it is generally easier to construct them using isospin and U-spin arguments, than to start with the complete list of reactions and apply (6.50). For example, the three reactions $\pi^+p \to \pi^{\circ}\Delta^{++}$, $\pi^+p \to \eta^{\circ}\Delta^{++}$, and $K^+p \to K^{\circ}\Delta^{++}$ all are related by U-spin. A simple SU(2) calculation leads to

$$\mathcal{T}(\pi^+p \to \pi^{\circ}\Delta^{++}) + \sqrt{3}\,\mathcal{T}(\pi^+p \to \eta^{\circ}\Delta^{++}) = \sqrt{2}\,\mathcal{T}(K^+p \to K^{\circ}\Delta^{++}). \tag{6.52}$$

A number of other relations can be obtained similarly.

Since $10 \otimes \overline{10} = 64 \oplus 27 \oplus 8 \oplus 1$, reactions of the type $B\overline{B} \to D\overline{D}$ are also given in terms of four reduced amplitudes $\mathcal{T}^{(27)}$, $\mathcal{T}^{(8f)}$, $\mathcal{T}^{(8d)}$ and $\mathcal{T}^{(1)}$. By using U-spin, or by explicit calculation using the Clebsches, one can find relations of the same kind as (6.52).

Finally there are reactions leading to two octets in the final state, including elastic scattering. Here each representation in the reduction of $8 \otimes 8$ can couple to itself; in addition, 8_f in the initial state can couple to 8_d in the final state, and *vice versa*. Thus there are a total of eight reduced amplitudes here, namely $\mathcal{T}^{(27)}$, $\mathcal{T}^{(10)}$, $\mathcal{T}^{(\overline{10})}$, $\mathcal{T}^{(8ff)}$, $\mathcal{T}^{(8dd)}$, $\mathcal{T}^{(8fd)}$, $\mathcal{T}^{(8df)}$ and $\mathcal{T}^{(1)}$. Having so many linearly independent terms makes it difficult to find any relations among different scattering processes.

A simplification can be introduced by the assumption of "octet dominance". In considering (3.25) we used the argument that the amplitude for $\pi^-p \to K^+\Sigma^-$ should be negligible because, in the one-particle exchange model, it requires a

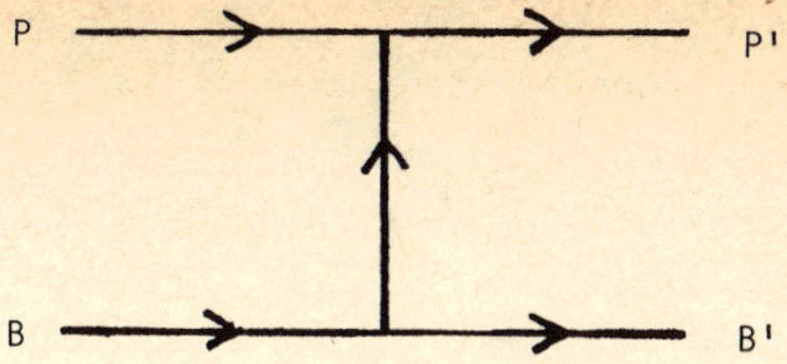

Fig.(6-3) PB scattering.

non-existent I = 2 meson. The SU(3) generalization of this idea is that there are no known mesons classified in 10, $\overline{10}$, or 27, so the contribution of these irreducible representations via particle exchange should be small.

For example, consider meson–baryon scattering $PB \to P'B'$ as a one-particle exchange process, as shown in Fig.(6-3). Looking at the exchanged particle in these processes is equivalent to looking at the intermediate states of $B\overline{B}' \to P\overline{P}'$.

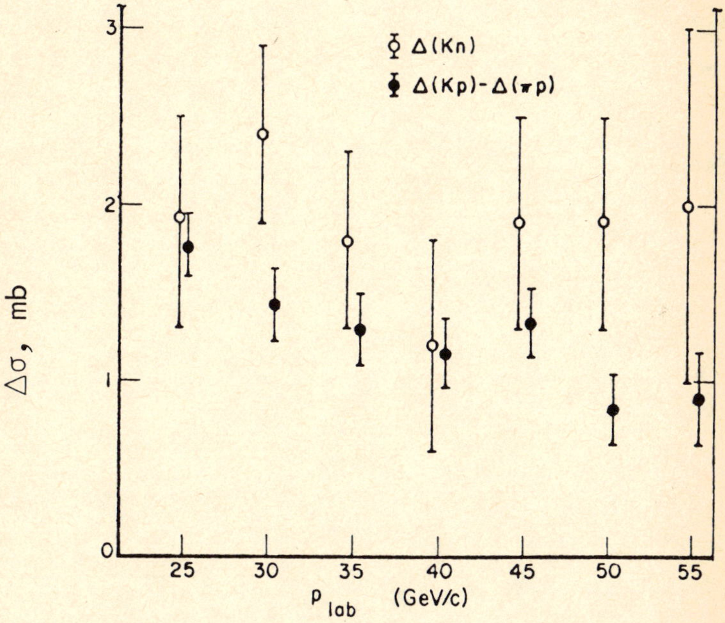

Fig.(6-4) Comparison of the Barger–Rubin relation (6.54) with experiment, with $\Delta(MN) = \sigma(MN) - \sigma(\overline{M}N)$ denoting differences of particle and antiparticle cross sections. (From S. Meshkov, *Phys. Rev.* **D6**, 399 (1972).)

Assuming that the contributions of 10, $\overline{10}$, and $\overline{27}$ are negligible here leads to the Barger–Rubin relation among elastic scattering amplitudes

$$\mathcal{T}_e(\pi^+p) - \mathcal{T}_e(\pi^-p) + \mathcal{T}_e(K^+p) - \mathcal{T}_e(K^+n) + \mathcal{T}_e(K^-n) +$$

$$\mathcal{T}_e(K^-p) = 0. \tag{6.53}$$

This relation is easily compared with experiment for forward scattering, where the optical theorem relates the imaginary part of the amplitude to the total cross section.

The prediction of (6.53) may then be written

$$\sigma(\pi^-p) - \sigma(\pi^+p) + \sigma(K^+p) - \sigma(K^-p) - \sigma(K^+n) + \sigma(K^-n).$$

A comparison of these two quantities is shown in Fig.(6-4). The agreement is quite good, particularly when one realizes that the cross sections involved are individually in the 15–25 mb range.

Chapter 7

ADDING SPIN

It is a striking fact that neither mesons with half-integral spin, nor baryons with integral spin, are known to exist. The parallel appearance of only integral or half-integral isospin values is neatly accounted for in SU(2) as a result of the isospin assignments of the non-strange quarks. In the same way, the assumption that the quarks have spin ½ will lead to precisely the angular momenta observed. Mesons, being quark–antiquark states, will have integral spin, while the three quarks composing a baryon will lead to half-integral spin.

Thus it seems that an extension of the unitary symmetry schemes to include a spin degree of freedom in the basis states may lead to a successful description of the hadrons' angular momenta, as well as their isospin and hypercharge values. If we describe quarks of spin ½ by writing $\mathscr{P}_{\pm}$, $\mathscr{N}_{\pm}$, and $\lambda_{\pm}$, the resulting basis state analogous to (4.1) will be six-dimensional,

$$|\psi_6\rangle = \begin{pmatrix} \psi_{\mathscr{P}_+} \\ \psi_{\mathscr{P}_-} \\ \psi_{\mathscr{N}_+} \\ \psi_{\mathscr{N}_-} \\ \psi_{\lambda_+} \\ \psi_{\lambda_-} \end{pmatrix} \tag{7.1}$$

and the group obtained by considering transformations on it will be SU(6). (A more ambitious scheme would be to represent each quark by a four-component Dirac spinor; then we obtain SU(12). It seems, unfortunately, that the higher the group, the less successful it is.)

SU(4)

In order to introduce the procedures involved in going from SU(3) to SU(6), and the difficulties that arise, let us first forget about the strange particles and study the simpler case of combining spin and isospin. That is, we consider only the group of transformations on the first four components of (7.1),

$$|\psi_4\rangle = \begin{pmatrix} \psi_{\mathscr{P}_+} \\ \psi_{\mathscr{P}_-} \\ \psi_{\mathscr{N}_+} \\ \psi_{\mathscr{N}_-} \end{pmatrix}, \tag{7.2}$$

which is SU(4). The structure of this group was first investigated long ago by Wigner, who used the four nucleon states to provide the basis and classified nuclei into spin–isospin "supermultiplets" corresponding to its irreducible representations.

There are fifteen generators of SU(4), corresponding to the fifteen linearly independent traceless hermitian 4-by-4 matrices. Three of them can be simultaneously diagonal. In order to analyze this rank-3 group by the Cartan techniques, we first note the presence of four SU(3) subgroups containing transformations which leave one component of (7.2) unaffected, such as

$$U_4 = \begin{pmatrix} U_{11} & U_{12} & U_{13} & 0 \\ U_{21} & U_{22} & U_{23} & 0 \\ U_{31} & U_{32} & U_{33} & 0 \\ 0 & 0 & 0 & 1 \end{pmatrix}. \tag{7.3}$$

Transformations of this form produce SU(3) for the first three states, and SU(1) – i.e. no transformation at all – for the fourth state, and for that reason this is called the SU(3) ⊗ SU(1) decomposition of SU(4). These four subgroups are analogous to the isospin, U-spin, and V-spin SU(2) subgroups of SU(3). Any one of them may be conserved; that is, the diagonal generators may be taken so that two of them are the appropriate SU(3) generators, and the third commutes with the SU(3) subgroup. For (7.3), this means choosing

$$D_1 = \begin{pmatrix} 1 & & & \\ & -1 & & \\ & & 0 & \\ & & & 0 \end{pmatrix}, \qquad D_2 = \begin{pmatrix} 1 & & & \\ & 1 & & \\ & & -2 & \\ & & & 0 \end{pmatrix}.$$

$$D_3 = \begin{pmatrix} 1 & & & \\ & 1 & & \\ & & 1 & \\ & & & -3 \end{pmatrix}, \tag{7.4}$$

Six of the non-diagonal generators will be those of the SU(3) subgroup; the remaining six are easily constructed. The commutation relations of SU(4) can then be established from this four-dimensional elementary representation in the same way as for SU(3).

Since we are conserving the SU(3) subgroup (7.3), its Casimir operators will be diagonal; they commute with D_1' and D_2 by definition, and with D_3 by construction, so they can be diagonalized simultaneously with them. Furthermore, the SU(3) subgroup itself contains a conserved SU(2) subgroup – in this case, the spin group of the $\mathscr{P}$ quark – and its Casimir operator can also be diagonalized. Thus the analysis we are using here produces six quantum numbers.

The detailed analysis of the group in this way leads to the determination of its irreducible representations as in SU(3). Since SU(4) is of rank 3, the representation space is three-dimensional, and there are three Casimir invariants. Unfortunately, however, the mathematically elegant procedures of Cartan are physically meaningless here. None of the three-dimensional subspaces defined as in (7.3) has any particular significance in terms of spin–isospin states. Indeed, to proceed intuitively from our definition of the states (7.2), we would look for the two SU(2) subgroups corresponding to spin and isospin, rather than the SU(3) decomposition given. These are given by the direct product matrices

$$S_1 = \tfrac{1}{2}\begin{pmatrix} 0 & 1 & 0 & 0 \\ 1 & 0 & 0 & 0 \\ 0 & 0 & 0 & 1 \\ 0 & 0 & 1 & 0 \end{pmatrix} \qquad S_2 = \tfrac{1}{2}\begin{pmatrix} 0 & -i & 0 & 0 \\ i & 0 & 0 & 0 \\ 0 & 0 & 0 & -i \\ 0 & 0 & i & 0 \end{pmatrix}$$

$$S_3 = \tfrac{1}{2}\begin{pmatrix} 1 & & & \\ & -1 & & \\ & & 1 & \\ & & & -1 \end{pmatrix}$$

$$I_1 = \tfrac{1}{2}\begin{pmatrix} 0 & 0 & 1 & 0 \\ 0 & 0 & 0 & 1 \\ 1 & 0 & 0 & 0 \\ 0 & 1 & 0 & 0 \end{pmatrix} \qquad I_2 = \tfrac{1}{2}\begin{pmatrix} 0 & 0 & -i & 0 \\ 0 & 0 & 0 & -i \\ i & 0 & 0 & 0 \\ 0 & i & 0 & 0 \end{pmatrix}$$

$$I_3 = \begin{pmatrix} 1 & & & \\ & 1 & & \\ & & -1 & \\ & & & -1 \end{pmatrix} \tag{7.5}$$

It is easily verified that these matrices are representations satisfying

$$[S_i, S_j] = i\epsilon_{ijk}S_k \tag{7.6a}$$

$$[I_i, I_j] = i\epsilon_{ijk}I_k \tag{7.6b}$$

$$[S_i, I_j] = 0, \tag{7.6c}$$

so that **S** and **I** do generate two commuting SU(2) subgroups, as expected. They are therefore chosen to represent six of the generators. The other nine can be constructed in this representation (but not generally) as products of them

$$D_{ij} = S_i \otimes I_j$$

from which it follows that the remaining commutation relations are

$$[S_i, D_{jk}] = i\epsilon_{ijl}D_{lk} \tag{7.6d}$$

$$[I_i, D_{jk}] = i\epsilon_{ik\ell}D_{j\ell} \tag{7.6e}$$

$$[D_{ij}, D_{kl}] = \tfrac{1}{4}i(\delta_{ik}\epsilon_{jlm}I_m + \delta_{jl}\epsilon_{ikm}S_m). \tag{7.6f}$$

In this SU(2) ⊗ SU(2) decomposition, the three diagonal generators are

$$S_3 = \tfrac{1}{2}\begin{pmatrix} 1 & & & \\ & -1 & & \\ & & 1 & \\ & & & -1 \end{pmatrix} = \tfrac{1}{2}[D_1 - \tfrac{1}{3}(D_2 - D_3)]$$

$$I_3 = \tfrac{1}{2}\begin{pmatrix} 1 & & & \\ & 1 & & \\ & & -1 & \\ & & & -1 \end{pmatrix} = \tfrac{1}{6}(2D_2 + D_3)$$

$$D_{33} = \tfrac{1}{4}\begin{pmatrix} 1 & & & \\ & -1 & & \\ & & -1 & \\ & & & 1 \end{pmatrix} = \tfrac{1}{4}[D_1 + \tfrac{1}{3}(D_2 - D_3)]. \qquad (7.7)$$

Now suppose we wish to classify the eigenstates of SU(4) in a spin–isospin "supermultiplet" decomposition, maintaining the SU(2) ⊗ SU(2) structure to the fullest extent possible. Then we expect them to be eigenstates of $\mathbf{I}^2$ and $\mathbf{S}^2$. But the commutation relations imply that in general

$$[\mathbf{I}^2, D_{33}] = i\epsilon_{i3k}(I_i D_{3k} + D_{3k} I_i) \neq 0$$

$$[\mathbf{S}^2, D_{33}] = i\epsilon_{i3k}(S_i D_{k3} + D_{k3} S_i) \neq 0. \qquad (7.8)$$

Thus $\mathbf{I}^2$ and $\mathbf{S}^2$ do not commute with D_{33}, so they cannot be diagonalized simultaneously with it. This means that eigenstates of $\mathbf{I}^2$ and $\mathbf{S}^2$ will not, in general, be eigenstates of D_{33}, and *vice versa.* In fact, it is impossible to construct three diagonal generators which will commute with $\mathbf{I}^2$ and $\mathbf{S}^2$. Since the states in representation space are chosen to be eigenstates of three generators, it follows that eigenstates of $\mathbf{I}^2$ and $\mathbf{S}^2$ cannot be plotted in representation space.

Thus the physically pertinent SU(2) ⊗ SU(2) decomposition of SU(4) differs considerably from the simpler SU(3) ⊗ SU(1) procedure. The structure of the group can be analyzed in either way, of course; but the Cartan methods described earlier, which apply only to the SU(3) ⊗ SU(1) decomposition since they use the representation space, are much more straightforward. Once the irreducible representations have been found in this way, their spin–isospin multiplet structure is easily determined by inspecting the eigenvalue spectra of S_3 and I_3.

In Table 7-A we list some of the smaller irreducible representations of SU(4). As in SU(3), we identify them by giving their dimensionalities, complex conjugate representations being indicated by bars. There are actually five independent 20-dimensional representations. One of these, labeled simply 20, is self-conjugate, but the other two, called 20′ and 20″, are distinct from their conjugates $\overline{20}'$ and $\overline{20}''$. The spin and isospin multiplets into which the eigenstates of these representations may be classified using the SU(2) ⊗ SU(2) decomposition are given by writing their sub-dimensions $(2S + 1, 2I + 1)$, S and I being the total spin and iso-

TABLE 7-A Some smaller irreducible representations of SU(4), showing their labelings [analogous to (λ, μ) in SU(3)] and their spin–isospin decompositions

	(λ, μ, ν)	$SU(2) \otimes SU(2)$ $(2S + 1, 2I + 1)$
1	(0, 0, 0)	(1, 1)
4	(1, 0, 0)	(2, 2)
$\overline{4}$	(0, 0, 1)	(2, 2)
$6 = \overline{6}$	(0, 1, 0)	(1, 3) (3, 1)
10	(2, 0, 0)	(3, 3) (1, 1)
$15 = \overline{15}$	(1, 0, 1)	(3, 3) (3, 1) (1, 3)
$20 = \overline{20}$	(0, 1, 0)	(5, 1) (1, 5) (3, 3) (1, 1)
$20'$	(1, 1, 0)	(4, 2) (2, 4) (2, 2)
$20''$	(3, 0, 0)	(4, 4) (2, 2)
84	(2, 0, 2)	(5, 5), (5, 3), (3, 5), (5, 1), (1, 5), (3, 3), (3, 3), (1, 1)

spin respectively. Thus the 10 representation contains (3,3), an isotriplet with spin one, and (1,1), a spinless isosinglet.

The non-strange mesons will be classified in SU(4) as quark–antiquark bound states. We have defined the quarks as the elementary representation 4; the antiquarks belong correspondingly to $\overline{4}$. The reduction formula for $4 \otimes \overline{4}$ can be shown to be

$$4 \otimes \overline{4} = 15 \oplus 1, \tag{7.9}$$

which tells us that the mesons should appear as singlets or 15-plets. In the 15, which is the regular representation of SU(4), we expect an $S = 1$ isotriplet, an $S = 0$ isotriplet and an $S = 1$ isosinglet. This multiplet therefore accounts very conveniently for the ρ's, the π's, and some combination of ω and ϕ. In a similar way, the baryons are classified according to

$$4 \otimes 4 \otimes 4 = 20'' \oplus 20' \oplus 20' \oplus \overline{4}. \tag{7.10}$$

The nucleons could be classified in any of these representations, but the Δ must belong to the $20''$, since that is the only representation containing the necessary (4,4) state.

We shall not go further into SU(4) classifications, since our principal concern in this chapter is SU(6). Before leaving the smaller group, however, it is interesting to consider the possibility of defining a mass splitting operator in SU(4). We expect all members of a spin–isospin multiplet to have the same mass. If the mass splitting is described by an irreducible tensor operator, as in SU(3), the operator must be scalar and isoscalar, i.e. it must have the transformation properties of the (1, 1) member of some irreducible representation of SU(4). Among the representations listed in Table 7-A only the 10, 20, and 84 contain such a state. Evaluating the mass of a particle belonging to the irreducible representation N of SU(4) will require knowing the Clebsch–Gordan coefficients

$$\begin{pmatrix} N & M & N \\ \nu & \mu & \nu \end{pmatrix}$$

for M = 10, 20, or 84. It is possible to prove, however, that for M = 10 the Clebsches vanish; that is, $10 \otimes N$ does not contain N. Thus the simplest acceptible mass splitting operator in SU(4) transforms as an irreducible tensor corresponding to the 20 representation. This is a more complicated result than in SU(3), where the mass splitting operator could be assigned to the regular representation.

SU(6)

Spin and SU(3) may be combined by a very similar procedure. The elementary representation, constructed using the basis states ($\mathscr{P}_\pm$, $\mathscr{N}_\pm$, $\lambda_\pm$), is six-dimensional. There are thirty-five generators, corresponding to the thirty-five linearly independent traceless hermitian 6-by-6 matrices, of which five can be chosen diagonal. Therefore the rank of SU(6) is 5, and the representation space is five-dimensional.

The commutation relations of the generators may be obtained in the $SU(3) \otimes SU(2)$ direct product representation we are using analogously to (7.6). We label eight of the generators F_a, as in SU(3), and assume that they reproduce that subgroup with respect to the SU(3) indices without affecting the spin indices. In the basis defined by (7.1), these are represented by

$$F_1 = \tfrac{1}{2}\begin{pmatrix} 0 & 0 & 1 & 0 & 0 & 0 \\ 0 & 0 & 0 & 1 & 0 & 0 \\ 1 & 0 & 0 & 0 & 0 & 0 \\ 0 & 1 & 0 & 0 & 0 & 0 \\ 0 & 0 & 0 & 0 & 0 & 0 \\ 0 & 0 & 0 & 0 & 0 & 0 \end{pmatrix} \quad F_2 = \tfrac{1}{2}\begin{pmatrix} 0 & 0 & -i & 0 & 0 & 0 \\ 0 & 0 & 0 & -i & 0 & 0 \\ i & 0 & 0 & 0 & 0 & 0 \\ 0 & i & 0 & 0 & 0 & 0 \\ 0 & 0 & 0 & 0 & 0 & 0 \\ 0 & 0 & 0 & 0 & 0 & 0 \end{pmatrix}$$

$$F_3 = \tfrac{1}{2}\begin{pmatrix} 1 & 0 & 0 & 0 & 0 & 0 \\ 0 & 1 & 0 & 0 & 0 & 0 \\ 0 & 0 & -1 & 0 & 0 & 0 \\ 0 & 0 & 0 & -1 & 0 & 0 \\ 0 & 0 & 0 & 0 & 0 & 0 \\ 0 & 0 & 0 & 0 & 0 & 0 \end{pmatrix} \quad F_4 = \tfrac{1}{2}\begin{pmatrix} 0 & 0 & 0 & 0 & 1 & 0 \\ 0 & 0 & 0 & 0 & 0 & 1 \\ 0 & 0 & 0 & 0 & 0 & 0 \\ 0 & 0 & 0 & 0 & 0 & 0 \\ 1 & 0 & 0 & 0 & 0 & 0 \\ 0 & 1 & 0 & 0 & 0 & 0 \end{pmatrix}$$

$$F_5 = \tfrac{1}{2}\begin{pmatrix} 0 & 0 & 0 & 0 & -i & 0 \\ 0 & 0 & 0 & 0 & 0 & -i \\ 0 & 0 & 0 & 0 & 0 & 0 \\ 0 & 0 & 0 & 0 & 0 & 0 \\ i & 0 & 0 & 0 & 0 & 0 \\ 0 & i & 0 & 0 & 0 & 0 \end{pmatrix} \quad F_6 = \tfrac{1}{2}\begin{pmatrix} 0 & 0 & 0 & 0 & 0 & 0 \\ 0 & 0 & 0 & 0 & 0 & 0 \\ 0 & 0 & 0 & 0 & 1 & 0 \\ 0 & 0 & 0 & 0 & 0 & 1 \\ 0 & 0 & 1 & 0 & 0 & 0 \\ 0 & 0 & 0 & 1 & 0 & 0 \end{pmatrix}$$

$$F_7 = \tfrac{1}{2}\begin{pmatrix} 0 & 0 & 0 & 0 & 0 & 0 \\ 0 & 0 & 0 & 0 & 0 & 0 \\ 0 & 0 & 0 & 0 & -i & 0 \\ 0 & 0 & 0 & 0 & 0 & -i \\ i & 0 & 0 & 0 & 0 & 0 \\ 0 & i & 0 & 0 & 0 & 0 \end{pmatrix} \quad F_8 = \frac{1}{2\sqrt{3}}\begin{pmatrix} 1 & & & & & \\ & 1 & & & & \\ & & 1 & & & \\ & & & 1 & & \\ & & & & -2 & \\ & & & & & -2 \end{pmatrix} \tag{7.11}$$

Similarly, the spin operators affect only the spin indices:

$$S_1 = \tfrac{1}{2}\begin{pmatrix} 0 & 1 & 0 & 0 & 0 & 0 \\ 1 & 0 & 0 & 0 & 0 & 0 \\ 0 & 0 & 0 & 1 & 0 & 0 \\ 0 & 0 & 1 & 0 & 0 & 0 \\ 0 & 0 & 0 & 0 & 0 & 1 \\ 0 & 0 & 0 & 0 & 1 & 0 \end{pmatrix} \quad S_2 = \tfrac{1}{2}\begin{pmatrix} 0 & -i & 0 & 0 & 0 & 0 \\ i & 0 & 0 & 0 & 0 & 0 \\ 0 & 0 & 0 & -i & 0 & 0 \\ 0 & 0 & i & 0 & 0 & 0 \\ 0 & 0 & 0 & 0 & 0 & -i \\ 0 & 0 & 0 & 0 & i & 0 \end{pmatrix}$$

$$S_3 = \tfrac{1}{2}\begin{pmatrix} 1 & & & & & \\ & -1 & & & & \\ & & 1 & & & \\ & & & -1 & & \\ & & & & 1 & \\ & & & & & -1 \end{pmatrix} \tag{7.12}$$

In the elementary representation the remaining twenty-four generators are given by products of these,

$$D_{i\alpha} = S_i F_\alpha. \tag{7.13}$$

Using (7.11), (7.12), and (7.13) we obtain the SU(6) commutation relations

$$[F_\alpha, S_i] = 0$$

$$[F_\alpha, F_\beta] = if_{\alpha\beta\gamma}F_\gamma$$

$$[S_i, S_j] = i\epsilon_{ijk}S_k$$

$$[F_\alpha, D_{i\beta}] = if_{\alpha\beta\gamma}D_{i\gamma}$$

$$[S_i, D_{j\alpha}] = i\epsilon_{ijk}D_{k\alpha}$$

$$[D_{i\alpha}, D_{j\beta}] = \tfrac{1}{4}i\ \delta_{ij}f_{\alpha\beta\gamma}F_\gamma + \tfrac{1}{6}i\ \delta_{\alpha\beta}\epsilon_{ijk}S_k + \tfrac{1}{2}\ i\epsilon_{ijk}d_{\alpha\beta\gamma}D_{k\gamma} \tag{7.14}$$

where $f_{\alpha\beta\gamma}$ and $d_{\alpha\beta\gamma}$ are given in (4.21) and (4.23). The five diagonal generators in this SU(3) ⊗ SU(2) notation are F_3, F_8, S_3, D_{33}, and D_{38}.

As in SU(4), however, the physically motivated SU(3) ⊗ SU(2) decomposition of SU(6) is not amenable to the Cartan-type analysis. The actual construction of the irreducible representations, and their classification into direct product states of SU(3) and spin, is a long and complicated process, even in the SU(5) chain decomposition. We shall not go into the details here; the important representations of SU(6), and their SU(3) ⊗ SU(2) content, are listed in Table 7-B.

There is a third decomposition of SU(6), in addition to SU(3) ⊗ SU(2) and the SU(5) chain, which has some physical significance. Suppose we maintain the SU(4) subgroup structure we have already studied of the non-strange quark states, plus the SU(2) spin algebra of the λ quarks. This is accomplished by choosing fifteen of the generators in the form

TABLE 7-B Some smaller irreducible representations of SU(6), with their SU(3) ⊗ SU(2) and SU(4) ⊗ SU(2) decompositions. For the latter, the entries are ordered in lines corresponding to decreasing values of hypercharge, with N_λ showing the number of λ quarks in the state (numbers of antiquarks indicated by bars)

	SU(3) ⊗ SU(2) $(N_3, 2S+1)$	SU(4) ⊗ SU(2) (N_4, N_2)	(N_λ)
6	(3, 2)	(4, 1)	(0)
		(1, 2)	(1)
15	$(\bar{3}, 3)$	(6, 1)	(0)
	(6, 1)	(4, 2)	(1)
		(1, 1)	(2)
20	(8, 2)	$(\bar{4}, 2)$	(0)
	(1, 4)	(6, 2)	(1)
		(4, 1)	(2)
21	(6, 3)	(10, 1)	(0)
	(3, 1)	(4, 2)	(1)
		(1, 3)	(2)
35	(8, 3)	(4, 2)	$(\bar{1})$
	(8, 1)	(15, 1), (1, 3), (1, 1)	$(1 + \bar{1})$
	(1, 1)	$(\bar{4}, 2)$	(1)
56	(10, 4)	(20″, 1)	(0)
	(8, 2)	(10, 2)	(1)
		(4, 3)	(2)
		(1, 4)	(3)
70	(10, 2)	(20, 1)	(0)
	(8, 4)	(10, 2), (6, 2)	(1)
	(8, 2)	(4, 1), (4, 3)	(2)
	(1, 2)	(1, 2)	(3)

$$H_i = \left(\begin{array}{cccc|cc} & & & & 0 & 0 \\ & & U_i & & 0 & 0 \\ & & & & 0 & 0 \\ & & & & 0 & 0 \\ \hline 0 & 0 & 0 & 0 & 0 & 0 \\ 0 & 0 & 0 & 0 & 0 & 0 \end{array}\right), \tag{7.15a}$$

where U_i are the appropriate four-dimensional matrix representations of the SU(4) generators, and three more as

$$S_i^{(\lambda)} = \left(\begin{array}{cccc|cc} & & & & 0 & 0 \\ & & 0 & & 0 & 0 \\ & & & & 0 & 0 \\ & & & & 0 & 0 \\ \hline 0 & 0 & 0 & 0 & \multicolumn{2}{c}{\multirow{2}{*}{τ_i}} \\ 0 & 0 & 0 & 0 & & \end{array}\right) \tag{7.15b}$$

with τ_i the Pauli spin matrices. Four diagonal generators are included in (7.15a); the fifth will commute with all of the H_i and S_i if we take it to be the hypercharge,

$$Y = \tfrac{1}{3}\begin{pmatrix} 1 & & & & & \\ & 1 & & & & \\ & & 1 & & & \\ & & & 1 & & \\ & & & & -2 & \\ & & & & & -2 \end{pmatrix} \tag{7.15c}$$

The remaining sixteen generators, corresponding to the matrix elements which vanish in both H_i and S_i, are easily constructed.

Although the eigenstates appropriate to this SU(4) ⊗ SU(2) decomposition, like those of SU(3) ⊗ SU(2), cannot be plotted in representation space, it is still useful to know the SU(4) ⊗ SU(2) content of an irreducible representation of SU(6). This is given also in Table 7-B, where for each value of Y we show (N_4, N_2), N_4 and N_2 being the dimensionalities of the SU(4) and SU(2) representations. Thus N_4 indicates the classification of the non-strange quarks, while N_2 shows how the spins of the strange quarks are coupled.

For example, let us inspect the structure of the 15-dimensional representation of SU(6). With respect to the SU(3) ⊗ SU(2) decomposition, there are two "submultiplets", (6, 1) and $(\bar{3}, 3)$; thus the 15 contains a spin-zero sextet and a spin-one antitriplet. Alternatively, we may look at the SU(4) ⊗ SU(2) structure. At Y = 2/3, there is a 6 of SU(4), i.e. a spin-zero isotriplet and a spin-one isosinglet. (The former state occurs in the (6, 1) of the SU(3) ⊗ SU(2) decomposition, the latter in the $(\bar{3}, 3)$.) Similarly a 4 of SU(4) occurs at Y = –2/3, accompanied by a single λ quark. At Y = –2/3 there is an SU(4) singlet; this value of the hypercharge implies the presence of two λ quarks, their spins being coupled to J = 1.

SU(6) Assignments

As in SU(3) and SU(4), the mesons are assigned in SU(6) to quark–antiquark states, i.e. to $6 \otimes \overline{6}$. The reduction formula

$$6 \otimes \overline{6} = 35 \oplus 1 \tag{7.16}$$

leads once again to meson classification in the regular representation, in this case the 35. Table 7-B shows that its SU(3) ⊗ SU(2) decomposition

$$35 = (8, 1) + (8, 3) + (1, 3)$$

contains an octet with spin zero and an octet plus a singlet with spin one. These thirty-five states are in exact agreement with the observed spectrum of pseudoscalar and vector mesons, including the singlet state needed for ω–ϕ mixing. The pseudoscalar octet, however, is not accompanied here by a singlet state; this apparently explains the fact that singlet–octet mixing was not necessary for it. The pseudoscalar and vector mesons may therefore be represented by quark–antiquark bound states according to the wave functions given in Table 7-C.

The baryons are to be classified as three-quark states. The reduction formula

$$6 \otimes 6 \otimes 6 = 20 \oplus 70 \oplus 70 \oplus 56 \tag{7.17}$$

is the SU(6) analogue of (5.11b) and (7.10). The SU(3) ⊗ SU(2) reductions of these representations are

$$20 = (8, 2) + (1, 4) \tag{7.18a}$$

$$70 = (10, 2) + (8, 4) + (8, 2) + (1, 2) \tag{7.18b}$$

$$56 = (10, 4) + (8, 2). \tag{7.18c}$$

In contrast to the mesons, we have here considerable latitude in making particle assignments. The $J^P = 3/2^+$ baryon decuplet can only be accommodated in the 56, since neither 70 nor 20 contains (10, 4); but all three representations contain (8, 2), an octet of spin ½ baryons. To which of them should the B octet be assigned?

If we expect that all states of an irreducible representation will be filled, the 70 is immediately ruled out by the absence of a $\frac{1}{2}^+$ decuplet among the known particles. Similarly, the 20 requires an SU(3) singlet with $J^P = 3/2^+$ in addition to the octet. No such particle has been observed. The 56, however, is exactly filled

TABLE 7-C SU(6) wave functions of the meson 35

Vector mesons

$\rho^+_{+1} = \bar{\mathscr{N}}_+ \mathscr{P}_+$ $\rho^\circ_{+1} = 1/\sqrt{2}\,(\bar{\mathscr{P}}_+ \mathscr{P}_+ - \bar{\mathscr{N}}_+ \mathscr{N}_+)$

$\rho^-_{+1} = \bar{\mathscr{P}}_+ \mathscr{N}_+]$

$K^{*+}_{+1} = \bar{\lambda}_+ \mathscr{P}_+$ $K^{*\circ}_{+1} = \bar{\lambda}_+ \mathscr{N}_+$

$\bar{K}^{*\circ}_{+1} = \bar{\mathscr{N}}_+ \lambda_+$ $K^{*-}_{+1} = \bar{\mathscr{P}}_+ \lambda_+$

$\phi^{(8)}_{+1} = 1/\sqrt{6}\,(2\bar{\lambda}_+ \lambda_+ - \bar{\mathscr{P}}_+ \mathscr{P}_+ - \bar{\mathscr{N}}_+ \mathscr{N}_+)$

$\omega^{(1)}_{+1} = 1/\sqrt{3}\,(\bar{\lambda}_+ \lambda_+ + \bar{\mathscr{P}}_+ \mathscr{P}_+ + \bar{\mathscr{N}}_+ \mathscr{N}_+)$

To get $S_3 = 0$ states, change $\bar{a}_+ b_+$ to $1/\sqrt{2}\,(\bar{a}_+ b_- + \bar{a}_- b_+)$

To get $S_3 = -1$ states, change $\bar{a}_+ b_+$ to $\bar{a}_- b_-$.

Pseudoscalar mesons

$\pi^+ = (\bar{\mathscr{N}} \mathscr{P})$ $\pi^\circ = 1/\sqrt{2}\,((\bar{\mathscr{P}} \mathscr{P}) - (\bar{\mathscr{N}} \mathscr{N}))$ $\pi^- = (\bar{\mathscr{P}} \mathscr{N})$

$K^+ = (\bar{\lambda} \mathscr{P})$ $K^\circ = (\bar{\lambda} \mathscr{N})$

$\bar{K}^\circ = (\bar{\mathscr{N}} \lambda)$ $K^- = (\bar{\mathscr{P}} \lambda)$

$\eta^\circ = 1/\sqrt{6}\,(2(\bar{\lambda}\lambda) - (\bar{\mathscr{P}} \mathscr{P}) - (\bar{\mathscr{N}} \mathscr{N}))$

The quark spin states are denoted here by $(\bar{a}b) = 1/\sqrt{2}\,(\bar{a}_+ b_- - \bar{a}_- b_+)$.

by the decuplet plus the octet. Thus the two most prominent SU(3) baryon multiplets can be elegantly classified in a single irreducible representation of SU(6).

There is, furthermore, some experimental evidence for the 56 assignment of the B octet. Suppose we consider the PBB coupling constants defined in (6.43). In SU(3), there were two independent couplings of $8 \otimes 8$ to 8, leading us to define the D/F ratio. In SU(6), however, these couplings involve a 35 meson and two

56 baryons. Picturing the process as a meson decaying to a baryon–antibaryon pair, we must couple a 35 to

$$56 \otimes \overline{56} = 1 \oplus 35 \oplus 405 \oplus 2695. \tag{7.19}$$

Since (7.19) contains a single 35, there is a unique coupling; that is, in SU(6) the D/F ratio is determined. We shall not go into the calculational details here; the result is that

TABLE 7-D SU(6) wave functions of the baryon 56

Decuplet baryons

$$\Delta^{++}_{++} = \mathscr{P}_+ \mathscr{P}_+ \mathscr{P}_+$$

$$\Delta^{+}_{++} = 1/3\,(\mathscr{P}_+ \mathscr{P}_+ \mathscr{N}_+ + \mathscr{P}_+ \mathscr{N}_+ \mathscr{P}_+ + \mathscr{N}_+ \mathscr{P}_+ \mathscr{P}_+)$$

$$\Delta^{\circ}_{++} = 1/3\,(\mathscr{P}_+ \mathscr{N}_+ \mathscr{N}_+ + \mathscr{N}_+ \mathscr{P}_+ \mathscr{N}_+ + \mathscr{N}_+ \mathscr{N}_+ \mathscr{P}_+)$$

$$\Delta^{-}_{++} = \mathscr{N}_+ \mathscr{N}_+ \mathscr{N}_+$$

$$Y^{*-}_{++} = 1/3\,(\lambda_+ \mathscr{P}_+ \mathscr{P}_+ + \mathscr{P}_+ \lambda_+ \mathscr{P}_+ + \mathscr{P}_+ \mathscr{P}_+ \lambda_+)$$

$$Y^{*+}_{++} = 1/6\,(\lambda'_+ \mathscr{P}_+ \mathscr{N}_+ + \lambda_+ \mathscr{N}_+ \mathscr{P}_+ + \mathscr{P}_+ \lambda_+ \mathscr{N}_+ + \mathscr{P}_+ \mathscr{N}_+ \lambda_+ +$$
$$+ \mathscr{N}_+ \lambda_+ \mathscr{P}_+ + \mathscr{N}_+ \mathscr{P}_+ \lambda_+)$$

$$Y^{*-}_{++} = 1/3\,(\lambda_+ \mathscr{N}_+ \mathscr{N}_+ + \mathscr{N}_+ \lambda_+ \mathscr{N}_+ + \mathscr{N}_+ \mathscr{N}_+ \lambda_+)$$

$$\Xi^{*\circ}_{++} = 1/3\,(\lambda_+ \lambda_+ \mathscr{P}_+ + \lambda_+ \mathscr{P}_+ \lambda_+ + \mathscr{P}_+ \lambda_+ \lambda_+)$$

$$\Xi^{*\circ}_{++} = 1/3\,(\lambda_+ \lambda_+ \mathscr{N}_+ + \lambda_+ \mathscr{N}_+ \lambda_+ + \mathscr{N}_+ \lambda_+ \lambda_+)$$

$$\Omega^{-}_{++} = \lambda_+ \lambda_+ \lambda_+$$

To get $S_3 = +\frac{1}{2}$ states, change $a_+ b_+ c_+$ to $(a_+ b_+ c_- + a_+ b_- c_+ + a_- b_+ c_+)/\sqrt{3}$.

To get $S_3 = -\frac{1}{2}$ states, change $a_+ b_+ c_+$ to $(a_+ b_- c_- + a_- b_+ c_- + a_- b_- c_+)/\sqrt{3}$.

To get $S_3 = -3/2$ states, change $a_+ b_+ c_+$ to $a_- b_- c_-$.

$$\frac{D}{D+F} = \frac{3}{5}\ , \qquad (7.20)$$

which is consistent with the observed values of ~0.7. If we assigned the B octet to the 20, (7.19) would be replaced by

$$20 \otimes \overline{20} = 1 \oplus 35 \oplus 189 \oplus 175,$$

leading again to a unique determination of D/F. In this case it turns out that only D-type coupling is allowed, i.e.

$$\frac{D}{D+F} = 1.$$

TABLE 7-D (continued)

Octet baryons

$$p = 1/\sqrt{3}\ [(\mathcal{P}\mathcal{P}\mathcal{N})_m + (\mathcal{P}\mathcal{N}\mathcal{P})_m + (\mathcal{N}\mathcal{P}\mathcal{P})_m]$$

$$n = 1/\sqrt{3}\ [(\mathcal{N}\mathcal{N}\mathcal{P})_m + (\mathcal{N}\mathcal{P}\mathcal{N})_m + (\mathcal{P}\mathcal{N}\mathcal{N})_m]$$

$$\Sigma^+ = 1/\sqrt{3}\ [(\mathcal{P}\mathcal{P}\lambda)_m + (\mathcal{P}\lambda\mathcal{P})_m + (\lambda\mathcal{P}\mathcal{P})_m]$$

$$\Sigma^\circ = 1/\sqrt{6}\ [(\mathcal{P}\mathcal{N}\lambda)_m + (\mathcal{P}\lambda\mathcal{N})_m + (\lambda\mathcal{P}\mathcal{N})_m + (\lambda\mathcal{N}\mathcal{P})_m + \\ + (\mathcal{N}\mathcal{P}\lambda)_m + (\mathcal{N}\lambda\mathcal{P})_m]$$

$$\Sigma^- = -1/\sqrt{3}\ [(\mathcal{N}\mathcal{N}\lambda)_m + (\mathcal{N}\lambda\mathcal{N})_m + (\lambda\mathcal{N}\mathcal{N})_m]$$

$$\Xi^\circ = -1/\sqrt{3}\ [(\lambda\lambda\mathcal{P})_m + (\lambda\mathcal{P}\lambda)_m + (\mathcal{P}\lambda\lambda)_m]$$

$$\Xi^- = 1/\sqrt{3}\ [(\lambda\lambda\mathcal{N})_m + (\lambda\mathcal{N}\lambda)_m + (\mathcal{N}\lambda\lambda)_m]$$

For $S_3 = \pm\frac{1}{2}$ states, $(abc)_{\pm 1/2} = (2a_\pm b_\pm c_\mp - a_\pm b_\mp c_\pm - a_\mp b_\pm c_\mp)/\sqrt{6}$.

$$\Lambda^\circ_{\pm 1/2} = 1/\sqrt{12}\ [\mathcal{P}_+\mathcal{N}_-\lambda_\pm - \mathcal{P}_-\mathcal{N}_+\lambda_\pm - \mathcal{N}_+\mathcal{P}_-\lambda + \mathcal{N}_+\mathcal{P}_-\lambda_\pm \\ + \lambda_\pm\mathcal{P}_+\mathcal{N}_- - \lambda_\pm\mathcal{P}_-\mathcal{N}_+ - \lambda_\pm\mathcal{N}_+\mathcal{P}_- + \lambda_\pm\mathcal{N}_-\mathcal{P}_+ \\ + \mathcal{P}_+\lambda_\pm\mathcal{N}_- - \mathcal{P}_-\lambda_\pm\mathcal{N}_+ - \mathcal{N}_+\lambda_\pm\mathcal{P}_- + \mathcal{N}_-\lambda_\pm\mathcal{P}_+]$$

Thus the D/F ratio indicates that the 56 classification is the correct one.

The ½$^+$ octet and the 3/2$^+$ decuplet may therefore be assigned three-quark wave functions appropriate to the 56. These are given in Table 7-D. One significant puzzle associated with this classification scheme should be pointed out here. The 56 is symmetric under permutations of the quark labels. This fact, shown in detail in Appendix A, is obvious in the wave functions in Table 7-D; interchanging any pair of quarks does not alter the wave function. But if the quarks have spin ½, they should be fermions, and the overall baryon wave function should be antisymmetric. If the SU(6) part of the wave function is symmetric, then the spatial part of the wave function must be antisymmetric.

Ordinarily, however, the ground state wave function of a many-particle system turns out to be symmetric. Since the B octet is certainly the "ground state" of the baryons, it should therefore have a symmetric spatial wave function. Furthermore, it is generally true that an antisymmetric spatial wave function leads to a form factor $F(q^2) = \int d^3r|\psi(\mathbf{r})|^2 e^{i\mathbf{q}\cdot\mathbf{r}}$ which has a zero for some value of q^2, and the known proton form factor does not vanish out to quite large values of q^2. Therefore it seems that the B octet should have a symmetric spatial wave function and an antisymmetric SU(6) wave function. This means that it should belong to the 20, which is the totally antisymmetric three-quark state, rather than to the 56. Since quarks have proved so elusive, however, arguments based on their statistics may be of questionable validity; in any case, the 56 classification is at present accepted.

Consequences of SU(6)

Attempts to use SU(6) in studying the properties of hadrons and of their interactions have, in general, led to more difficulties than successes. It would be expected that the non-relativistic treatment of spin might lead to problems in relativistic situations. At any energy, however, spin is a dynamical quantity, so SU(6) inevitably mixes physical space with isospin-hypercharge space. Perhaps for this reason, even static SU(6) (i.e. the zero-energy limit) has not been notably successful. To illustrate the difficulty, we need only observe that in SU(6) there is no easy way to account for orbital angular momentum. The decay of a vector meson into two pseudoscalar mesons, $\rho^\circ \to \pi^+\pi^-$ for example, is forbidden in SU(6) because the total spin must be conserved. Such decays are observed, of course, since the spin of the ρ° becomes orbital angular momentum in the two-pion system. A number of attempts to extend SU(6) further and avoid such problems have been made; to discuss them here would carry us beyond the scope of this book, however. We shall therefore mention only two of the relatively successful results of SU(6); its failures are amply documented elsewhere.

The mass splitting operator in SU(3) was satisfactorily identified as a tensor operator transforming like the hypercharge. We may attempt a similar identification in SU(6) by assuming that the mass splitting still transforms like the appropriate member Y of the regular representation; that is, we assign it to the 35. It turns out, unfortunately, that this assignment leads to the results of the Gell-Mann–Okubo formula (6.15) for the entire SU(6) multiplet, so the mass of the Δ is not split from that of the nucleon, or the ρ from the π, etc. The reason for this failure is easily seen in the light of our discussion of mass operators in SU(4). In order to separate states with different eigenvalues of S^2, such as N and Δ, the SU(4) $\otimes$ SU(2) decomposition of the SU(6) mass operator must contain either (20, 1) or (84, 1). The 35 does not contain either of these; its content for Y = 0 is (15, 1) + (1, 3) + (1, 1), of which only the (1, 1) is acceptable for an SU(4) mass operator. Therefore the 35 will not introduce any splitting of the SU(4) multiplets in Table 7-B.

Thus the mass operator in SU(6) must contain higher-order tensor contributions. The conditions we would impose on it may be summarized as follows:

a) it must be diagonal;

b) it should be a scalar with respect to the spin part of the SU(3) $\otimes$ SU(2) decomposition, since the mass does not depend on the eigenvalue of S_3;

c) it should transform under the SU(3) subgroup like singlet or octet, as suggested by the Gell-Mann–Okubo formula;

d) With respect to the SU(4) $\oplus$ SU(2) decomposition, it must belong to (1, 1), (20, 1), or (84, 1).

In addition to the 35, only two representations with dimensionality less than 1000 satisfy these conditions. These are the 189 and the 405; both contain two acceptable terms, namely (8, 1) and (1, 1) under SU(3) $\otimes$ SU(2). A satisfactory form for a general SU(6) mass operator is therefore

$$M = M_1^1 + M_{35}^8 + M_{189}^1 + M_{189}^8 + M_{405}^1 + M_{405}^8, \qquad (7.21)$$

where the subscript denotes the SU(6) representation and the superscript its SU(3) content.

The implications of (7.21) for the 56 baryons follow directly from evaluating the matrix element of M. Since

$$56 \otimes \overline{56} = 1 \oplus 35 \oplus 405 \oplus 2695,$$

the 189 terms do not contribute here. It follows that all of the baryon masses are given in terms of four invariant amplitudes and the SU(6) Clebsch–Gordan coefficients. In fact, it can be shown that in this case (7.21) is equivalent to a simple extension of the Gell-Mann–Okubo formula,

$$M = m_0 + aY + b(Y^2 - \tfrac{1}{4}I(I+1)) + cS(S+1). \tag{7.22}$$

Thus the successful results of SU(3) are reproduced. In addition, however, (7.22) predicts that the same values of a and b hold in both octet and decuplet, i.e. that

$$M(\Xi) - M(\Sigma) = M(\Xi^*) - M(Y_1^*). \tag{7.23}$$

Experimentally, (7.23) yields 130 MeV versus 145 MeV, which is reasonably good agreement.

For the 35, however, we have

$$35 \otimes 35 = 1 \oplus 35 \oplus 35 \oplus 189 \oplus 280 \oplus \overline{280} \oplus 405$$

so that all six terms of (7.21) contribute. Furthermore, there are two independent 35 terms, the SU(6) analogue of D and F coupling in $8 \otimes 8$. Therefore there are seven independent reduced matrix elements for the meson masses. Because of this complexity, only the familiar equation

$$4m^2(K) = m^2(\pi) + 3m^2(\eta^\circ)$$

is obtained; no other relations follow from SU(6).

Thus the easy elegance of the Gell-Mann–Okubo mass formula in SU(3) is lost when we go to SU(6). Even though (7.22) is close to it, contributions from the 405 were necessary to obtain this relation, whereas in SU(3) no representation larger than the regular representation was required.

As a second example of SU(6) we consider briefly meson–baryon scattering. If both the initial and the final states consist of a 35 meson plus a 56 baryon, the reduced amplitudes describing the scattering process correspond to the irreducible representations in the expansion

$$35 \otimes 56 = 56 \oplus 70 \oplus 700 \oplus 1134.$$

If we ignore dynamical complications, then, all of these scattering amplitudes are given in terms of only four independent ones. There are six elastic amplitudes among these which are easily measured, for $K^\pm p$, $K^\pm n$, and $\pi^\pm p$ scattering. Expressing these six in terms of the four independent amplitudes leads to two relations between them, known as the *Johnson–Treiman relations*:

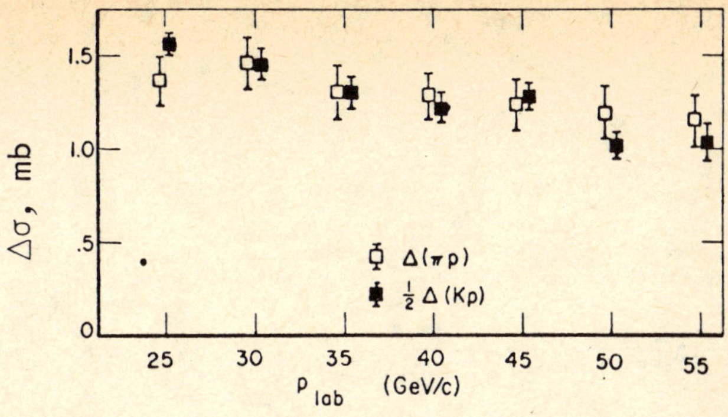

Fig.(7-1) Comparison with experiment of the Johnson–Treiman relation $\Delta(\pi p) = \frac{1}{2}\Delta(Kp)$, where $\Delta(Mp) = \sigma(Mp) - \sigma(\overline{M}p)$. A similar comparison with $\Delta(Kn)$ indicates that the neutron differences, with much larger error bars, are somewhat too large. (From S. Meshkov, *Phys. Rev.* **D6**, 399 (1972).)

$$\tfrac{1}{2}(\mathscr{T}_e(K^-p) - \mathscr{T}_e(K^+p)) = \mathscr{T}_e(\pi^-p) - \mathscr{T}_e(\pi^+p) = \mathscr{T}_e(K^-n) - \mathscr{T}_e(K^+n).$$

Experimentally these equations are valid to about 15%, as shown in Fig.(7-1). By combining them appropriately, however, we obtain the antisymmetric sum rule derived by Barger and Rubin using SU(3) plus octet dominance,

$$\mathscr{T}_e(\pi^+p) - \mathscr{T}_e(\pi^-p) + \mathscr{T}_e(K^-p) - \mathscr{T}_e(K^-n) + \mathscr{T}_e(K^+n) - \mathscr{T}_e(K^+p) = 0,$$

which is in considerably better agreement with the data.

A number of other relations can be obtained in this way, involving inelastic meson–baryon scattering as well as baryon–baryon interactions. While there are some successful predictions among these, there are also many that are mediocre and a number that are poor. The difficulties involved in combining SU(3) with dynamics are probably responsible, but as yet no clear way of avoiding these failures has emerged.

Chapter 8

THE INDEPENDENT QUARK MODEL

One of the most frustrating facts in fundamental particle physics is that the free quark has proved so elusive. When Gell-Mann and Zweig, prompted by the success of the eightfold way, suggested these elementary objects as a physical source of the observed symmetries, the way seemed open to an understanding of the strong interactions as a newer and deeper form of nuclear physics. The principal proviso on this conjecture was the necessity of observing the free quark, the "hydrogen nucleus" of the system, which would be distinguished by its fractional electric charge. When none were found, the emphasis in research turned quickly to a more recondite mathematical viewpoint which regarded quarks as a group theoretic abstraction rather than as a physical entity. In other words, physical quarks became mathematical quirks. It was assumed that SU(3) is, for some unknown reason, a symmetry of the strong interactions Hamiltonian allowing only those representations with integral triality. Attempts were made to enlarge the theory by building larger groups containing spin and by making it fully relativistic. As we have indicated, however, little of this work was successful.

When the higher group theories proved unsatisfactory, a reaction set in which led to a more literal physical picture of the quark model. The observation which prompted most of this work is a simple one: if mesons are made up of two quarks, while baryons consist of three, then the ratio of mesonic to baryonic cross sections should be about 2/3. The experimental data on total cross sections at high energies seem compatible with this result; for example, above 10 GeV/c laboratory momentum $\sigma(\pi^- p)/\sigma(pp)$ is about 0.62.

In this chapter we shall study the independent quark model, which has grown from the success of that startlingly simple observation. The basic idea here is that everything happens to the hadrons because it happens to the quarks. Two areas of investigation have developed, dealing respectively with low-energy and high-energy quark phenomena. We shall describe some of the important results of each of these.

The Low-Energy Quark Model

The idea that the quarks are indeed physically present inside the hadrons allows us to follow a more intuitive logic in discussing their properties. For example, we consider first the description of the mass splitting in SU(3) as approached from the independent quark model.

The mass of a hadron is clearly equal to the sum of the masses of the quarks inside it, less the binding energy. Suppose first that the binding energy is approximately independent of which quarks are involved. Then the mass differences of the hadrons reflect the mass differences of the quarks; since mass is independent of I_3, we have $m(\mathscr{P}) = m(\mathscr{N})$, but $m(\lambda)$ may be different from these. Indeed, the equal spacing law of the baryon decuplet is easily explained in this way, and implies $m(\lambda) - m(\mathscr{P}) \approx 145$ MeV.

A similar explanation can be made of the $\omega - \phi$ splitting. If $m(\mathscr{P}) = m(\mathscr{N}) \neq m(\lambda)$, then the mass eigenstates of the quark–antiquark system corresponding to ω and ϕ will be

$$\omega = 1/\sqrt{2}\,(\mathscr{P}\bar{\mathscr{P}} + \mathscr{N}\bar{\mathscr{N}})$$

$$\phi = \lambda\bar{\lambda} \tag{8.1}$$

since these states contain only equally massive quarks. Comparison of (8.1) with (6.21) shows that $\cos\theta_V = \sqrt{2/3}$, i.e. that the mixing angle $\theta_V = 35.5°$, in reasonably good agreement with the value obtained earlier.

This simplistic picture also tells us that the mass differences in the vector meson nonet are due to the heavier λ quark. Suppose $m(\mathscr{P}) = m(\mathscr{N}) = m$, $m(\lambda) = M$, and the binding energy B is constant throughout the nonet. Then we have

$$m(\omega) = m(\rho) = 2m + B$$

$$m(K^*) = m(\bar{K}^*) = m + M + B$$

$$m(\phi) = 2M + B, \tag{8.2}$$

implying that

$$2m(K^*) = m(\rho) + m(\phi) = m(\omega) + m(\phi). \tag{8.3}$$

Equation (8.2) predicts the equality of the ω and ρ masses, which differ experimentally by less than 3%. Using the measured K^* and ϕ masses in (8.3) yields $m(\omega) = m(\rho) \approx 760$ MeV, again in good agreement. Finally, we note that here

$M - m \approx 130$ MeV, which is not incompatible with the value estimated from the baryon decuplet.

This simple picture fails to describe the pseudoscalar mesons, however; since $m(\eta^\circ) \gg m(\pi)$, we must identify the η° as the $\lambda\bar{\lambda}$ state, and (8.3) then becomes

$$2m(K) = m(\eta^\circ) + m(\pi),$$

which disagrees with experiment. It is also less than satisfactory for the baryon octet, where it predicts $m(\Lambda^\circ) = m(\Sigma^\circ)$. On the whole, however, the model works surprisingly well considering the naivety of its assumptions.

A more detailed version can be constructed in which the requirement that the binding energy is SU(3)-symmetric is relaxed. If we assume that only two-body forces are present, then the mass will be given by

$$M = \sum_i m(i) - \sum_{i \neq j} B(i, j), \tag{8.4}$$

where m(i) denotes the mass of quark i and B(i, j) the binding energy resulting from the interaction between quarks i and j. The latter term may depend on the relative spin orientations or the quarks as well as on their SU(3) labels. If we neglect electromagnetic effects and assume that isosymmetry is valid, (8.4) contains two independent masses and eight binding energy terms. Explicit evaluation using the SU(6) wave functions for the 56 baryons leads to the following relations:

$$m(\Xi) - m(\Sigma) = m(\Xi^*) - m(Y_1^*) \tag{8.5a}$$

$$m(\Omega^-) - m(\Delta) = 3\,[m(Y_1^*) - m(\Delta)] \tag{8.5b}$$

$$m(\Sigma) - m(\Lambda) = \tfrac{2}{3}[m(\Delta) - m(N) - m(\Xi^*) + m(\Xi)]\,. \tag{8.5c}$$

Equation (8.5a) says, in effect, that the mass splittings in the octet and decuplet are identical; it is the same as the result (7.23) of the SU(6) mass formula, and is reasonably well satisfied. The second result is a more general form of the equal spacing of the decuplet masses. The third equation is one we have not seen before; the measured masses yield a value of 58 MeV for the right hand side, versus a $\Sigma - \Lambda$ mass difference of 74 MeV.

Electromagnetic mass differences may also be considered in this model by removing the isosymmetry restrictions. Then (8.4) contains more terms, but many more masses are available for comparison. The calculation leads to nine relations among the physical masses:

$$m(\Delta^{++}) - m(\Delta^-) = 3[m(\Delta^+) - m(\Delta^\circ)]$$

$$m(\Delta^+) - m(\Delta^\circ) = m(Y_1^{*+}) - m(Y_1^{*-}) - m(\Xi^{*\circ}) + m(\Xi^{*-})$$

$$m(\Delta^+) + m(\Delta^-) - 2m(\Delta^\circ) = m(Y_1^{*+}) + m(Y_1^{*-}) - 2m(Y_1^{*\circ})$$

$$m(\Delta^+) - m(\Delta^\circ) = m(p) - m(n)$$

$$m(Y_1^{*+}) + m(Y_1^{*-}) - 2m(Y_1^{*\circ}) = m(\Sigma^+) + m(\Sigma^-) - 2m(\Sigma^\circ)$$

$$m(p) - m(n) = m(\Sigma^+) - m(\Sigma^-) - m(\Xi^\circ) + m(\Xi^-)$$

$$m(\Omega^-) - m(\Delta^-) = 3[m(\Xi^{*-}) - m(Y_1^{*-})]$$

$$m(\Xi^{*-}) - m(Y_1^{*-}) = m(\Xi^-) - m(\Sigma^-)$$

$$-\bar{m}(\Delta) + \bar{m}(Y_1^*) + \bar{m}(\Xi^*) - m(\Omega^-) = 3m(\Lambda^\circ) + \bar{m}(\Sigma)$$

$$- 2\bar{m}(N) - 2\bar{m}(\Xi), \tag{8.6}$$

where $\bar{m}$ denotes the average mass of the isomultiplet.

A second electromagnetic application of the low-energy quark model deals with the magnetic moments of the baryons. If we assume that the quarks are in s-wave bound states, so that no orbital angular momentum is present, then the magnetic moment of the bound state is the sum of the three quark moments,

$$\mu(B)\boldsymbol{\sigma}(B) = \langle B| \sum_i \mu(i)\boldsymbol{\sigma}(i)|B\rangle \tag{8.7}$$

where $\boldsymbol{\sigma}(i)$ is the quark's spin, $\mu(i)$ its magnetic moment, and $|B\rangle$ the SU(6) baryon wave function. Under SU(3) symmetry, the quarks, like the 10, would have magnetic moments proportional to their charges,

$$\mu(i) = Q\mu_Q \tag{8.8}$$

with μ_Q the "quark magneton". The magnitude of the magnetic moment (8.7) is then obtained from

$$\mu(B)\sigma_3(B) = \langle B| \sum_i Q(i)\sigma_3(i)|B\rangle\mu_Q. \tag{8.9}$$

Evaluating (8.9) for the proton and neutron leads to the prediction

$$\mu(p)/\mu(n) = -\tfrac{3}{2}, \tag{8.10}$$

which is very well satisfied by the experimental values, $\mu(p)/\mu(n) = -1.46$. In addition (8.9) leads to the familiar SU(3) results

$$\mu(\Sigma^+) = \mu(p) \tag{6.27a}$$

$$\mu(\Lambda) = \tfrac{1}{2}\mu(n), \tag{6.31}$$

which we discussed in Chapter 6. The breaking of SU(3) can be accounted for by allowing the magnetic moment of the λ quark to differ from that of the $\mathscr{N}$. Then (8.10) is nonetheless obtained, since the nucleons contain no λ's. Instead of (6.27a) and (6.31), we find a more complicated result,

$$\mu(\Lambda^\circ) + 3\mu(\Sigma^+) = \tfrac{8}{3}\mu(p), \tag{8.11}$$

which is in fair agreement with the measured values. A number of other predictions may be derived by similar arguments.

The basic idea behind all of this work is that the known hadrons are the lowest-lying bound states of a quark–antiquark or three-quark system. If that is literally true, then the study of strong interactions is confronted with the same problem that has been met many times already, in atoms as well as in nuclei. To conclude this description of the low-energy quark model, we therefore turn to the most literal calculations of all, first proposed by Greenberg and subsequently studied intensively by Dalitz.

Here we assume that the baryons can be described as "three-quark nuclei" in which each quark sees only an effective central potential. The spatial wave function can be written in an independent-particle "shell model" analogously to that of the three-electron atom or the three-nucleon nucleus. The state of each quark is denoted by the usual spectroscopic labeling (nl), where n gives an energy level and l = s, p, d, . . . indicates the quark's orbital angular momentum. Thus the wave function corresponding to all three quarks in the ground state (1s) is $(1s)^3$, etc. Since the $(1s)^3$ state should be the ground state of the system, we assign the 56 baryons to it. That is, we suppose that their wave functions are

$$\Psi(1,2,3) = \psi_{1s}(1)\psi_{1s}(2)\psi_{1s}(3)\chi_{56}(1,2,3), \tag{8.12}$$

where $\psi_{1s}(i)$ is the 1s spatial wave function for quark i, and $\chi_{56}(1,2,3)$ is the SU(6) wave function. The wave function (8.12) is totally symmetric under interchange of the quark labels, so it assumes that the quarks are bosons, in violation of the usual spin-statistics theorems.

If the 56 baryons are described by (8.12), it seems natural to try to describe the higher baryon resonances as orbital excitations in this model. The first excited state will be $(1s)^2(1p)$, with one quark excited into a higher angular momentum state. The spatial wave function $\psi(r_1, r_2, r_3) = \psi_{1s}(r_1)\psi_{1s}(r_2)\psi_{1p}(r_3)$ will then have negative parity, corresponding to total orbital angular momentum L = 1. It is neither symmetric nor antisymmetric, but has mixed symmetry; in order to obtain a total wave function that is symmetric, it must be multiplied by an SU(6) wave function which also has mixed symmetry. As shown in Appendix A, this means that the particles assigned to this state must belong to a 70 of SU(6).

The SU(3) ⊗ SU(2) decomposition of the 70 has been given as

$$70 = (10, 2) + (8, 4) + (8, 2) + (1, 2), \tag{7.18b}$$

i.e. a spin-½ decuplet and singlet plus octets with spin ½ and spin 3/2. The spin must, of course, be combined with the orbital angular momentum to produce an eigenstate of total angular momentum. For example, the only unambiguous assignment resulting from (7.18b) involves the Δ states of the (10, 2) decuplet. Combining the spin of the SU(6) 70 wave function with the orbital angular momentum will produce particles with angular momentum $J = 1 \pm \frac{1}{2} = \frac{1}{2}, \frac{3}{2}$. The 70 should therefore contain two $I = \frac{3}{2}$, $Y = 1$ states, with $J^P = \frac{1}{2}^-$ and $\frac{3}{2}^-$; the Δ(1640) and the Δ(1690) have precisely these values.

Similarly, one might identify the nucleonic states of (8, 2) and (8, 4) with $J^P = \frac{1}{2}^-, \frac{3}{2}^-, \frac{3}{2}^-$, and $\frac{5}{2}^-$ nucleon resonances. The remaining states of the 70 may also be considered in the same way, as well as the higher excited states resulting from other configurations such as $(1s)^2(1d)$, $(1s)(1p)^2$, etc. In general the baryon spectrum is still very confused, but in many cases it seems to yield substantially successful agreement for this model.

Additivity

Next we consider the implications of the quark model for high-energy scattering. The basic assumption, mentioned earlier, is that the total cross section for scattering of an arbitrary hadron is proportional to the number of quarks within that hadron. This is essentially an "optical" assumption; it would be expected to hold when the dimensions of the hadrons are much larger than the wavelength of the probing particle or radiation. Therefore it should be tested in high-energy scattering, where it predicts that mesonic cross sections should be ⅔ as large as the corresponding baryonic ones. This ratio is verified in pion–proton and proton–proton scattering.

That the model is far too crude is easily seen, however, since it predicts that all mesonic total cross sections are equal. Experimentally the kaon cross sections are 25% smaller than those of pions. This discrepancy is easily explained by saying that the λ quark is "smaller" than the $\mathscr{P}$ and $\mathscr{N}$. A more appropriate way to state this idea is provided by the *additivity* hypothesis: the scattering amplitude for a given process is the sum of amplitudes corresponding to each quark involved. For example, the π^+ contains $\bar{\mathscr{N}}$ and $\mathscr{P}$ quarks, so the elastic scattering amplitude describing the process $\pi^+A \to \pi^+A$, where A is an arbitrary scatterer, is the sum of an $\bar{\mathscr{N}}$ amplitude and a $\mathscr{P}$ amplitude:

$$F(\pi^+A) = f_{\pi^+}(\bar{\mathscr{N}}A) + f_{\pi^+}(\mathscr{P}A). \tag{8.13}$$

This hypothesis is not a new idea; it has long been known in atomic and nuclear physics as the impulse approximation. It is not difficult to derive using potentials describing the π^+A interaction in terms of two-body forces involving the quarks. The result of this derivation is that the quark amplitudes $f_{\pi^+}(\bar{\mathscr{N}}A)$ and $f_{\pi^+}(\mathscr{P}A)$ in (8.13) are separable into the product form

$$f_{\pi^+}(QA) = S_{\pi^+}(\tfrac{1}{2}\mathbf{\Delta})f(QA), \tag{8.14}$$

where $\mathbf{\Delta}$ is the momentum transferred in the scattering, $S_{\pi^+}(q) \equiv \int d^3r|\psi_{\pi^+}(r)|^2 e^{i\mathbf{q}\cdot\mathbf{r}}$ is the "form factor" of the pion (i.e. the Fourier transform of its spatial density $|\psi_{\pi^+}(r)|^2$), and f(QA) describes the scattering of a free quark on A.

Suppose we initially consider only forward scattering, i.e. $\mathbf{\Delta} = 0$. The forward scattering amplitude is particularly appropriate, since the optical theorem gives the total cross section in terms of it; furthermore, the definition of the form factor implies that $S(\mathbf{0}) = 1$ for *any* bound state. It follows then from (8.13) that

$$F(\pi^+A) = f(\bar{\mathscr{N}}A) + f(\mathscr{P}A). \tag{8.15}$$

A similar expansion can be made for other meson scattering processes, yielding

$$F(\pi^-A) = f(\bar{\mathscr{P}}A) + f(\mathscr{N}A)$$

$$F(K^+A) = f(\bar{\lambda}A) + f(\mathscr{P}A)$$

$$F(K^\circ A) = f(\bar{\lambda}A) + f(\mathcal{N}A)$$

$$F(\bar{K}^- A) = f(\bar{\mathcal{P}}A) + f(\lambda A)$$

$$F(\bar{K}^\circ A) = f(\bar{\mathcal{N}}A) + f(\lambda A). \tag{8.16}$$

The quark amplitudes can be eliminated from (8.15) and (8.16) to yield a linear relation among the amplitudes

$$F(\pi^+ A) - F(\pi^- A) + F(K^\circ A) - F(K^+ A) + F(\bar{K}^- A) - F(\bar{K}^\circ A) = 0. \tag{8.17}$$

If we take A to be a proton and use the isospin relations

$$F(K^\circ p) = F(K^+ n), \; F(\bar{K}^\circ p) = F(\bar{K}^- n),$$

(8.17) becomes the very successful Barger–Rubin relation

$$F(\pi^+ p) - F(\pi^- p) + F(K^+ n) - F(K^+ p) + F(\bar{K}^- p) - F(\bar{K}^- n) = 0, \tag{6.53}$$

which we obtained earlier from SU(3) with octet dominance and again from SU(6).

A similar calculation can be made for nucleon and antinucleon scattering, yielding

$$F(pA) = 2f(\mathcal{P}A) + f(\mathcal{N}A)$$

$$F(nA) = f(\mathcal{P}A) + 2f(\mathcal{N}A)$$

$$F(\bar{p}A) = 2f(\bar{\mathcal{P}}A) + f(\bar{\mathcal{N}}A)$$

$$F(\bar{n}A) = f(\bar{\mathcal{P}}A) + 2f(\bar{\mathcal{N}}A). \tag{8.18}$$

Comparing (8.18) with (8.17) leads to the predictions

$$F(pA) - F(nA) = F(K^+ A) - F(K^\circ A) \tag{8.19a}$$

$$F(\bar{p}A) - F(\bar{n}A) = F(\bar{K}^- A) - F(\bar{K}^\circ A) \tag{8.19b}$$

$$F(pA) + F(\bar{n}A) = 2F(\pi^+ A) + F(\pi^- A) \tag{8.20a}$$

$$F(nA) + F(\bar{p}A) = F(\pi^+ A) + 2F(\pi^- A). \tag{8.20b}$$

These also may be tested by letting A be a proton and using isosymmetry.

Equations (8.19) imply that

$$\sigma(pp) - \sigma(pn) - \sigma(K^+p) + \sigma(K^+n) = 0$$

$$\sigma(\bar{p}p) - \sigma(\bar{p}n) - \sigma(\bar{K}^-p) + \sigma(\bar{K}^-n) = 0;$$

the experimental values of these sums, shown in Fig.(8-1), are in reasonably good agreement with zero. For (8.20), however, the results

$$\sigma(pp) + \sigma(\bar{p}n) - 2\sigma(\pi^+p) - \sigma(\pi^-p) = 0$$

$$\sigma(pn) + \sigma(\bar{p}p) - \sigma(\pi^+p) - 2\sigma(\pi^-p) = 0$$

disagree violently with the data [Fig.(8-2)].

It is not hard to guess a reason for this failure. The amplitudes involved in (8.20) are all varying fairly rapidly as a function of the momentum of the incident particle. In making the comparison shown, we have taken the same value for the nucleon and meson incident momenta. But the result depends on the equality of the quark amplitudes, which presumably depend on the quark's incident momentum. Thus we have assumed, in fact, that this quantity does not depend on whether the quark belongs to baryon or meson. That assumption is certainly questionable; it seems far more logical that the quarks should share

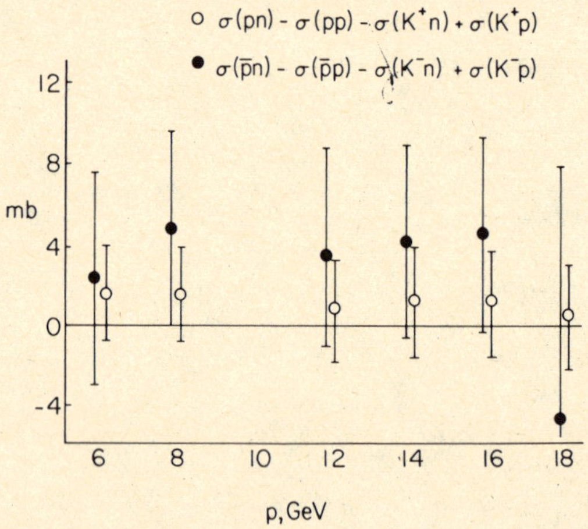

Fig.(8-1) Comparison of equations (8.19) with experiment.

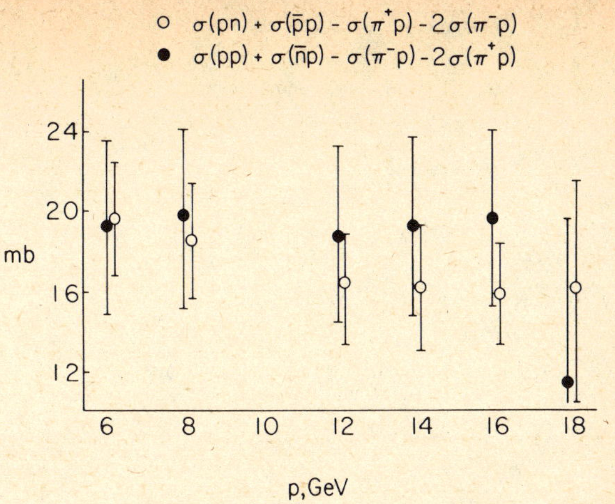

Fig.(8-2) Comparison of equations (8.20) with experiment.

equally in the total momentum. Comparing at equal quark momenta would then mean choosing the meson data at a momentum 2/3 that of the baryon. A third, and perhaps simpler, possibility is to assume that these relations hold in the rest frame of the meson or baryon, with the A particle considered the incident beam. None of these proposals is completely successful in accounting for the failure of (8.20), however; quark-model predictions relating mesonic and baryonic amplitudes must therefore be viewed with suspicion.

It should be noted that none of these results has required any detailed knowledge of the scatterer A. If we take it to be a nucleon, as above, a further application of additivity can be made to express the quark–nucleon amplitudes in terms of quark–quark amplitudes. Then we have

$$f(\mathscr{N}p) = f(\mathscr{P}n) = 2f(\mathscr{N}\mathscr{P}) + f(\mathscr{P}\mathscr{P})$$

$$f(\mathscr{P}p) = f(\mathscr{N}n) = 2f(\mathscr{P}\mathscr{P}) + f(\mathscr{N}\mathscr{P}) \tag{8.21}$$

since isosymmetry implies that $f(\mathscr{P}\mathscr{P}) = f(\mathscr{N}\mathscr{N})$. Making this decomposition will not alter the results, of course; it will merely replace each quark–nucleon amplitude by the appropriate sum of terms.

Inelastic scattering can also be treated by means of additivity. Indeed, isosymmetry tells us that charge-exchange amplitudes are related to elastic ones. For quark–quark scattering, it is easily seen that

$$f(\mathscr{P}\mathscr{N} \to \mathscr{N}\mathscr{P}) = f(\mathscr{P}\mathscr{P}) - f(\mathscr{P}\mathscr{N})$$

$$f(\bar{\mathscr{P}}\mathscr{P} \to \bar{\mathscr{N}}\mathscr{N}) = f(\bar{\mathscr{N}}\mathscr{P}) - f(\bar{\mathscr{P}}\mathscr{P}),$$

which implies also that, as expected,

$$f(\mathscr{P}n \to \mathscr{N}p) = f(\mathscr{P}p) - f(\mathscr{P}n)$$

$$f(\mathscr{P}p \to \mathscr{N}n) = f(\mathscr{N}p) - f(\mathscr{P}p).$$

The result of (8.15) and (8.16) can thus be shown consistent with

$$\sqrt{2}F(\pi^- p \to \pi^\circ n) = F(K^+ n \to K^\circ p) - F(K^- p \to \bar{K}^\circ n), \tag{8.22}$$

the inelastic form of (6.53). A similar calculation for the reaction $\pi^- p \to \eta^\circ n$ leads to

$$\sqrt{6}F(\pi^- p \to \eta^\circ n) = F(K^+ n \to K^\circ p) + F(K^- p \to \bar{K}^\circ n). \tag{8.23}$$

From (8.22) and (8.23) follows a relation among the differential cross sections

$$\frac{d\sigma}{d\Omega}(\pi^- p \to \pi^\circ n) + 3\frac{d\sigma}{d\Omega}(\pi^- p \to \eta^\circ n) = \frac{d\sigma}{d\Omega}(K^+ n \to K^\circ p) + \frac{d\sigma}{d\Omega}(K^- p \to \bar{K}^\circ n), \tag{8.24}$$

which seems to agree reasonably well with experiment near the forward direction. It should be remembered, of course, that the variation of the form factor in (8.14) for $\mathbf{\Delta} \neq 0$ will vitiate the quark model results.

A second example of the use of additivity in inelastic reactions deals with the spin states of the Δ produced by a meson–nucleon scattering, e.g. $\pi^+ p \to \pi^\circ \Delta^{++}$. This process involves changing the spin dependence of the proton's SU(6) wave function. The result of assuming additivity here is that an $m = \pm\frac{1}{2}$ proton can yield only $m = \pm\frac{3}{2}$ or $m = \mp\frac{1}{2}$ for the final spin state of the Δ^{++}, and yields the former value three times as often as the latter. Thus three-fourths of the Δ's produced should have $|m| = \frac{3}{2}$.

Many other calculations can be made within the framework of additivity. We shall not attempt here to give any further details. Instead we may summarize the additive quark model of high-energy scattering by saying that the model works surprisingly well in a surprisingly large number of cases. Where it fails, furthermore, a simple and intuitive explanation of the failure is often available.

The combined successes of low-energy and high-energy quark models do not, of course, provide any real proof of the existence of free quarks. Indeed, most of the best quark-model results can also be explained by other theories; SU(3) with octet dominance yields the Barger–Rubin relation, and many of the mass relations follow from SU(3) or SU(6) mass formulas. It does seem, however, that the quark model unifies a large number of otherwise unrelated predictions. That unification is persuasive evidence that, whether quarks exist or not, the model has some physical sensibility.

A1. Irreducible Representations of SU(N)

The irreducible representations of a Lie group can all be built up by considering direct product states of the elementary representation. We consider an n-fold direct product state

$$|n\rangle = |a_1\rangle\, |a_2\rangle\, |a_3\rangle \ldots |a_n\rangle \tag{A.1}$$

with the corresponding definition of the generators

$$X_a^{(n)} = \sum_{i=1}^{n} X_a(i), \tag{A.2}$$

where $X_a(i)$ stands for the operator X applied to the ith state, times a unit operator on all other states. Here a_i stands for the full set of quantum numbers identifying a given state in the elementary representation.

Now let us define a permutation operator P_{ij} which interchanges the labels of the states in (A.1), e.g.

$$P_{12}|n\rangle = |a_2\rangle\, |a_1\rangle\, |a_3\rangle \ldots |a_n\rangle$$

$$P_{13}|n\rangle = |a_3\rangle\, |a_2\rangle\, |a_1\rangle \ldots |a_n\rangle \tag{A.3}$$

etc. Any more complicated permutation

$$P\begin{pmatrix} 1 & 2 & . & . & . & n \\ i_1 & i_2 & . & . & . & i_n \end{pmatrix},$$

defined by

$$P\begin{pmatrix} 1 & 2 & . & . & . & n \\ i_1 & i_2 & . & . & . & i_n \end{pmatrix} |n\rangle = |a_{i_1}\rangle\, |a_{i_2}\rangle \ldots |a_{i_n}\rangle, \tag{A.4}$$

can be written as a product of the P_{ij}. The full set of such permutations, plus the unit operator, forms the permutation group S(n) on n objects. Unlike the unitary groups, S(n) is a finite group.

It is nonetheless possible to define irreducible matrix representations of the permutation group. The most familiar case is that of N = 2. Taking as basis states $|a_1\rangle\, |a_2\rangle$ and $|a_2\rangle\, |a_1\rangle$, we find a matrix representation of P_{12} given by

$$P_{12} = \begin{pmatrix} 0 & 1 \\ 1 & 0 \end{pmatrix}, \tag{A.5a}$$

which is diagonalized by transforming to the symmetric basis states $(|a_1\rangle\, |a_2\rangle \pm |a_2\rangle\, |a_1\rangle)/\sqrt{2}$, yielding

$$P'_{12} = \begin{pmatrix} 1 & 0 \\ 0 & -1 \end{pmatrix}. \tag{A.5b}$$

Thus $S^{(2)}$ has two one-dimensional irreducible representations, $P_{12} = 1$ and $P_{12} = -1$.

For n > 2, however, life is much more complicated. With n = 3, for example, there are six basis states (the permutations of $|a_1\rangle\, |a_2\rangle\, |a_3\rangle$) and three operators P_{12}, P_{13}, and P_{23}. Six-dimensional matrix representations of these operators can be defined as in (A.5a), and by appropriate transformation can be reduced. If P_{12} is chosen diagonal, we have

$$P_{12} = \begin{pmatrix} 1 &&&&& \\ & -1 &&&& \\ && 1 &&& \\ &&& -1 && \\ &&&& 1 & \\ &&&&& -1 \end{pmatrix} \quad P_{13} = \tfrac{1}{2}\begin{pmatrix} 2 &&&&& \\ & -2 &&&& \\ && -1 & \sqrt{3} && \\ && \sqrt{3} & 1 && \\ &&&& -1 & -\sqrt{3} \\ &&&& -\sqrt{3} & 1 \end{pmatrix}$$

$$P_{23} = \tfrac{1}{2}\begin{pmatrix} 2 &&&&& \\ & -2 &&&& \\ && -1 & -\sqrt{3} && \\ && -\sqrt{3} & 1 && \\ &&&& -1 & -\sqrt{3} \\ &&&& -\sqrt{3} & 1 \end{pmatrix}; \tag{A.6}$$

in addition to the two one-dimensional representations corresponding to the symmetric and antisymmetric combinations of states, there are two equivalent two-dimensional irreducible representations.

A similar analysis can be carried out for S(n). The general result for the irreducible representations is most easily expressed in terms of *Young diagrams.* Let us represent each of the states in (A.1) by a box, □. For n = 2, we have two boxes, which can be joined horizontally (□□) or vertically (). Suppose we define the symmetric combination to correspond to □□, and the antisymmetric to ; thus

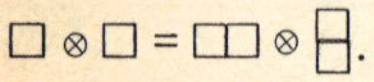 (A.7)

Adding a third box to get n = 3, we have

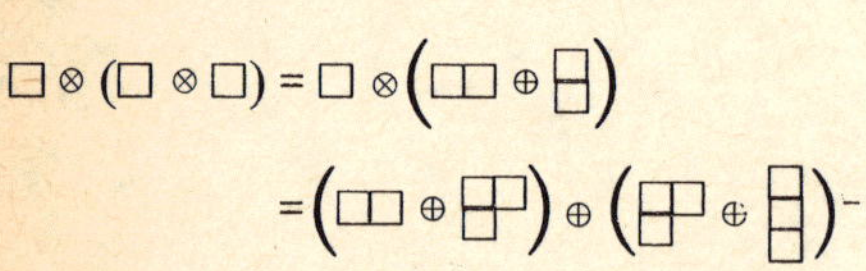 (A.8)

The □□□ clearly corresponds to the symmetric one-dimensional representation, and the

to the totally antisymmetric. It can be shown, although we shall not go into the details here, that the two  diagrams correspond exactly with the two two-dimensional representations. Similarly, the irreducible representations of S(n) can be shown to be in one-to-one correspondence with the possible arrangements of n boxes subject to the rules defining a Young diagram. These are simply (a) the number of boxes in a horizontal row does not increase downward; (b) the number of boxes in a vertical column does not increase moving to the right. Thus for n = 5, for example, the Young diagrams are

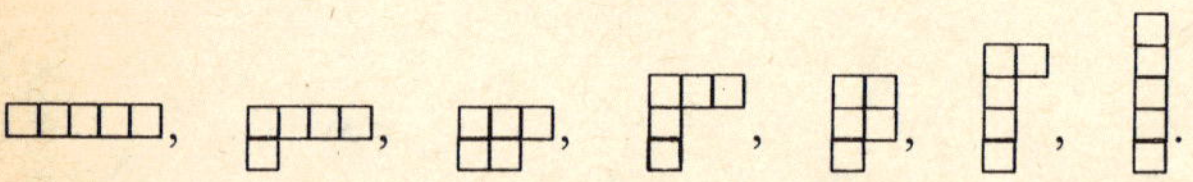

Since any permutation only interchanges the order of terms in (A.2), the permutation group commutes with all of the generators,

$$P_{ij} X_\alpha^{(n)} P_{ij} = X_\alpha^{(n)}. \tag{A.9}$$

If $|n\rangle$ is chosen to be an eigenstate of one of the Lie group generators, then so is any permutation of $|n\rangle$, with the same eigenvalue. These states may be separated according to the irreducible representations of S(n). Applying any of the generators $X_a^{(n)}$ will not mix these irreducible representations of the permutation group, so they provide invariant subspaces. It follows that the irreducible representations of the Lie group are in one-to-one correspondence with those of S(n).

The number of independent states in an irreducible representation of SU(N) may thus be calculated using Young diagrams. In SU(2), for example, the elementary representation is two-dimensional; thus each of the a_i in (A.1) is two-valued, say $a_i = 1, 2$ corresponding to $I_3 = \pm\frac{1}{2}$. A particular state may be identified by putting a value for a_i inside the box, $|+\frac{1}{2}\rangle = \boxed{1}$ and $|-\frac{1}{2}\rangle = \boxed{2}$. Then the symmetric n = 2 states are

$$\boxed{1}\boxed{1} = |\tfrac{1}{2}\rangle\,|\tfrac{1}{2}\rangle$$

$$\boxed{1}\boxed{2} = \frac{1}{\sqrt{2}}\left(|\tfrac{1}{2}\rangle\,|-\tfrac{1}{2}\rangle + |-\tfrac{1}{2}\rangle\,|\tfrac{1}{2}\rangle\right) = \boxed{2}\boxed{1}$$

$$\boxed{2}\boxed{2} = |-\tfrac{1}{2}\rangle\,|-\tfrac{1}{2}\rangle. \tag{A.10}$$

The antisymmetric state is

$$\begin{array}{|c|}\hline 1\\\hline 2\\\hline\end{array} = \frac{1}{\sqrt{2}}\left(|\tfrac{1}{2}\rangle\,|-\tfrac{1}{2}\rangle - |-\tfrac{1}{2}\rangle\,|\tfrac{1}{2}\rangle\right); \tag{A.11}$$

notice that

$$\begin{array}{|c|}\hline 1\\\hline 1\\\hline\end{array} = \begin{array}{|c|}\hline 2\\\hline 2\\\hline\end{array} = 0,$$

since the antisymmetric combination of two identical states vanishes. It follows that for SU(2) $\begin{array}{|c|}\hline \ \\\hline \ \\\hline\end{array}$ is invariably a singlet state, while any column of three or more boxes vanishes, e.g.

$$\begin{array}{|c|}\hline 1\\\hline 1\\\hline 1\\\hline\end{array} = \begin{array}{|c|}\hline 1\\\hline 1\\\hline 2\\\hline\end{array} = \begin{array}{|c|}\hline 1\\\hline 2\\\hline 2\\\hline\end{array} = \begin{array}{|c|}\hline 2\\\hline 2\\\hline 2\\\hline\end{array} = 0. \tag{A.12}$$

Therefore the most general allowed Young diagram for SU(2) is of the form

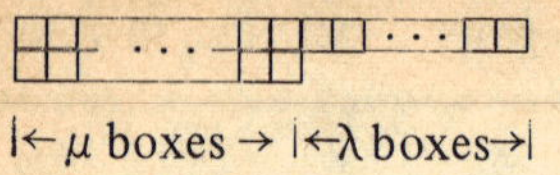

containing μ two-box columns and λ single boxes. Since the former must be filled as $\begin{smallmatrix}\boxed{1}\\\boxed{2}\end{smallmatrix}$, the only freedom in assigning definite states is in the λ single boxes. Furthermore, the symmetrization condition sets all labellings with λ_1 1's and $(\lambda - \lambda_1)$ 2's equal; thus the only non-identical states are those with different values of λ_1. The possible values of λ_1 are 0, 1, . . . , λ, so the dimensionality of the representation is $\lambda + 1$, independent of μ. The irreducible representations of SU(2) can thus be labeled by single integer λ, the number of unpaired boxes in the Young diagram. This result is the same as that obtained from the rotation group, with $\lambda = 2J$.

For SU(3), we may follow the same procedure. There are now three possible values of the a_i, say $a_i = 1, 2, 3$ for $\mathscr{P}$, $\mathscr{N}$, λ. The dimensionality of an irreducible representation is therefore given by the number of independent ways of filling the boxes of its Young diagram with $a_i = 1, 2, 3$. To eliminate identical labelings we require that the a_i increase vertically and do not decrease horizontally. Thus the independent $\square\square$ states are $\boxed{1}\boxed{1}$, $\boxed{1}\boxed{2}$, $\boxed{1}\boxed{3}$, $\boxed{2}\boxed{2}$, $\boxed{2}\boxed{3}$, and $\boxed{3}\boxed{3}$, so $\square\square$ is six-dimensional, while $\begin{smallmatrix}\square\\\square\end{smallmatrix}$ can only be filled with $\begin{smallmatrix}\boxed{1}\\\boxed{2}\end{smallmatrix}$, $\begin{smallmatrix}\boxed{1}\\\boxed{3}\end{smallmatrix}$, and $\begin{smallmatrix}\boxed{2}\\\boxed{3}\end{smallmatrix}$ and is three-dimensional. Thus

$$\square \otimes \square = \square\square \oplus \begin{smallmatrix}\square\\\square\end{smallmatrix}$$

implies that

$$3 \otimes 3 = 6 \oplus \bar{3}. \tag{A.13}$$

The $\begin{smallmatrix}\square\\\square\end{smallmatrix}$ state is labeled $\bar{3}$ to distinguish it from $\square$. This labeling correctly implies that it corresponds to the complex conjugate representation. In SU(3) the singlet representation is

$$\begin{smallmatrix}\square\\\square\\\square\end{smallmatrix},$$

and to obtain it we must combine $\square$ with $\begin{smallmatrix}\square\\\square\end{smallmatrix}$; thus $\begin{smallmatrix}\square\\\square\end{smallmatrix}$ is effectively the "antiparticle" of $\square$, since combining them leads to the "vacuum" singlet state.

Any column with more than three boxes vanishes for SU(3). Therefore the irreducible representations of SU(3) correspond to Young diagrams of the form

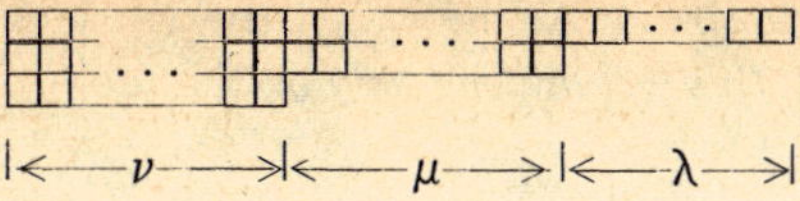

The three-box columns must be filled with

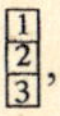

so the dimensionality of the representation results from the possible labelings of the $\mu + \lambda$ remaining columns, and is independent of ν. Therefore we may consider only $\nu = 0$.

It is not hard to see that λ and μ here are identical with the labels introduced in Chapter 5. To obtain the maximal value of Y, we want only $\mathscr{N}$ and $\mathscr{P}$ quarks, i.e. only $a_i = 1, 2$. Then we have effectively reverted to SU(2), and the number of such states is $(\lambda + 1)$. A parallel argument holds for μ. Counting up the possible ways of putting all three values of a_i into this diagram is relatively straightforward and leads to a dimensionality

$$N(\lambda, \mu) = \tfrac{1}{2}(\lambda + 1)(\mu + 1)(\lambda + \mu + 2). \tag{A.14}$$

More detailed arguments of this nature can be used to derive completely the structure we give in Chapter 5 for an SU(3) representation.

For higher groups the same procedure can again be followed, with the larger numbers of possible values of resulting in correspondingly more complicated counting. All of the results we have given for SU(4) and SU(6) can be derived in this way. For example, the three-quark SU(6) state is analyzed by writing

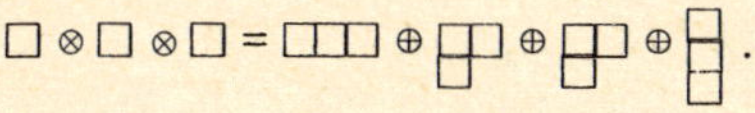

The totally symmetric ☐☐☐ diagram can be filled now with $a_i = N = 1, 2, 3, 4, 5, 6$. If we put N_1 in the first box, the second box must have $N_2 \geqslant N_1$, and the third $N_3 \geqslant N_2$. The number of possibilities for N_3 is therefore $6 - N_2$, for each N_2. The total is therefore

$$\sum_{N_1=1}^{6} \sum_{N_2=N_1}^{6} \sum_{N_3=N_2}^{6} = 56,$$

which is simply the number of symmetric combinations of three six-state objects. In more complicated diagrams, however, this technique is very useful; for , for example, we find

$$N(1, 1) = \sum_{N_1=1}^{6} \sum_{N_2=N_1}^{6} \sum_{N_3=N_1+1}^{6} 1 = 70.$$

Young diagrams for SU(N) can have no more than N boxes in a vertical column, since they must be filled antisymmetrically with the N basis states. Furthermore, the N-box columns are always singlets, and can be disregarded; the structure of the irreducible representations results only from the remaining columns of the diagram. Thus an irreducible representation of SU(N) corresponding to a Young diagram

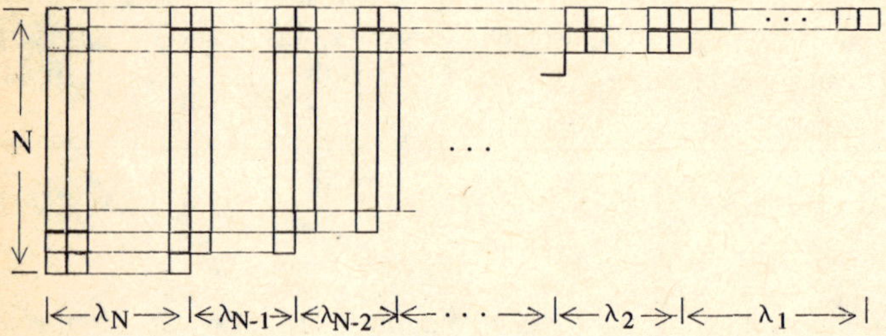

is completely described by giving the lengths $\lambda_1, \lambda_2, \ldots, \lambda_{N-1}$ of the non-singlet columns. These values can be used to describe the dimensions and shape of an irreducible representation in representation space, as we have seen in SU(3); the number of them is identical with the number of Casimir operators, so they may be used to identify the different irreducible representations.

A2 Table of the Hadrons

The particles listed in this table include all of those to which reference is made in the text, as well as most of those which are relatively well established. Since the experimental situation changes constantly, however, more precise information should, when needed, be obtained from the more up-to-date and complete listings appearing annually in Particle Data Tables. The data below have been taken from the 1974 edition of that list, but are intended more as a summary of basic properties than as a compendium of experimental detail. For each particle, we list in addition to its mass and width, the isospin I and the spin-parity J^P;

eigenvalues of G-parity and C-parity are also shown for eigenstates of those operations. Some of these quantities not yet definite, of course, and in such cases we give the currently accepted most probable value.

	Particle	I^G	J^{PC}	Mass (MeV)	Width (MeV)
	π	1^-	0^{-+}	$\pi^{\pm}$: 139.57	≈0
				π°: 134.96	(7.2 ± 1.2) eV
	η (549)	0^+	0^{-+}	548.8 ± 0.6	(2.63 ± 0.64) keV
	ρ (765)	1^+	1^{--}	770 ± 10	150 ± 10
	ω (784)	0^-	1^{--}	782.7 ± 0.6	10.0 ± 0.4
	η (958)	0^+	0^{-+}	957.6 ± 0.3	⩽1
	ϕ (1020)	0^-	1^{--}	1019.7 ± 0.3	4.2 ± 0.2
	A_1 (1100)	1^-	1^{++}	~1100	~300
	B (1235)	1^+	1^{+-}	1237 ± 10	120 ± 20
S = 0	f (1270)	0^+	2^{++}	1270 ± 10	170 ± 30
Mesons	A_2 (1310)	1^-	2^{++}	1310 ± 10	100 ± 10
	E (1420)	0^+	0^{-+}	1416 ± 10	60 ± 20
	f (1516)	0^+	2^{++}	1516 ± 3	40 ± 10
	g (1680)	1^+	3^{--}	1686 ± 20	180 ± 30
	S (1930)	$1^?$	?	~1930	?
	T (2200)	$1^?$	?	~2200	?
	K	½	0^-	K^+: 493.70	≈0
				K°: 497.70	
	K* (890)	½	1^-	892.2 ± 0.5	49.8 ± 1.1
S = 1	K_A (1240)	½	1^+	1242 ± 10	127 ± 25
Mesons	K_V (1420)	½	2^+	1421 ± 5	100 ± 10
	(S = −1 mesons are the antiparticles of these)				

	Particle	I	J^P	Mass (MeV)	Width (MeV)
	N	½	$\frac{1}{2}^+$	p 938.28 n 939.57	
	N (1520)	½	$\frac{3}{2}^-$	1510–1540	105–150[a]
	N (1670)	½	$\frac{5}{2}^-$	1670–1685	115–175
	N (1688)	½	$\frac{5}{2}^+$	1680–1690	105–180
	N (1700)	½	$\frac{1}{2}^-$	1665–1765	100–300
	N (1780)	½	$\frac{1}{2}^+$	1650–1860	50–350
	N (1810)	½	$\frac{3}{2}^+$	1770–1860	180–330
	N (1990)	½	$\frac{7}{2}^+$	1980–2000	120–300
S = 0	N (2040)	½	$\frac{3}{2}^-$	2030–2060	120–300
Baryons	N (2190)	½	$\frac{7}{2}^-$	2000–2260	150–325
	Δ (1236)	$\frac{3}{2}$	$\frac{3}{2}^+$	1230–1236	110–122
	Δ (1650)	$\frac{3}{2}$	$\frac{1}{2}^-$	1615–1695	140–200
	Δ (1670)	$\frac{3}{2}$	$\frac{3}{2}^-$	1650–1720	190–270
	Δ (1890)	$\frac{3}{2}$	$\frac{5}{2}^+$	1840–1920	140–350
	Δ (1910)	$\frac{3}{2}$	$\frac{1}{2}^+$	1780–1935	200–340
	Δ (1950)	$\frac{3}{2}$	$\frac{7}{2}^+$	1930–1980	170–270

[a] The nucleon resonance spectrum is too complicated for the precise masses and width to be easily determined.

	Particle	I	J^P	Mass (MeV)	Width (MeV)
	Λ	0	$\frac{1}{2}^+$	1115.6	≈0
	Λ (1405)	0	$\frac{1}{2}^-$	1405 ± 5	40 ± 10
	Λ (1520)	0	$\frac{3}{2}^-$	1518 ± 2	16 ± 2
S = −1	Λ (1670)	0	$\frac{1}{2}^-$	1660–1680	23–40
Baryons	Λ (1690)	0	$\frac{3}{2}^-$	1690 ± 10	30–70
	Λ (1815)	0	$\frac{5}{2}^+$	1820 ± 5	70–100
($Y_0^* = \Lambda$)	Λ (1830)	0	$\frac{5}{2}^-$	1810–1840	70–120
	Λ (2100)	0	$\frac{7}{2}^-$	2090–2120	80–140

	Particle	I	J^P	Mass (MeV)	Width (MeV)
	Σ	1	$\frac{1}{2}^+$	+: 1189.4 0: 1192.5 –: 1197.4	≈0
S = –1 Baryons	Σ (1385)	1	$\frac{3}{2}^+$	+: 1383 ± 1 –: 1387 ± 1	35 ± 2 42 ± 5
	Σ (1670)	1	$\frac{3}{2}^-$	1670 ± 10	35–60
$(Y_1^* = \Sigma)$	Σ (1750)	1	$\frac{1}{2}^-$	1700–1790	50–100
	Σ (1765)	1	$\frac{5}{2}^-$	1765 ± 5	~120
	Σ (1915)	1	$\frac{5}{2}^+$	1900–1930	50–120
	Σ (2030)	1	$\frac{7}{2}^+$	2020–2040	120–170
	Ξ	$\frac{1}{2}$	$\frac{1}{2}^+$	0: 1314.9 ± 0.6 –: 1321.3 ± 0.14	
S = –2 Baryons	Ξ (1530)	$\frac{1}{2}$	$\frac{3}{2}^+$	0: 1531.8 ± 0.3 –: 1535.1 ± 0.7	9.1 ± 0.5 10.6 ± 2.6
	Ξ (1820)	$\frac{1}{2}$	$\frac{3}{2}^-$?	1795–1870	12–100
	Ξ (1930)	$\frac{1}{2}$	$\frac{5}{2}^-$?	1920–1960	40–140
	Ξ (2030)	$\frac{1}{2}$	$\frac{5}{2}^+$?	2030	50
S = –3 Baryon	Ω	0	$\frac{3}{2}^+$	1672.2 ± 0.4	≈0

Reading List for Group Theory

The two principal references for the material covered in Part I are:

1. Harry J. Lipkin, *Lie Groups for Pedestrians* New York: Interscience, 1965), and
2. Peter A. Curruthers, *Introduction to Unitary Symmetry* (New York: Interscience, 1966).

The first is a readable and interesting introduction to the unitary groups, approached by means of creation and annihilation operators. The isospin group and SU(3) are constructed in this manner and their important features are nicely shown. The higher groups SU(4), SU(6), and SU(12) are also treated briefly. Although some parts of this book are of more interest to nuclear than to particle physicists, the basic ideas are well covered and the experimental consequences of SU(3) are given in some detail.

The second is a more detailed exposition of the structure of SU(3) and its application to particle physics. The topics we treated in Chapters 4, 5, and 6 are addressed in greater depth here. The student wishing to go further into the unitary symmetry schemes will gain greatly from a careful study of this book.

A third general reference is:

3. Murray Gell-Mann and Yuval Ne'eman, *The Eightfold Way* (New York: Benjamin, 1965).

This book is primarily a collection of reprints, and as such is more of historic than pedagogic interest. It does, however, contain papers which are the definitive treatments of unitary symmetry in particle physics, notably that by J. H. deSwart (*Rev. Mod. Phys.* **35**, 916 (1963)), in which the basic structure of SU(3) is derived, the irreducible representations are constructed, and tables of the isoscalar factors are given. The early experimental vindications of unitary symmetry are also included in this book, as are the applications of SU(3) in the weak interactions.

A second volume of reprints which includes material on the advent of the higher symmetry schemes as well as their decline and fall is:

4. Freeman J. Dyson, *Symmetry Groups in Nuclear and Particle Physics* (New York: Benjamin, 1966).

An excellent review of dynamical aspects of higher symmetry schemes was also given by

5. A. Pais, *Rev. Mod. Phys.* **38**, 215 (1966);

the details of SU(6) mass splitting are derived by

6. M. A. B. Beg and V. Singh, *Phys. Rev. Letters* **13**, 418, 601(E) (1964).

The independent quark model described in Chapter 8 is reviewed in some detail by:

7. J. J. J. Kokkedee, *The Quark Model* (New York: Benjamin, 1969).

The details of the quark shell model are given by:

8. R. H. Dalitz, *Proceedings of the International Conference on Symmetries and Quark Models,* Wayne State University, 1969 (New York: Gordon and Breach, 1970).

A review of mass and electromagnetic properties in the quark model is also given by:

9. H. R. Rubenstein, *Lectures on Quark Model Results at Low Energies,* Summer Institute of the Niels Bohr Institute, 1967.

For the details of constructing the irreducible representations of SU(3) and reducing their direct products,

10. S. Coleman, "Fun with SU(3)", in *High Energy Physics and Elementary Particles* (Vienna: IAEA, 1965)

should be studied.

Finally, the student wishing to study the more formal aspects of group theory in particle physics can turn to any of a number of more mathematical treatments; one of the most lucid is:

11. P. A. Rowlatt, *Group Theory and Elementary Particles* (New York: American Elsevier, 1966).

Part II
Analyticity

Chapter 9

SINGULARITIES IN POTENTIAL THEORY

In the first part of this book we have dealt with the discrete quantum numbers which characterize the fundamental particles. Although the unitary symmetry scheme seems to provide some clues to relationships between these quantities, we have absolutely no idea *why* they appear or what physical principle they represent. The existence of half-integral spin was a similar mystery in the early years of quantum mechanics; we now recognize it as a manifestation of the structure of representations of the rotation group. Perhaps a similar insight into the origin of hypercharge and isospin will lead to, or follow from, a deeper understanding of the dynamics of strong interactions. Until then, however, they must remain mere labels by which we classify the symmetries of hadrons.

The continuous variables with which we must deal, on the other hand, present no such problem. We have a good idea of the meaning of energy and momentum; the mystery here is the way in which hadron dynamics depends on them. Our intuition is far better equipped, therefore, to delve into the details of how particles interact than to understand the particles themselves. We can invent models, non-relativistic quantum mechanics for example, which seem "reasonable" as descriptions of the physics we observe, and compare them with the experimental data. In such models it is important to separate the grain from the chaff, to recognize ideas and results which may be more general than the model itself, and to attempt to use these ideas to the fullest extent without depending on the specific model from which we guessed them. It is this task with which we shall be occupied for the remainder of the book.

The models will be descriptions of how systems of two or more hadrons interact, i.e., of collision processes. To see the interaction itself we must separate the *kinematics* from the *dynamics*; that is, we must distinguish between how conservation of energy and momentum limits the possible final states (kinematics) and how the colliding particles actually choose among them (dynamics). To put it more simply, kinematics tells us what *can* happen, dynamics what *does* happen. Kinematical calculations are straightforward, if sometimes laborious, but we can only guess at the dynamics.

Although non-relativistic quantum mechanics is not adequate to describe the strong interactions, it is by far the best guide for any excusion into the hadron jungle. We shall therefore begin by reviewing briefly the formalism of the Schrödinger equation with a central potential and the theory of scattering. In order to make the discussion more precise we shall include a familiar example, the spherical square well potential.

Scattering by a Central Potential

The three-dimensional Schrödinger equation

$$\left[-\frac{1}{2m}\nabla^2 + V(r)\right]\psi(\mathbf{r}) = E\psi(\mathbf{r}) \tag{9.1}$$

can be separated if the potential is spherically symmetric by substituting the partial wave decomposition

$$\psi(\mathbf{r}) = \sum_{\ell,m} a_{\ell m} R_\ell(r) Y_{\ell m}(\theta,\phi), \tag{9.2}$$

where the $Y_{\ell m}(\theta,\phi)$ are the spherical harmonics. In general, axial symmetry will always be maintained, so that all ϕ-dependence can be eliminated. Therefore we replace (9.2) immediately by

$$\psi(\mathbf{r}) = \psi(r,\theta) = \sum_{\ell} a_\ell R_\ell(r) P_\ell(\cos\theta), \tag{9.3}$$

$P_\ell(\cos\theta)$ being the Legendre polynomial of order 1. The resulting radial equation is

$$\frac{1}{r}\frac{d^2}{dr^2}(rR_\ell(r)) + \left[2m(E - V(r)) - \frac{\ell(\ell+1)}{r^2}\right]R_\ell(r) = 0. \tag{9.4}$$

The mathematical solution of (9.4) must be subjected to appropriate boundary conditions. First of all, the requirement that the wave function be continuous and have a continuous first derivative is the mathematical equivalent in this case of the assumption that nature is "smooth". The physical situation to be described then supplies the remaining boundary conditions needed to determine $\psi(r,\theta)$.

We are interested specifically in the physics of two types of system, which correspond to the two principal observations available to experiment: bound states and the scattering of an incident beams of particles. Provided that the energy scale is so defined that $V(r) \to 0$ as $r \to \infty$, these two describe the phenomena occurring for negative and positive energies, respectively. When $E < 0$, solutions of (9.4) will increase too rapidly for large r to permit normalizability except for certain energy values, and we obtain a discrete spectrum of bound states for each ℓ. If $E > 0$, on the other hand, solutions corresponding to scattering by the potential can be found for any energy.

The boundary condition describing the scattering situation requires that far from the scattering center the wave function should represent only the incident beam plus a superposition of diverging spherical waves which compose the scattered beam. If the incident beam is represented by a monoenergetic plane wave, we expect that

$$\psi(r,\theta) \approx e^{i\mathbf{k}\cdot\mathbf{r}} + f(k,\theta)\,\frac{e^{ikr}}{r} \tag{9.5}$$

where the polar axis is defined by the vector **k**. The function $f(k,\theta)$ defined by (9.5) is usually referred to in potential theory as the scattering amplitude. It determines the relative intensity of the beam scattered into a solid angle element $d\Omega$, i.e. the differential cross section, as

$$\frac{d\sigma}{d\Omega} = |f(k,\theta)|^2. \tag{9.6}$$

For ease in recalling some basic facts about $f(k,\theta)$ let us assume that the potential vanishes identically beyond some radius $r = a$. Then in this region the radial equation can be written

$$\frac{1}{r}\,\frac{d}{dr^2}(rR_\ell(r)) + \left[k^2 - \frac{\ell(\ell+1)}{r^2}\right]R_\ell(r) = 0, \tag{9.7}$$

with $k^2 = 2mE$. Equation (9.7) is a form of one of the standard differential equations of physics; its two independent solutions are the spherical Bessel functions of the first and second kinds, denoted by j_ℓ and n_ℓ respectively. In terms of them the most general solution of (9.7) is given by

$$R_\ell(r) = b_\ell j_\ell(kr) + c_\ell n_\ell(kr), \tag{9.8}$$

with b_ℓ and c_ℓ arbitrary constants.

In order to impose the boundary conditions we shall need to know the behaviour of these functions for large and small values of r. As $r \to \infty$, we have

$$\begin{aligned} j_\ell(kr) &\approx \frac{1}{kr}\cos\left(kr - \tfrac{1}{2}(\ell+1)\pi\right) \\ n_\ell(kr) &\approx \frac{1}{kr}\sin\left(kr - \tfrac{1}{2}(\ell+1)\pi\right), \end{aligned} \tag{9.9}$$

whereas when $r \to 0$,

$$\begin{aligned} j_\ell(kr) &\approx (kr)^\ell/(2\ell+1)!! \\ n_\ell(kr) &\approx -(2\ell-1)!!(kr)^{-\ell} \\ (2\ell+1)!! &\equiv 1\,.\,3\,.\,5\,.\,.\,.\,(2\ell+1). \end{aligned} \tag{9.10}$$

Because (9.5) contains e^{ikr}/r rather than the trigonometric forms in (9.9), it is appropriate to replace $j_\ell(kr)$ and $n_\ell(kr)$ by linear combinations

$$\begin{aligned} h_\ell^{(1)}(kr) &= j_\ell(kr) + in_\ell(kr) \\ h_\ell^{(2)}(kr) &= j_\ell(kr) - in_\ell(kr), \end{aligned} \tag{9.11}$$

which are known as the spherical Hankel functions of the first and second kinds. These functions have the asymptotic behavior

$$\begin{aligned} h_\ell^{(1)}(kr) &\approx \frac{1}{kr}\, e^{i(kr - \frac{1}{2}(\ell+1)\pi)} \\ h_\ell^{(2)}(kr) &\approx \frac{1}{kr}\, e^{-i(kr - \frac{1}{2}(\ell+1)\pi)}. \end{aligned} \tag{9.12}$$

If the potential is real, it is possible to show that b_ℓ and c_ℓ in (9.8) must be chosen relatively real. Since we are considering a stationary state, the probability density must be constant; therefore the probability current through any closed surface must vanish. Considering a spherical surface at $r_0 > a$ shows that this implies the vanishing at $r = r_0$ of

$$R_\ell^* \frac{d}{dr} R_\ell - \left(\frac{d}{dr} R_\ell^*\right) R_\ell = (b_\ell^* c_\ell - c_\ell^* b_\ell)\left[j_\ell \frac{d}{dr} n_\ell - n_\ell \frac{d}{dr} j_\ell\right],$$

which means that

$$b_\ell^* c_\ell - c_\ell^* b_\ell = 2i\mathrm{Im}(b_\ell^* c_\ell) = 0.$$

It follows that b_ℓ and c_ℓ have the same phase. Therefore we write

$$R_\ell(r) = d_\ell[\cos\delta_\ell j_\ell(kr) - \sin\delta_\ell n_\ell(kr)] \tag{9.13}$$

with $|d_\ell|^2 = |b_\ell|^2 + |c_\ell|^2$ and a real value of $\delta_\ell = \arctan(-c_\ell/b_\ell)$. In terms of the Hankel functions,

$$R_\ell(r) = \tfrac{1}{2} d_\ell[e^{i\delta_\ell} h_\ell^{(1)}(kr) + e^{-i\delta_\ell} h_\ell^{(2)}(kr)]. \tag{9.14}$$

The wave function (9.3) is therefore of the form

$$\psi(r, \theta) = \tfrac{1}{2} \sum_\ell d_\ell[e^{i\delta_\ell} h_\ell^{(1)}(kr) + e^{-i\delta_\ell} h_\ell^{(2)}(kr)] P_\ell(\cos\theta). \tag{9.15}$$

If the scattered part of the wave function contains only outgoing spherical waves, it must be free of any contributions from $h_\ell^{(2)}(kr)$ in the asymptotic region; therefore the boundary condition (9.5) is equivalent to

$$\psi(r, \theta) \approx e^{i\mathbf{k}\cdot\mathbf{r}} + \sum_\ell g_\ell h_\ell^{(1)}(kr) P_\ell(\cos\theta). \tag{9.16}$$

Expressing the incident plane wave by means of a standard formula

$$e^{i\mathbf{k}\cdot\mathbf{r}} = \sum_\ell (2\ell + 1) i^\ell j_\ell(kr) P_\ell(\cos\theta)$$

allows us to compare (9.15) with (9.16). The two are consistent only if

$$\begin{aligned} d_\ell &= (2\ell + 1) i^\ell e^{i\delta_\ell} \\ g_\ell &= (2\ell + 1) i^{\ell+1} e^{i\delta_\ell} \sin\delta_\ell. \end{aligned} \tag{9.17}$$

The solution of the Schrödinger equation corresponding to scattering is therefore

$$\psi(r,\theta) = e^{i\mathbf{k}\cdot\mathbf{r}} + \sum_{\ell} (2\ell+1) i^{\ell+1} e^{i\delta_\ell} \sin\delta_\ell h_\ell^{(1)}(kr) P_\ell(\cos\theta). \qquad (9.18)$$

Using the asymptotic form of the Hankel function, we have

$$\psi(r,\theta) \approx e^{i\mathbf{k}\cdot\mathbf{r}} + \sum_{\ell} (2\ell+1) e^{i\delta_\ell} \sin\delta_\ell \frac{e^{ikr}}{kr} P_\ell(\cos\theta), \qquad (9.19)$$

from which the scattering amplitude can be identified as

$$f(k,\theta) = \frac{1}{k} \sum_{\ell} (2\ell+1) e^{i\delta_\ell} \sin\delta_\ell P_\ell(\cos\theta). \qquad (9.20)$$

Equation (9.20) defines the *partial wave expansion* of $f(k,\theta)$. The quantities δ_ℓ, which determine the relative phases of the scattered spherical waves in (9.18), are known as the *phase shifts.* Since the $P_\ell(\cos\theta)$ provide a complete orthogonal set of functions, any function of θ can be expanded in terms of them; therefore we can write generally

$$f(k,\theta) = \frac{1}{k} \sum_{\ell=0}^{\infty} (2\ell+1) f_\ell(k) P_\ell(\cos\theta) \qquad (9.21)$$

with the partial wave amplitudes

$$f_\ell(k) = k \int_{-1}^{1} d(\cos\theta) f(k,\theta) P_\ell(\cos\theta), \qquad (9.22a)$$

which we shall also denote by

$$a_\ell(k) = f_\ell(k)/k. \qquad (9.22b)$$

It follows from (9.20) that the amplitudes $f_\ell(k)$ are related to the phase shifts by

$$f_\ell(k) = e^{i\delta_\ell}\sin\delta_\ell = \frac{1}{2i}(e^{2i\delta_\ell} - 1). \tag{9.22c}$$

The reality of δ_ℓ then implies that

$$\mathrm{Im} f_\ell = |f_\ell|^2. \tag{9.23a}$$

This result is a consequence of the reality of the potential; more fundamentally, it is a result of the unitarity of the scattering matrix, which guarantees the conservation of probability. For this reason we refer to it as the partial-wave elastic unitarity relation. (We shall see in Chapter 12 that in a more general situation which includes the possibility of inelastic scattering, (9.23a) is replaced by

$$\mathrm{Im} f_\ell = |f_\ell|^2 + I_\ell, \tag{9.23b}$$

with $0 \leqslant I_\ell \leqslant \frac{1}{4}$.)

From (9.20) we can easily obtain the total cross section,

$$\sigma_T = \int d\Omega\, |f(k,\theta)|^2.$$

The integration can be carried out after writing $|f(k,\theta)|^2$ explicity as a product of two summations to yield

$$\sigma_T = \frac{4\pi}{k^2}\sum_\ell (2\ell + 1)\sin^2\delta_\ell. \tag{9.24}$$

Since $P_\ell(1) = 1$ for all ℓ, a comparison of (9.20) with (9.24) establishes the *optical theorem,*

$$\sigma_T = \frac{4\pi}{k}\,\mathrm{Im} f(k,0). \tag{9.25}$$

One other consequence of (9.24) which should be noted here is the "unitarity limit" on the contribution of the ℓth partial wave to the total cross section

$$\sigma_\ell = \frac{4\pi}{k^2}(2\ell + 1)\sin^2\delta_\ell \leqslant \frac{4\pi}{k^2}(2\ell + 1). \tag{9.26}$$

The Square Well

In order to provide a simple illustration of potential theory for present as well as future reference, let us consider the spherical square well potential defined by

$$\begin{aligned} V(r) &= -V_0, \quad r < a \\ &= 0, \qquad r > a. \end{aligned} \tag{9.27}$$

For $r > a$, the radial Schrödinger equation is identical with (9.7), so the solution is a superposition of Hankel functions; for $r < a$, the same equation is obtained if we replace k^2 by $K^2 = 2m(E + V_0)$. Since the interior region contains the origin, the wave function there must not have any contribution from the diverging Bessel function of the second kind $n_\ell(Kr)$; thus we must have

$$\psi(r < a, \theta) = \sum_\ell a_\ell j_\ell(Kr) P_\ell(\cos\theta). \tag{9.28}$$

The form of the wave function for $r > a$ depends on whether it is to describe scattering or a bound state.

In the bound state situation, $-V_0 < E < 0$, we have $k^2 = 2mE < 0$, so k must be purely imaginary, i.e. $k = i\kappa$. The asymptotic form (9.12) of the Hankel functions is still valid for imaginary arguments,

$$h_\ell^{(1)}(i\kappa r) \approx \frac{1}{i\kappa r}\, e^{-\kappa r - \frac{1}{2}i(\ell+1)\pi}$$

$$h_\ell^{(2)}(i\kappa r) \approx \frac{1}{i\kappa r}\, e^{\kappa r - \frac{1}{2}i(\ell+1)\pi}.$$

Clearly the bound state wave function cannot depend on $h_\ell^{(2)}(i\kappa r)$, so

$$\psi(r > a, \theta) = \sum_\ell A_\ell h_\ell^{(1)}(i\kappa r) P_\ell(\cos\theta). \tag{9.29}$$

Matching (9.28) with (9.29) in the usual way leads to an equation determining the bound state energies,

$$i\kappa h_\ell^{(1)\prime}(i\kappa a) j_\ell(Ka) - K h_\ell^{(1)}(i\kappa a) j_\ell'(Ka) = 0. \tag{9.30}$$

The spectrum of bound states for a given value of ℓ thus corresponds to those energies $E < 0$ providing a solution of (9.30); it can be shown that these solutions are real and non-degenerate.

For positive energies k is real rather than imaginary, and the wave function for $r > a$ can be written in the scattering form (9.15). Applying again the matching conditions at $r = a$ leads to a phase shift δ_ℓ given by

$$\tan\delta_\ell = \frac{kj_\ell'(ka)j_\ell(Ka) - Kj_\ell(ka)j_\ell'(Ka)}{kn_\ell'(ka)j_\ell(Ka) - Kn_\ell(ka)j_\ell'(Ka)}. \tag{9.31}$$

The phase shift is real, as expected, since $j_\ell(ka)$ and $n_\ell(ka)$ are real functions. Using this result for the phase shift, the partial wave expansion of $f(k, \theta)$ can be constructed by noting that

$$e^{i\delta_\ell}\sin\delta_\ell = \frac{\tan\delta_\ell}{1 - i\tan\delta_\ell}$$

which leads to a partial wave amplitude $f_\ell(k)$ given by

$$f_\ell(k) = i\,\frac{kj_\ell'(ka)j_\ell(Ka) - Kj_\ell(ka)j_\ell'(Ka)}{kh_\ell^{(1)\prime}(ka)j_\ell(Ka) - Kh_\ell^{(1)}(ka)j_\ell'(Ka)}. \tag{9.32}$$

Much more can be said about this potential, but in deriving the bound state energies and the scattering amplitude we have obtained all the information we need. Suppose now that we think of k not as a physical quantity but as a complex variable, of which $a_\ell(k)$ and $f(k,\theta)$ are functions. Then $a_\ell(k)$ may be evaluated anywhere in the complex k-plane, and in particular on the imaginary axis, the region which corresponds to negative energy and bound states. Comparing (9.32) with (9.30), we make a rather surprising discovery: the bound state energy values are those for which the denominator of $a_\ell(k)$ vanishes! Consequently we surmise that *a bound state corresponds to a pole of the scattering amplitude.*

Bound State Poles

The correspondence between a bound state and a pole of the scattering amplitude is not, of course, a result peculiar to the square well potential. It is built into the theory of scattering, and comes about through the requirement that the boundary condition

$$\psi(r,\theta) \approx e^{i\mathbf{k}\cdot\mathbf{r}} + f(k,\theta)\,\frac{e^{ikr}}{r} \tag{9.33}$$

of the scattering problem, which defines $f(k,\theta)$, can be extrapolated so that it describes the bound state situation as well. If the components of the vector k become imaginary, $\mathbf{k} = i\boldsymbol{\kappa}$, as they must for $k^2 < 0$, then the plane wave term $e^{i\mathbf{k}\cdot\mathbf{r}} = e^{-\kappa\cdot\mathbf{r}}$ diverges as $r \to \infty$ whenever $\kappa \cdot \mathbf{r} < 0$. This form is therefore incompatible with the characteristics of a bound state unless the coefficient of the plane wave vanishes. Since an overall normalization constant has been omitted in writing (9.33), the vanishing of the plane wave is effectively the same as making the scattered term infinitely more important than the incident beam contribution. Thus for a bound state with energy $E_b = -k_b^2/2m$, we must require that $f(k,\theta) \to \infty$ as $E \to E_b$, and that it do so independently of the value of θ. In the partial wave expansion it follows that one of the $f_\ell(k)$ must be infinite at $k^2 = k_b^2$. Ordinarily the infinity will occur in a single partial wave, producing a bound state with definite angular momentum ℓ.

Equivalently we may simply observe that as long as $V(r) \to 0$ as $r \to \infty$, the solution of the radial equation must become asymptotically a combination of spherical Bessel functions,

$$R_\ell(r) \approx e^{i\delta_\ell} h_\ell^{(1)}(kr) + e^{-i\delta_\ell} h_\ell^{(2)}(kr). \tag{9.34}$$

This behavior defines the phase shifts for $E > 0$, and allows a bound state for $E < 0$ only if $e^{-i\delta_\ell} = 0$. Therefore the partial wave amplitude

$$f_\ell = \frac{\sin\delta_\ell}{e^{-i\delta_\ell}}$$

must become infinite for a bound state.

What sort of infinity will occur? The easiest to imagine is a simple pole, and in fact that is generally assumed to be the case. In a later chapter we shall investigate more thoroughly the allowed singularities of the partial wave amplitude and learn how knowing about them helps us actually to determine $f(k,\theta)$. It seems strange that the scattering amplitude, defined originally only for $E > 0$, is also able to tell us the bound state spectrum; but when we assume that physics is described by a solution of the Schrödinger equation, we are accepting that nature possesses the mathematical characteristics of that solution as well as the physical ones. One of these is analyticity as a function of k, and for that reason a smooth and meaningful transition from one value of k to another is guaranteed.

A knowledge of the scattering amplitude is thus enough to describe completely the bound state spectrum, since the poles of the scattering amplitude in the region $E < 0$ must be in one-to-one correspondence with the bound states. The next question we must ask is whether there are any other singularities in $f(k,\theta)$, and if so what physical significance they have. For real positive values of E, there can be no poles in $a_\ell(k)$, for in that case the total cross section would diverge. There may, however, be other poles of $a_\ell(k)$ which occur for *complex* values of the energy. Of course, such values are not physically attainable, but they may nonetheless have a physical significance which we recognize from the effects they produce. These effects are called resonances.

Resonance Poles

What is a resonance? There are several definitions, all of which are similar if not entirely equivalent. (Indeed, a physicist's answer to this question not only distinguishes theorist from experimentalist, but even separates various subgroups within those two categories!) In Part I, we considered a resonance "as good as a particle" and postponed a precise definition. Let us now proceed by giving the historical definition; then we shall show how others have sprung from it.

When a scattering system seems to have formed an unstable but relatively long-lasting compound state, we call that state a resonance. In one-dimensional quantum mechanics this idea is often illustrated by introducing the concept of a "time delay"; the same argument can also be used in three dimensions. We write the wave function in the form (9.18):

$$\psi_{\mathbf{k}}(r,\theta) = e^{i\mathbf{k}\cdot\mathbf{r}} + \sum_{\ell} (2\ell+1)i^{\ell+1}e^{i\delta_\ell} \sin\delta_\ell h_\ell^{(1)}(\kappa r)P_\ell(\cos\theta).$$

The incident particle is of finite extent, however, and should therefore be represented by a localized wave packet rather than by a simple plane wave. Since

$$\Psi(\mathbf{r},t) = \int d^3k\Phi(\mathbf{k})\,\psi_{\mathbf{k}}(r,\theta)e^{-iE(\mathbf{k})t} \tag{9.35}$$

is a solution of the time-dependent Schrödinger equation for any $\Phi(\mathbf{k})$, we may identify an incident and a scattered wave packet by substituting (9.18) into (9.35), obtaining

$$\Psi_{in}(\mathbf{r},t) = \int d^3k\Phi(\mathbf{k})e^{i\mathbf{k}\cdot\mathbf{r}-iE(\mathbf{k})t} \tag{9.36}$$

$$\Psi_{sc}(\mathbf{r}, t) = \int d^3k\Phi(\mathbf{k}) \sum_{\ell} (2\ell + 1) i^{\ell+1} e^{i\delta_\ell} \sin\delta_\ell h_\ell^{(1)}(kv) P_\ell(\cos\theta) e^{-iE(k)t}$$

Now the motion of the incident and scattered wave packets can be studied by assuming that $\Phi(\mathbf{k})$ is sharply peaked about some $\mathbf{k} = \mathbf{k}_0$. In the incident wave we expand E as a function of $\mathbf{k}$,

$$E(\mathbf{k}) = E(\mathbf{k}_0) + (\mathbf{k} - \mathbf{k}_0) \cdot \nabla E(\mathbf{k}_0), \tag{9.37}$$

where $\nabla E(\mathbf{k}_0)$ denotes the gradient of $E(\mathbf{k})$ with respect to $\mathbf{k}$, evaluated at $\mathbf{k} = \mathbf{k}_0$. Keeping only the leading terms, we have

$$\Psi_{in}(\mathbf{r}, t) \approx e^{i\lambda(t)} \int d^3k\Phi(\mathbf{k}) e^{i\mathbf{k}\cdot(\mathbf{r}-\nabla E(\mathbf{k}_0)t)}, \tag{9.38}$$

where the phase $\lambda(t)$ is independent of $\mathbf{k}$, and can be taken outside the integral. It follows immediately from (9.38) that

$$\Psi_{in}(\mathbf{r}, t) = e^{i\lambda(t)} \Psi_{in}(\mathbf{r} - \mathbf{v}t, 0) \tag{9.39}$$

where the velocity $\mathbf{v}$ is given by $\mathbf{v} = \nabla E(\mathbf{k}_0)$. Because $E = k^2/2m$, we have the expected result that $\mathbf{v} = \mathbf{k}_0/m$. In other words, during the time t the incident wave packet has been translated by the vector $\mathbf{v}t$; it has acquired a phase factor, but its *shape* remains unchanged. Consequently it does provide a reasonable representation of the incident particle.

The scattered wave can be treated in an analogous fashion. For convenience we write (9.36) as

$$\Psi_{sc}(\mathbf{r}, t) = \sum_{\ell} (2\ell + 1) R_\ell(r, t) P_\ell(\cos\theta) \tag{9.40}$$

and examine the behavior of an individual partial wave,

$$R_\ell(r, t) = \int d^3k\Phi(k) i^{\ell+1} e^{i\delta_\ell(k)} \sin\delta_\ell(k) h_\ell^{(1)}(kr) e^{-iE(k)t} \tag{9.41}$$

For large r, this becomes

$$R_\ell(r, t) \approx \int d^3k\Phi(\mathbf{k}) \sin\delta_\ell(k) \frac{1}{kr} e^{i[kr+\delta_\ell(k)-E(k)t]} \tag{9.42}$$

Since we know that both $\delta_\ell(k)$ and $E(k)$ are functions of only the magnitude $k = |\mathbf{k}|$, it is more convenient to expand them here simply as a function of this variable. The exponent in (9.42) is therefore replaced by

$$kr + \delta_\ell(k) - E(k)t \approx$$

$$kr + \delta_\ell(k_0) + (k - k_0)\delta_\ell'(k_0) - [E(k_0) + (k - k_0)E'(k_0)]\,t \tag{9.43}$$

where

$$\delta_\ell' = \frac{d\delta_\ell}{dk} \quad \text{and} \quad E'(k) = \frac{dE}{dk}.$$

Substituting (9.43) into (9.42), we find that

$$R_\ell(r, t) \approx e^{i\zeta_\ell(t)} \sum d^3k\Phi(k) \sin\delta_\ell(k) \frac{1}{kr}\, e^{ik[r+\delta_\ell'(k_0)-E'(k_0)t]} \tag{9.44}$$

where the phase $\zeta_\ell(t)$ is again independent of k. Thus, as in (9.39), we have an equation describing the time development of the wave packet for large r,

$$rR_\ell(r, t) = e^{i\zeta(t)}[r - v(t - \tau_\ell)]\, R_\ell(r - v(t - \tau_\ell), 0) \tag{9.45}$$

with $v = E'(k_0)$ and $\tau_\ell = \delta_\ell'(k_0)/v$.

The contribution of a given angular momentum state to the scattered wave packet therefore propagates outward with a radial velocity v. There is a phase change, as with the incident wave, but the *shape* of the packet does not change except for the usual attenuation factor $1/r$ associated with spherical waves; furthermore, it is the same shape $\Phi(\mathbf{k})$ as that of the incident particle. Consequently the ℓth partial wave is effectively the incident wave packet, transformed by the potential into spherical form. In passing through the potential region, however, the packet appears to have suffered a time delay $\tau_\ell = \delta_\ell'(k_0)/v$; when this time delay is large, we say that the system has formed a resonance with angular momentum ℓ.

We therefore associate resonances with large values of $\delta_\ell'(k)$, i.e. with a phase shift which varies rapidly as a function of k. Since $a_\ell(k) = e^{i\delta_\ell(k)} \sin\delta_\ell(k)/k$, this means that the phase of the partial wave amplitude changes rapidly at a resonance.

If we assume now that $a_\ell(k)$ has a pole corresponding to a complex energy

$E = k^2/2m = E_R - i\Gamma$, then we know from unitarity that near this pole $a_\ell(k)$ must be given by

$$a_\ell = -\frac{i}{k}\,\frac{\Gamma}{E - E_R + i\Gamma} \tag{9.46}$$

The phase of a_ℓ is $\delta_\ell = \arctan(\Gamma/(E - E_R))$, and it is easily seen that $\delta_\ell(k)$ varies most rapidly when $E = E_R$. Indeed, the time delay corresponding to (9.46) is given as a function of E by

$$\tau_\ell = \frac{\Gamma}{(E - E_R)^2 + \Gamma^2} \tag{9.47}$$

which has a maximum value of $1/\Gamma$ when $E = E_R$. We see therefore that a pole of the partial wave amplitude for a complex energy $E = E_R - i\Gamma$ corresponds to a resonance at $E = E_R$ with a lifetime $\tau_\ell = 1/\Gamma$.

Conversely, it is also true that every resonance must correspond to such a pole. It is known in the theory of functions of a complex variable that the phase of an analytic function can vary rapidly only if a singularity is nearby. That this singularity must actually be a pole if it is to represent a resonance is a reasonable assumption, and will be proved in potential theory in a later chapter. We may therefore give a new definition of a resonance as *a pole of the scattering amplitude at a complex value of the energy.*

This new definition is the mathematical equivalent of the physical one we gave earlier, and it is at the root of several others. For example, if we look at the contribution to the total cross section of a resonating partial wave, we find that

$$\sigma_\ell = \frac{2\pi}{mE}(2\ell + 1)\frac{\Gamma^2}{(E - E_R)^2 + \Gamma^2}\,. \tag{9.48}$$

This function is essentially the same as the time delay, and is shown in Fig.(9-1). If the pole is near the real axis, so that Γ is much smaller than E_R, the lifetime is relatively long and the bump will be a sharp one. Clearly, then, if one partial wave forms a resonance while the others are varying more smoothly, the total cross section itself will show a bump centered at the resonant energy, with a width depending on the lifetime of the resonance. In Fig.(9-2) we show the most famous example of this phenomenon, the total cross section for pion–nucleon scattering in the isospin $\frac{3}{2}$ $\pi^+ p$ system. The huge enhancement which occurs when the pion has a momentum of about 300 MeV in the laboratory reference

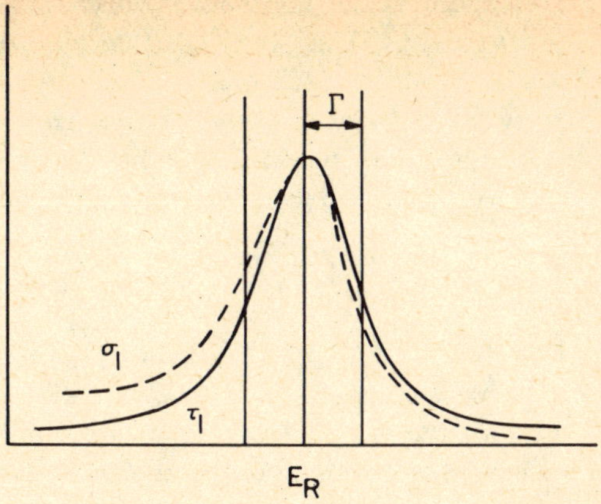

Fig.(9-1) The ℓ-wave total cross section and the time delay associated with a resonance.

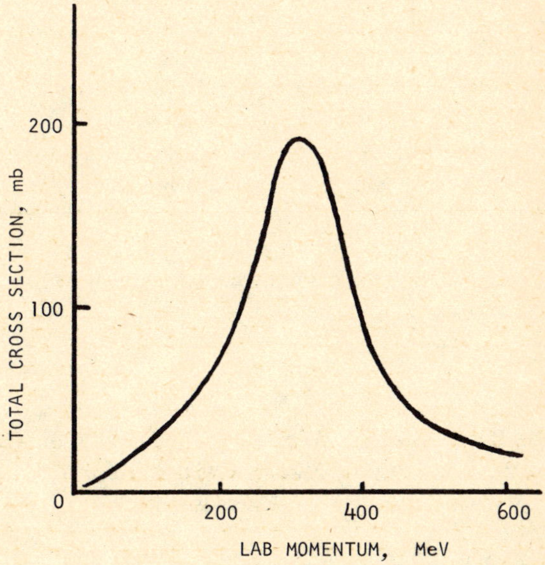

Fig.(9-2) The total cross section for π^+p scattering in the region of the Δ^{++} resonance.

frame is obviously a resonance; in fact, it is the Δ^{++}, which we have already met in the decuplet of spin-$\frac{3}{2}$ baryons. Thus we say that the bump represents the process $\pi^+p \Leftrightarrow \Delta^{++}$, i.e. the production followed by the decay of this resonance. It is common to identify the apparent "mass" of the resonance, rather than the incident particle's momentum or energy. This quantity is determined by conservation of energy-momentum, which yields

$$M(\Delta^{++}) = [m_\pi^2 + m_p^2 + 2m_p\sqrt{m_\pi^2 + k^2}]^{1/2}; \tag{9.49}$$

for the Δ^{++}, (9.49) leads to a value $M(\Delta^{++}) = (1236.0 \pm 0.6)$ MeV. The width is $\Gamma = (120 \pm 2)$ MeV, which corresponds to a lifetime of $(5.49 \pm 0.09) \times 10^{-23}$ seconds. Other data are necessary if we wish to know the spin, parity, etc., of the resonance; the total cross section tells us only the mass and width.

We might, therefore, define a resonance simply as a bump in the total cross section. This definition is an inferior one, unfortunately, because other effects can also cause similar enhancements in σ_T. Much more information is needed to be sure that we are not seeing, for example, the opening of a new inelastic reaction channel.

We can better judge whether a bump is a resonance or not if we isolate the contributions of different partial waves. For one thing, the angular distribution observed in the differential cross section should show a strong dependence on $P_\ell(\cos\theta)$. Also, we notice that in (9.45) the partial wave amplitude becomes purely imaginary when $E = E_R$, so that $\delta_\ell = \pi/2$. But the concept of a "time delay" is meaningful only if $\tau_\ell = 1/\Gamma$ is positive, so we must also have $\delta_\ell'(k) > 0$; therefore a resonance occurs only when the phase shift *increases* through $\pi/2$ with increasing energy. It also follows when $\delta_\ell = \pi/2$ that σ_ℓ is equal to the unitarity limit.

In the presence of inelasticity, however, all of these characteristics of resonance are altered because the elastic unitarity relation must, as indicated in (9.23b), be replaced by

$$\mathrm{Im} f_\ell = |f_\ell|^2 + I_\ell,$$

with $0 \leqslant I_\ell \leqslant \frac{1}{4}$. It then becomes necessary to look at both the real and the imaginary parts of a_ℓ. If we draw, as in Fig.(9-3), a plot of $\mathrm{Im} f_\ell$ versus $\mathrm{Re} f_\ell$ as a function of k, the inequality (9.23b) demands that

$$(\mathrm{Im} f_\ell - \tfrac{1}{2})^2 + (\mathrm{Re} f_\ell)^2 = \tfrac{1}{4} - I_\ell, \tag{9.50}$$

which implies that the resulting curve must lie on or inside a circle of radius $\frac{1}{2}$ centered at $f_\ell = i/2$. A resonance can no longer be identified with a purely

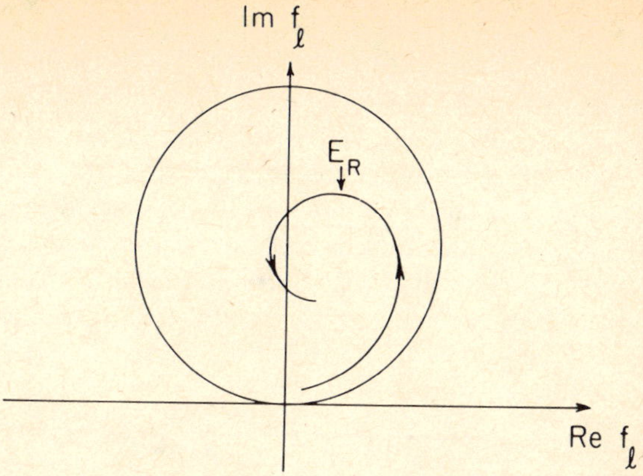

Fig.(9-3) A resonance appears as a counterclockwise loop in an Argand diagram, with the resonant energy at the top of the loop.

imaginary value of the partial wave amplitude because of the additional term I_ℓ in (9.23b). Provided that I_ℓ varies smoothly, however, we can show that there must be an equivalent effect, a loop in the curve, whenever the energy passes through a resonant value. An example of such a loop, which results directly from the presence of a pole in the complex k-plane, is shown in the figure. Such a plot, known as an Argand diagram, is a useful technique for spotting resonances when large inelastic effects might otherwise obscure them.

All of these methods of identification apply only when the resonance is formed by two particles involved in an elastic scattering process. Most of the resonances we discussed in Part I cannot be observed by this type of direct formation experiment; to see the K* in this way, for instance, would require a total cross section measurement of $K\pi$ scattering. We cannot as yet study these two-body processes directly, since we have no meson targets, but we can observe the $K\pi$ system in a less satisfactory way through inelastic processes such as $K^-p \rightarrow K^-\pi^+n$. The details of the $K^-\pi^+$ interaction are clouded by the presence of a neutron in the final state, with which both mesons will interact, but the implications of a resonance are fairly clear. The large total cross section of formation of a resonance will lead to a disproportionate number of $K^-\pi^+$ pairs in the final state having the apparent mass of the $K^{*\circ}$ meson. Furthermore, in the center-of-momentum frame of the $K^-\pi^+$ system the angular distribution of these particles should indicate its angular momentum. In Fig.(9-4) we show the results of such an experiment, in which there is a clear excess of events for which the mass of the $K^-\pi^+$ system is approximately 890 MeV, corresponding to the formation of the K^{*o} vector meson.

In searching for resonances as peaks in the mass spectrum of a many-body process, care must be taken to evaluate correctly effects due to the kinematics of the system. An appropriate way to do this is to compare the experimental data with the results obtained by assuming that the amplitude F for the process is simply a constant. Carrying out appropriate integrations then gives us a prediction for the shape of the mass spectrum in the absence of any resonances. The resulting distribution is called "phase space"; for the process $K^-p \to K^-\pi^+n$, it is the smooth curve shown in Fig.(9-4). When there are more particles in the

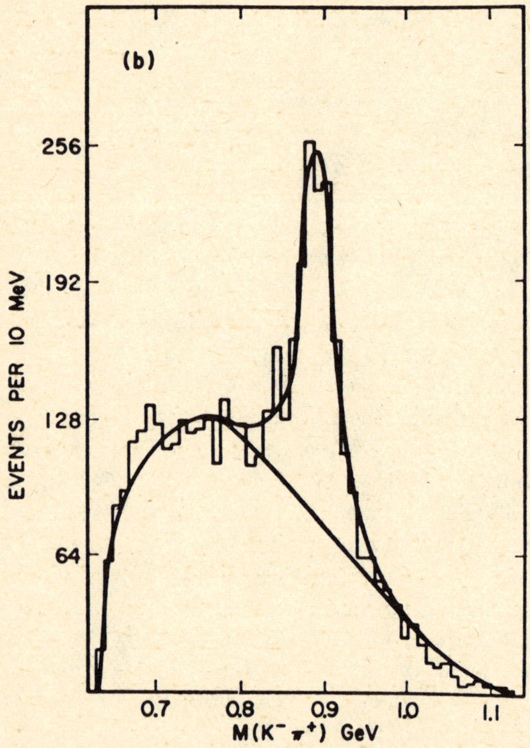

Fig.(9-4) Distribution of events as a function of the apparent mass of the $(K^-\pi^+)$ system in the final state of the reaction $K^-p \to K^-\pi^+\pi^-p$. (From P. M. Dauber, P. E. Schlein, W. E. Slater, and H. K. Ticho, *Phys. Rev.* 153, 1504 (1967).)

final state, an analogous calculation can be carried out for all of the different mass spectra which are possible in two-body states, three-body states, and so on. Since we shall be primarily concerned with two-body processes, however, we shall omit any further details of phase space methods.

High energy physicists are thus occupied with several different varieties of phenomena in the analysis of the experimental data, each of which may be considered evidence for the existence of a resonance. In all of these, however, the basic idea remains that every resonance corresponds to a pole of the partial wave amplitude for $E = E_R - i\Gamma$, and *vice versa*. In studying the analytic structure of the scattering amplitude we accept this correspondence as the fundamental definition of resonance. Therefore we may conceptually unite bound states and resonances by this common property.

That this definition is equivalent to the idea that a resonance is an "almost-bound" state becomes clear if we think of varying the strength of the potential. Let us return to the squre well potential for which

$$a_\ell = \frac{i}{k} \frac{k j_\ell'(ka) j_\ell(Ka) - K j_\ell(ka) j_\ell'(Ka)}{k h_\ell^{(1)\prime}(ka) j_\ell(Ka) - K h_\ell^{(1)}(ka) j_\ell'(Ka)}$$

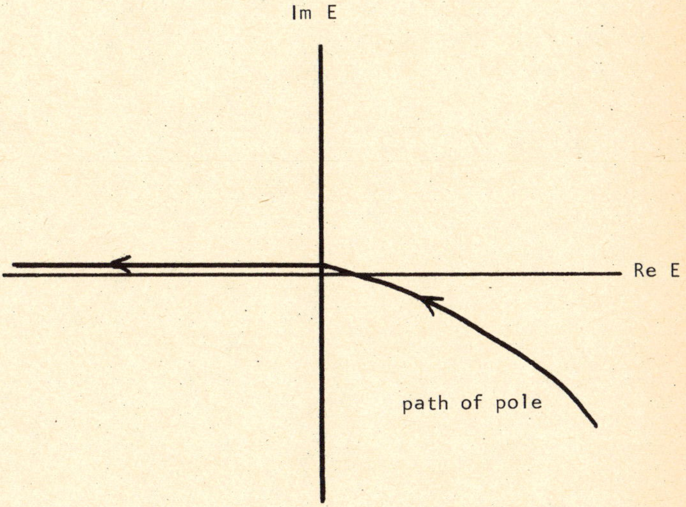

Fig.(9-5) Motion of a resonance pole as the potential becomes strong enough to convert it into a bound state.

Both bound states and resonances are determined by the poles of $a_\ell(k)$, which can come only from the vanishing of the denominator

$$kh_\ell^{(1)\prime}(ka)j_\ell(Ka) - Kh_\ell^{(1)}(ka)j_\ell{}'(Ka). \tag{9.51}$$

For a given value of V_0, the resonant energies are found by solving this equation with $k^2 = 2mE$ and $K^2 = 2m(E + V_0)$. There will be a certain number of solutions with real $E < 0$ which correspond to bound states, the rest being resonance poles. If V_0 is increased, however, it will eventually reach a value at which it is sufficiently strong to bind another state, and the number of bound states will increase by one. The new solution of (9.51) for real $E < 0$ has not appeared out of thin air, of course; one of the complex solutions has merely moved onto the real axis, as shown in Fig.(9-5). In other words, increasing the potential strength has transformed a resonance into a bound state.

Bound states and resonances are therefore simply different physical manifestations of a single cause, the presence of a pole in the scattering amplitude. If we believe that nature is analytic, these effects must be observed for every pole. Of course, a pole with a very large imaginary part Γ will not be apparent in the total cross section; but we believe nevertheless that it corresponds to a resonance, however short the lifetime may be. We can turn this argument around, and it takes on a significance far deeper than bump-hunting; the scattering amplitude is an analytic function having no singularities other than those specified by physics. If that is true, nature can keep few secrets from the physicist who understands the theory of functions of a complex variable.

Chapter 10

THE COMPLEX k-PLANE

The Schrödinger equation and the wave function obtained by solving it generally provide far more complete information about a quantum mechanical system than we are really interested in. Ordinarily the energy levels of the bound states and their angular momenta, plus the scattering amplitude, are enough to satisfy our needs. Occasionally the probability density is of interest, but a complete specification of the wave function hardly ever arises in studying strong interactions. We would be glad to know it, of course, since it contains the complete description of the system, but the price of that knowledge is very high – it requires a potential. Even if we were sure that strong interactions can be described by potential theory, the problem would be to construct the potential from scattering data, rather than the other way around. The added complications of inelastic processes, multiparticle production, etc., make invention of potentials a difficult and generally unrewarding venture.

In any case, our primary aim is to describe interactions. In non-relativistic quantum mechanics a potential provides an easy way of doing this, but for strong interactions it may be an unsatisfactory artifice. In the preceding chapter we have seen that the scattering amplitude $f(k, \theta)$ contained a description of the bound states of the system as well as of scattering. Therefore it may describe the interaction of a particle with a potential, or the interaction of two particles represented *via* a potential, just as completely as the potential itself does. Eliminating the direct study of wave functions and potentials in favor of the scattering amplitude was first suggested long ago by Heisenberg in his introduction of the scattering matrix, or S-matrix, which is the formal generalization of $f(k, \theta)$.

The crucial step in our analysis of $f(k, \theta)$ in Chapter 9 was the assumption that the momentum k could be treated as a complex variable in order to apply the theory of analytic functions. The idea that nature is smooth, and not obstreperous, is implicit in formulating the Schrödinger equation. When singularities arise, they do so for physical reasons. The same physics may produce these same singularities in situations where we do not know how to write a potential. Since knowing the singularities of an analytic function is equivalent to knowing the function, therefore, we may be able to formulate the scattering amplitude directly.

To be precise, let us consider an arbitrary partial wave amplitude $a_\ell(k)$ as a function of the complex variable k. Then by Cauchy's residue formula, we have

$$a_\ell(k) = \frac{1}{2\pi i}\oint \frac{a_\ell(k)}{k - k'} \qquad (10.1)$$

where c is a closed contour encircling the point k′ in the k-plane, as shown in Fig.(10-1), and we assume that $a_\ell(k)$ is analytic on and within c. The analyticity of $a_\ell(k)$ then allows us to deform the contour at will provided we avoid all singularities of the integrand. We have seen that there are poles of $a_\ell(k)$ corresponding to bound states and resonances; deforming the contour to that labeled c′ in Fig. (10-1), we pick up the contributions of these poles explicitly, plus the remainder of the contour. Thus

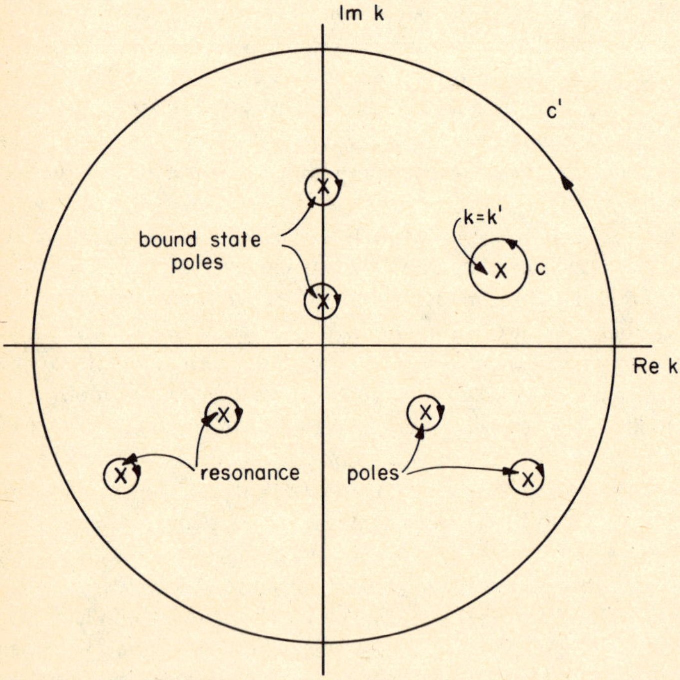

Fig.(10-1) Contours of integration for $a_\ell(k')$.

$$a_\ell(k') = \sum_i \frac{r_i}{k' - k_i} + \frac{1}{2\pi i} \oint_{c'} \frac{a_\ell(k)}{k - k'} dk \tag{10.2}$$

where the k_i's denote the pole positions of the bound states and resonances.

In the simplest imaginable case, we could hope that there are no singularities other than the poles, and that $|a_\ell(k)| \to 0$ as $|k| \to \infty$ in any direction. If we enlarge the contour c' to a circle of infinite radius, its contribution to (10.2) will then vanish, and the scattering amplitude will be given simply by the sum of the bound state and resonance poles. Knowing these energy levels and lifetimes would then enable us to predict the scattering amplitude unambiguously for *any* value of k. Unfortunately, life is not quite that simple; in (9.32), for example, it is clear that in addition to the pole structure there is a branch point in the k-plane resulting from the appearance of

$$K = \sqrt{k^2 + 2mV_0}.$$

Furthermore, the behavior as $|k| \to \infty$ is entirely unsatisfactory, leading to an essential singularity there.

It can be shown, however, that this essential singularity is purely a result of the discontinuous nature of the potential, and that it occurs in the first Born approximation. If we consider instead

$$\tilde{a}_\ell(k) = a_\ell(k) - a_\ell^B(k),$$

where $a_\ell^B(k)$ is the Born term, the essential singularity will be absent. The branch points at $k = \pm i\sqrt{2mV_0}$ will remain, however; they characterize the strength of the potential, and cannot be eliminated.

Thus a knowledge of the bound states and resonances alone is not enough to determine the scattering amplitude. Other singularities exist, and must be considered. In this chapter we shall therefore attempt a systematic study of the analytic properties of the partial wave amplitude in potential theory, in order to see if all of its singularities may be connected with a physical origin.

The Jost Function

In order to study the analytic properties of $a_\ell(k)$, we return to the radial wave equation (9.4), which we rewrite in the more convenient form

$$\frac{d^2}{dr^2}u_\ell(k, r) + \left[k^2 - U(r) - \frac{\ell(\ell + 1)}{r^2} \right] u_\ell(k, r) = 0, \tag{10.3}$$

with $u_\ell(k, r) = rR_\ell(r)$, $k^2 = 2mE$, and $U(r) = 2mV(r)$. Being a second-order differential equation, (10.3) has two linearly independent solutions. Assuming that $U(r)$ is less singular than r^{-2} near the origin, we find that generally as $kr \to 0$,

$$u_\ell(k, r) \approx A(k)(kr)^{\ell+1} + B(k)(kr)^{-\ell}.$$

Requiring that the wave function be integrable in the neighborhood of the origin leads to the *regular* solution $\phi_\ell(k, r)$, defined by the boundary condition

$$\phi_\ell(k, r) \approx \frac{(kr)^{\ell+1}}{(2\ell + 1)!!} \text{ as } (kr) \to 0. \tag{10.4}$$

(The second boundary condition needed to specify a solution of (10.3) uniquely is simply an overall normalization constant, which we may neglect.) Since $U(r) \to 0$ as $r \to \infty$, the asymptotic form of $\phi_\ell(k, r)$ is given by (9.14),

$$\phi_\ell(k, r) \approx C_\ell(k) r \left[e^{i\delta_\ell(k)} h_\ell^{(1)}(kr) - e^{-i\delta_\ell(k)} h_\ell^{(2)}(kr) \right], \tag{10.5}$$

where $C_\ell(k)$ and $\delta_\ell(k)$ must be determined by solving the equation. If we define $S_\ell(k) = e^{2i\delta_\ell(k)}$, this becomes

$$\phi_\ell(k, r) \approx C_\ell(k) e^{-i\delta_\ell(k)} \left[e^{-i(kr - \frac{1}{2}(\ell+1)\pi)} + S_\ell(k) e^{i(kr - \frac{1}{2}(\ell+1)\pi)} \right]. \tag{10.6}$$

Thus $S_\ell(k)$ is related to the ratio of outgoing to incoming spherical wave contributions in $\phi_\ell(k, r)$. It is related to the partial wave amplitude $a_\ell(k)$ by

$$S_\ell(k) = 1 + 2ika_\ell(k). \tag{10.7}$$

The analytic properties of $a_\ell(k)$ follow from those of $S_\ell(k)$. In general, these result from the fact that the boundary condition (10.4) is independent of k, and that the differential equation (10.3) depends explicitly on k in an analytic way. Thus small changes in k produce small changes in the differential equation, and none at all in the boundary condition, and we therefore expect only small changes to result in the solution as well. The formal statement of this idea is given in a theorem of Poincaré, which guarantees that solutions of a differential equation are analytic functions of a given parameter wherever the equation itself and the boundary condition are. Consequently we know that when solutions $\phi_\ell(k, r)$ of

(10.3) and (10.4) exist, they are analytic everywhere in the finite k-plane. Such functions are called entire functions of k.

It does not follow that $S_\ell(k)$ is analytic everywhere, however, since (10.6) determines $S_\ell(k)$ only as a ratio of terms in $\phi_\ell(k, r)$. Indeed, bound states and resonances require, as we have seen, the existence of poles in $S_\ell(k)$. In order to learn about $S_\ell(k)$ it is appropriate to study solutions of (10.3) which correspond to purely outgoing or incoming spherical waves, rather than the physical solutions $\phi_\ell(k, r)$. Thus we define $F_\ell(k, r)$ as the solution of (10.3) satisfying

$$F_\ell(k, r) \approx e^{-i(kr - \frac{1}{2}\ell\pi)}, \quad r \to \infty, \tag{10.8}$$

instead of (10.4). The two linearly independent solutions of (10.3) then correspond to $F_\ell(k, r)$ and $F_\ell(-k, r)$. Since $\phi_\ell(k, r)$ is also a solution of (10.3), it must be expressible as a combination of them, i.e.

$$\phi_\ell(k, r) = \alpha_\ell(k) F_\ell(k, r) + \beta_\ell(k) F_\ell(-k, r). \tag{10.9}$$

In order to satisfy (10.4), however, we must have

$$\phi_\ell(k, r) \sim F_\ell(-k, 0) F_\ell(k, r) - F_\ell(k, 0) F_\ell(-k, r). \tag{10.10}$$

(where $\sim$ means "proportional to"). Asymptotically, (10.10) implies that

$$\phi_\ell(k, r) \sim e^{-i(kr - \frac{1}{2}(\ell+1)\pi)} + \frac{F_\ell(k, 0)}{F_\ell(-k, 0)} e^{i(kr - \frac{1}{2}(\ell+1)\pi)}, \tag{10.11}$$

i.e., that

$$S_\ell(k) = \frac{F_\ell(k, 0)}{F_\ell(-k, 0)}. \tag{10.12}$$

This formulation of $S_\ell(k)$ was pioneered by R. Jost, and $F_\ell(k) \equiv F_\ell(k, 0)$ is therefore called the Jost function.

It is apparent from (10.12) that

$$S_\ell(-k) = \frac{1}{S_\ell(k)}. \tag{10.13a}$$

This symmetry could also have been derived directly by noting that both $\phi_\ell(k, r)$ and $\phi_\ell(-k, r)$ satisfy (10.3) and (10.4), so they must be proportional; (10.13a) then follows from (10.6). A more general symmetry of this type can be obtained by writing (10.3) using k* instead of k,

$$u_\ell''(k^*, r) + \left[k^{*2} - U(r) - \frac{\ell(\ell+1)}{r^2}\right]u_\ell(k^*, r) = 0,$$

and taking the complex conjugate of this equation to obtain

$$u_\ell^{*\prime\prime}(k^*, r) + \left[k^2 - U(r) - \frac{\ell(\ell+1)}{r^2}\right]u_\ell^*(k^*, r) = 0.$$

Similar treatment of the boundary condition (10.8) shows that

$$F_\ell^*(-k^*, r) \approx e^{-i(kr - \frac{1}{2}\ell\pi)}.$$

Thus $F_\ell^*(-k^*, r)$ satisfies the same differential equation and the same inhomogeneous boundary condition as $F_\ell(k, r)$; consequently they must be identical,

$$F_\ell^*(-k^*, r) = F_\ell(k, r).$$

Therefore

$$F_\ell^*(-k^*) = F_\ell(k), \tag{10.14}$$

$$S_\ell^*(-k^*) = S_\ell(k), \tag{10.13b}$$

and by combining (10.13a) and (10.13b), we also find that

$$S_\ell(k^*) = \frac{1}{S_\ell(k)} \tag{10.13c}$$

It should be noted here that $F_\ell^*(k^*)$ is a function of k, and not of k*; $F_\ell(k^*)$ is a function of k*, but taking its complex conjugate changes every k* back to k. This is an important point, since an analytic function of k may not depend on k*.

Analytic Properties of the Jost Function

To get more detailed information on the singularities of $S_\ell(k)$, we now study those of the Jost function $F_\ell(k)$. Poincaré's theorem applies also to $F_\ell(k, r)$, since the boundary condition (10.8) is analytic in its dependence of k. Therefore the Jost function is an analytic function of k for all k such that the solution $F_\ell(k, r)$ exists.

In considering the existence of solutions of (10.3) with certain boundary conditions, it is appropriate, for two reasons, to convert it into an integral equation. First, the integral equation automatically contains the boundary conditions. Second, the existence of the solution, and therefore its analyticity in k, can be established by proving the convergence of that equation. Thus it is not necessary to obtain the solution explicitly. Furthermore, the exact form of the potential need not be given, provided its asymptotic properties are specified.

For simplicity, we shall consider only the s-wave amplitude; with a little extra work, the same procedure can be applied to arbitrary ℓ. With $\ell = 0$, we have

$$\left(\frac{d^2}{dr^2} + k^2\right) u_0(k, r) = U(r) u_0(k, r). \tag{10.15}$$

This equation is easily converted to an integral equation satisfying the boundary condition (10.8) using Green's function techniques. Defining $G(r - r')$ as the solution of

$$\left(\frac{d^2}{dr^2} + k^2\right) G(r - r') = \delta(r - r') \tag{10.16a}$$

with the boundary condition

$$G(r - r') \to 0 \text{ as } r \to \infty, \tag{10.16b}$$

we find that $F_0(k, r)$ satisfies

$$F_0(k, r) = e^{-ikr} + \int dr' G(r - r') U(r') F_0(k, r'). \tag{10.17}$$

The determination of the Green's function $G(r - r')$ is a standard example of complex variable techniques. Writing $G(r - r')$ as a Fourier transform

$$G(r - r') = \frac{1}{2\pi} \int_{-\infty}^{\infty} e^{iK(r-r')} g(K) dK, \tag{10.18}$$

and using the Fourier representation of the delta function, we find from (10.16a) that

$$g(K) = \frac{1}{k^2 - K^2}. \tag{10.19}$$

Putting (10.19) back into (10.18) leads to an indefinite result because the integrand has poles at $K = \pm k$. A solution is obtained by moving the poles infinitesimally off the real axis; the form of the solution depends on whether they are moved above or below it. It is easily shown that the solution corresponding to the boundary condition (10.16b) is obtained if both poles are moved *below* the real axis, as shown in Fig.(10-2). Then if $r > r'$, the exponential term

$$e^{iK(r-r')} = e^{i(\mathrm{Re}K)(r-r')}e^{-(\mathrm{Im}K)(r-r')}$$

vanishes as $|k| \to \infty$ with $\mathrm{Im}K > 0$, so the integration contour can be closed in the upper half-plane. Since the closed contour contains no singularities, it follows that

$$G(r - r') = 0, \qquad r > r', \tag{10.20a}$$

satisfying (10.16b). For $r < r'$, the contour must instead be closed in the lower half-plane; then both poles contribute, yielding

$$G(r - r') = -\frac{1}{k} \sin k(r - r'), \qquad r < r'. \tag{10.20b}$$

Substituting (10.20) into (10.17), we find

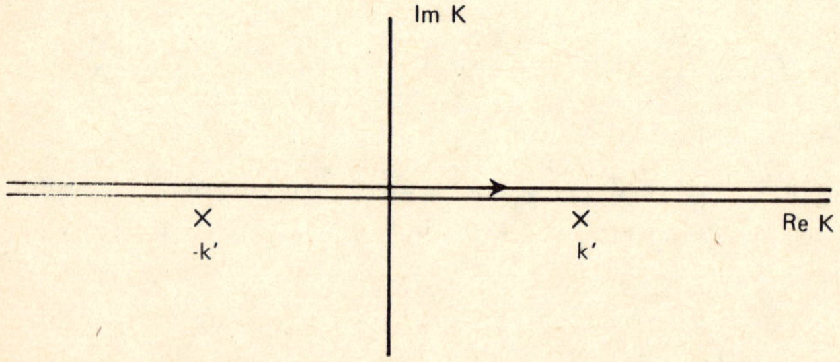

Fig.(10-2) Contour of integration for the Green's function (10.20).

$$F_0(k, r) = e^{-ikr} - \frac{1}{k}\int_r^\infty dr' \sin(k(r - r'))U(r')F_0(k, r'). \tag{10.21}$$

The analyticity of $F_0(k, r)$ as a function of k can now be established by proving that the integral in (10.21) converges. It is not necessary to *solve* the equation; general theorems on integral equations guarantee that it is soluble if it is convergent, and if it is soluble for a given k it must be analytic there. Consequently we need only study the convergence of the integral

$$I(k, r) = \int_r^\infty dr' \sin(k(r - r'))U(r')F_0(k, r'). \tag{10.22}$$

Since $U(r')$ is assumed to be finite for $r' > 0$, there is no divergence at the lower limit in $I(k, r)$ for $r > 0$. As $r \to 0$, we have

$$I(k, 0) = -\int_0^\infty dr' \sin kr'(U(r')F_0(k, r')$$

and convergence is still guaranteed provided $U(r')$ is no more singular than $1/r'$ as $r' \to 0$. (More singular potentials can sometimes be treated by more sophisticated techniques.)

To consider the convergence of $I(k, r)$ at the upper limit, we may use the known asymptotic behavior of $F_0(k, r)$ to write

$$\int_{r_0}^\infty dr' \sin(k(r - r'))U(r')F_0(k, r') \approx \int_{r_0}^\infty dr' \sin(k(r - r')U(r')e^{ikr'}$$

$$= \frac{1}{2i}\int_{r_0}^\infty dr' U(r')[e^{ikr}e^{-2ikr'} - e^{-ikr}] \tag{10.23}$$

for some large r_0. Assuming that $U(r')$ vanishes rapidly enough as $r' \to \infty$ that

$$\int^{\infty} dr' U(r')$$

exists, we see that the convergence of I(k, r) is guaranteed provided that

$$\int_{r_0}^{\infty} dr' e^{-2ikr'} U(r') < \infty.$$

Now we must make an assumption on the asymptotic behavior of $U(r')$ in order to proceed further. Suppose that $U(r')$ decreases faster than an exponential,

$$|U(r')| < Ce^{-\mu r'}, \qquad (10.24)$$

for some real, positive value of μ. Then for arbitrary complex k, we have

$$|e^{-2ikr'} U(r')| < Ce^{-(\mu - 2\mathrm{Im}k)r'}$$

and it follows that the integral converges for all values of k such that $(\mu - 2\mathrm{Im}k) > 0$, i.e., for

$$\mathrm{Im}k < \tfrac{1}{2}\mu. \qquad (10.25)$$

Consequently the integral is convergent, and therefore $F_0(k)$ is analytic, for all k satisfying (10.25). The domain of analyticity implied by (10.25) is shown in Fig.(10.3a). An ingenious method of enlarging this domain has been given by Regge, based on the substitution $r = Re^{i\alpha}$ in (10.15). We than have

$$e^{-2i\alpha} \frac{d^2}{dR^2} u_0(k, Re^{i\alpha}) + k^2 u(k, Re^{i\alpha}) = U(Re^{i\alpha}) u(k, Re^{i\alpha}),$$

and if we define $K = ke^{i\alpha}$, $\tilde{u}_0(K, R) = u_0(Ke^{-i\alpha}, Re^{i\alpha})$, and $U'(R) = e^{2i\alpha} U(Re^{i\alpha})$, this equation becomes

$$\frac{d^2}{dR^2} u_0(K, R) + K^2 \tilde{u}_0(K, R) = U'(R) \tilde{u}_0(K, R). \qquad (10.26)$$

The boundary condition (10.11), furthermore, becomes

$\tilde{u}_0(K, R) \approx e^{iKR}$ as $R \to \infty$.

Thus (10.26) is fully equivalent to the equation we have just analyzed with the replacement of U(r) by U′(R), so the same procedure is applicable. Let us assume that the asymptotic form of the potential is actually given by

$$U(r) \approx r^{\beta} e^{-\mu r} \text{ as } r \to \infty, \tag{10.27}$$

so that

$$U'(R) \approx e^{2i\alpha}(Re^{i\alpha})^{\beta} e^{-\mu(\cos\alpha + i \sin\alpha)}.$$

In order to maintain the convergence of

$$\int^{\infty} U'(R')dR',$$

we must have $\cos\alpha > 0$, i.e. $-\frac{1}{2}\pi < \alpha < \frac{1}{2}\pi$. The convergence criterion (10.24) becomes

$$|e^{-2iKR}U'(R)| \lesssim CR^{\beta}e^{-(\mu\cos\alpha - 2\operatorname{Im}K)R}.$$

As $R \to \infty$, the exponential dominates any power of R, so analyticity of the solution is guaranteed for

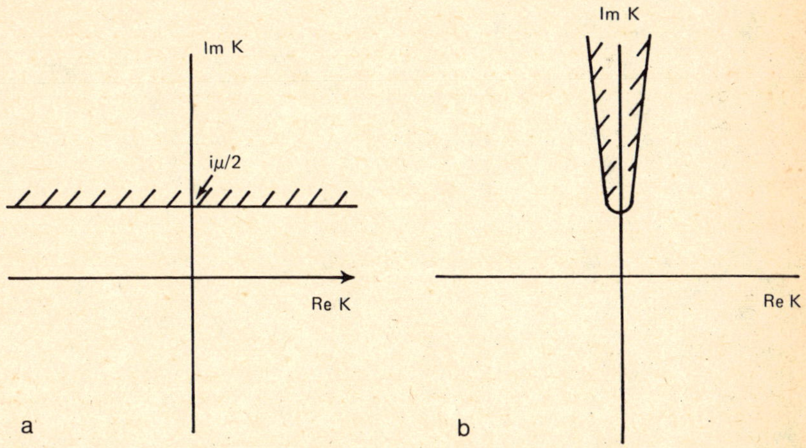

Fig.(10-3) (a) Domain of analyticity of $F_0(k)$ resulting from (10.25), and (b) its extension.

$$\text{Im}K = \text{Im}(ke^{i\alpha}) < \tfrac{1}{2}\mu \cos\alpha,$$

that is, for

$$\text{Im}k < \tfrac{1}{2}\mu - (\text{Re}k)\tan\alpha. \tag{10.27}$$

Since α is arbitrary within the limits $-\frac{1}{2}\pi < \alpha < \frac{1}{2}\pi$, it follows that (10.27) excludes only an arbitrarily narrow wedge of the k-plane, as shown in Fig. (10-3b).

The behavior of $F_0(k)$ as $|k| \to \infty$ can also be established from (10.23). The Riemann–Lebesgue lemma tells us that

$$\lim_{|k|\to\infty} \int_R^\infty dr' U(r') e^{-2ikr'} = 0$$

provided the integral is convergent. Consequently

$$\lim_{|k|\to\infty} I(k, r) = e^{-ikr} \int_r^\infty dr' U(r'),$$

and it follows from (10.21) that

$$\lim_{|k|\to\infty} F_0(k, r) = e^{-ikr} \left(1 + \frac{1}{k} \int_r^\infty U(r') dr' \right), \tag{10.28}$$

that is,

$$|F_0(k) - 1| \sim \frac{1}{|k|} \text{ as } |k| \to \infty \tag{10.29a}$$

provided the integral in (10.28) exists for $r = 0$. The possibility that $U(r') \sim 1/r'$ as $r' \to 0$ requires special treatment here; it can be shown, however, that in this case

$$|F_0(k) - 1| \sim \frac{\ln|k|}{|k|} \text{ as } |k| \to \infty \qquad (10.29b)$$

Both (10.29a) and (10.29b) maintain the desired result that

$$\lim_{|k|\to\infty} |f_0(k) - 1| = 0 \qquad (10.30a)$$

for which it follows that

$$\lim_{|k|\to\infty} |S_0(k) - 1| = 0, \qquad (10.30b)$$

so that dispersion relations can be written for the partial-wave amplitude.

The extension of these arguments to arbitrary ℓ is complicated considerably by the centrifugal potential term $\ell(\ell + 1)/r^2$, which fails to satisfy the necessary convergence criteria at either $r = 0$ or $r = \infty$. Thus the straightforward procedure of replacing U(r) by

$$U'(r) = U(r) + \frac{\ell(\ell + 1)}{r^2}$$

cannot succeed. Instead, we must follow the Green's function procedure using solutions of

$$\left(\frac{d^2}{dr^2} + k^2 - \frac{\ell(\ell + 1)}{r^2}\right)G(r - r') = \delta(r - r'),$$

which can be expressed in terms of spherical Hankel functions, rather than (10.16a). It can then be proved that, under similar conditions, $F_\ell(k)$ is analytic in the same region we have shown for $F_0(k)$.

The Yukawa Potential

Consequently the Jost function $F_\ell(k)$ is analytic in the entire k-plane except for the portion of the imaginary axis for which $\mathrm{Im}\,k > \frac{1}{2}\mu$. Exactly what singularities may lie there cannot be said without detailed knowledge of the potential. It could be nothing worse than simple poles, or it could be an infinite set of branch points. The quantity which reflects the structure of this non-analytic region is the discontinuity across it,

$$\Delta_\ell(\kappa) = F_\ell(i\kappa + \epsilon) - F_\ell(i\kappa - \epsilon)$$

with ϵ infinitesimal. A knowledge of this discontinuity is as good as having a potential.

The most important potential in the strong interactions is the Yukawa potential, so let us investigate the results obtainable for this case in more detail. With $\ell = 0$ and

$$U(r) = \lambda \frac{e^{-\mu r}}{r}, \tag{10.31}$$

we have for the integral equation (10.21)

$$F_0(k, r) = e^{-ikr} - \frac{\lambda}{k} \int_r^\infty dr' \sin(k(r - r')) \frac{e^{-\mu r'}}{r'} F_0(k, r') \tag{10.32}$$

A solution of (10.32) is conveniently found by iteration. If we take

$$F_0^{(0)}(k, r) = e^{-ikr}$$

$$F_0^{(n+1)}(k, r) = -\frac{\lambda}{k} \int_r^\infty dr' \sin(k(r - r')) \frac{e^{-\mu r'}}{r'} F_0^{(n)}(k, r'), \tag{10.33}$$

then it is readily seen that

$$F_0(k, r) = \sum_{n=0}^\infty F_0^{(n)}(k, r) \tag{10.34}$$

satisfies the integral equation. The corresponding Jost function is simply

$$F_0(k) = \sum_{n=0}^\infty F_0^{(n)}(k, 0) \tag{10.35}$$

Since $F_0^{(n)}(k, 0)$ is proportional to λ^n, as a result of (10.33), this is a power series in the potential strength; in fact it is just the Born series for the Jost function.

Let us examine the first iteration in (10.35),

$$F_0^{(1)}(k) = \frac{\lambda}{k} \int_0^\infty dr' \sin kr' \frac{e^{-\mu r'}}{r'} e^{-ikr'}. \tag{10.36}$$

This integration can be performed if we write

$$\frac{e^{-\mu r'}}{r'} = \int_\mu^\infty d\alpha e^{-\alpha r'}, \tag{10.37}$$

yielding

$$F_0^{(1)}(k) = \lambda \int_\mu^\infty d\alpha \frac{1}{\alpha(\alpha + 2ik)} \tag{10.38a}$$

$$= \frac{\lambda}{2ik} \ell n \left(\frac{\mu}{\mu + 2ik} \right). \tag{10.38b}$$

It is clear that $F_0^{(1)}(k)$ has exactly the analytic structure derived above. There is a logarithmic branch point at $k = \frac{1}{2}i\mu$; when $k = iK$, $K > \frac{1}{2}\mu$, the argument of the logarithm becomes real and negative. An alternative way of seeing the singularity is to note that the integrand in (10.38a) has a pole which moves onto the contour of integration in that region.

The corresponding structure for further iterations can be studied similarly. We shall omit the details here; the result is that after n iterations, there are precisely n branch points, located at $2ik = \mu, 2\mu, \ldots n\mu$. A more difficult calculation shows the same to be true for arbitrary ℓ. Consequently the Yukawa potential leads to a Jost function $F_\ell(k)$ with an infinite number of branch points, evenly spaced along the positive imaginary axis. The same analysis can be carried through for any potential which can be expressed as a superposition of exponentials similarly to (10.37),

$$U(r') = \int_{\mu}^{\infty} da C(a) e^{-ar'}, \tag{10.39}$$

with precisely the same results.

Singularities of $S_\ell(k)$

Knowing the analytic structure of $F_\ell(k)$ leads immediately to that of

$$S_\ell(k) = \frac{F_\ell(k)}{F_\ell(-k)}.$$

Clearly $S_\ell(k)$ has a cut from $k = \frac{1}{2}i\mu$ to $i\infty$, due to $F_\ell(k)$, and one from $k = -\frac{1}{2}i\mu$ to $-i\infty$ because of $F_\ell(-k)$. In addition, however, $S_\ell(k)$ will have poles wherever $F_\ell(-k) = 0$. The singularities of $S_\ell(k)$ are therefore as shown in Fig.(10-4).

The poles of $S_\ell(k)$ must come from zeros of the denominator, since the numerator $F_\ell(k)$ is analytic except for the cut and therefore has no poles. As we have seen, they will have as their physical consequences the appearance of bound states and resonances.

When $F_\ell(-k) = 0$, equation (10.12b) shows that the asymptotic behavior of the wave function is

$$\phi_\ell(k, r) \sim F_\ell(k) e^{ikr}. \tag{10.40}$$

If (10.40) is to represent a bound state, the wave function must vanish as $r \to \infty$, so the pole position must have $\text{Im}\, k > 0$. But in (10.15) it is clear that k^2 is the eigenvalue of a Hermitian operator, and therefore must be real. Thus the bound state poles should only appear on the positive imaginary axis.

It is easily shown that no other poles are allowed in the upper half-plane. We combine (10.3) and its complex conjugate to obtain

$$u_\ell^*(k, r) u_\ell''(k, r) - u_\ell^{*''}(k, r) u_\ell(k, r) = (k^2 - k^{*2}) |u_\ell(k, r)|^2. \tag{10.41}$$

Integrating (10.41) over all r yields

$$\left(u_\ell^* u_\ell' - u_\ell^{*'} u_\ell\right)\Big|_0^\infty = (k^2 - k^{*2}) \int_0^\infty dr |u_\ell(k, r)|^2. \tag{10.42}$$

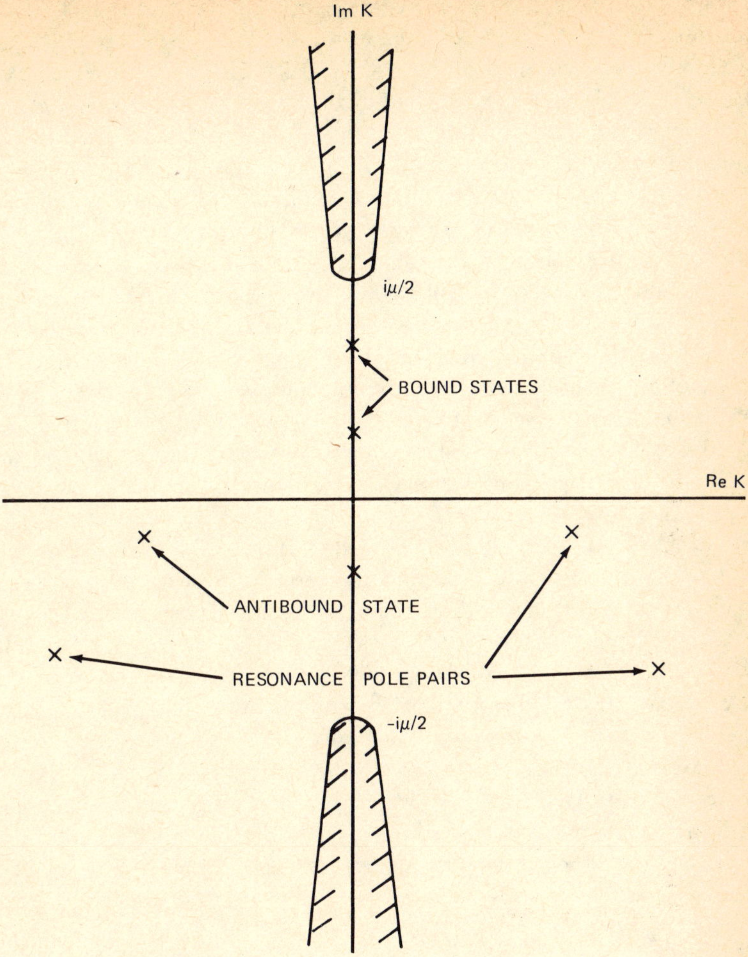

Fig.(10.4) Singularities of $S_\ell(k)$.

Since $u_\ell(k, 0) = 0$, the lower limit does not contribute in the left-hand side of (10.42), and if (10.40) holds as $r \to \infty$ with $\text{Im}k > 0$, then the upper limit also vanishes. It follows that $(k^2 - k^{*2}) = 0$, so k^2 is real, proving that all such poles are on the imaginary k-axis.

The resonance poles occur for $\text{Im } k < 0$. In general they will appear in pairs, since (10.14) implies that if $F_0(-k, 0) = 0$, then also $F_0(k^*, 0) = 0$; the pole positions are thus symmetric about the imaginary axis. The only poles that appear singly are those directly on the imaginary axis, which in this case are called "antibound states". None of these values of k correspond to physically allowed wave functions, since for Im $k < 0$ (10.40) leads to an increasing exponential behavior. In principle, we identify all of these poles as resonances. We have not proved, however, that there is a finite number of resonances, or that all of these poles are located near enough to the real k axis that they would be observed as resonances as described in Chapter 9. Indeed, it can be shown that the partial wave amplitude (9.32) resulting from the square well potential has an infinite number of resonance poles, some far from any physical value of k. In the next chapter, however, we shall see that this problem can be overcome.

Finally, we have the two cuts of $S_\ell(k)$. These result from the potential and reflect its properties. In particular, the cuts begin at $k = \pm\frac{1}{2}i\mu$, where μ describes the range of the potential. The discontinuities across these cuts depend on the detailed form of the potential. That is, they tell us about the interaction. Knowing the discontinuities is therefore essentially the same as knowing the potential, and evaluating the dispersion relation (10.2) is equivalent to solving the Schrödinger equation.

The advantage of writing dispersion relations, however, is that they do not depend on the convergence of a perturbative series. In solving (10.32), for example, we obtain for the Jost function a power series expansion in λ. We have indicated in Chapter 1 that if this Yukawa potential is used to represent exchange of a single pion between two nucleons, the resulting expansion fails to converge. It is clear from (10.40), however, that the location of the cut is independent of λ, depending only on the range μ of the forces. Thus we may hope that, even though perturbation theory cannot give us the scattering amplitude itself, it does give correctly its analytic structure. We may then do away with the need for a Schrödinger equation or a perturbative treatment of a potential, substituting a knowledge of the discontinuities for those problems.

Chapter 11

THE COMPLEX ENERGY PLANE

Although the analytic properties of the partial wave amplitude are more easily established by considering it as a function of the momentum k, they are more effectively put to use if we consider it as a function of the energy. We have shown that $a_\ell(k)$ is analytic in the k-plane except for the singularities shown in Fig.(10-4). Now let us define $A_\ell(E) = a_\ell(\sqrt{2mE})$, and ask what can be said about the analytic properties of $A_\ell(E)$ in the complex energy plane.

Since $E = k^2/2m$, both k and $-k$ correspond to the same value of E; thus we will need two full energy planes in order to represent a function of k. Suppose that z(k) is an arbitrary single-valued function of k; that is, for each value of k there is a unique value of z(k). Now let us map the values of z(k) onto the corresponding points in the E-plane, starting with positive real k, to define a new function Z(E). If $k = |k|e^{i\kappa}$, then $E = |E|e^{2i\kappa}$, so the values of z(k) corresponding to a given phase angle κ in the k-plane are mapped onto points with twice that phase angle in the E-plane; the z(k) along the positive imaginary k-axis wind up on the negative real E-axis, and the first quadrant of the k-plane is mapped onto the upper half of the E plane. Continuing this process, we find that the negative real k-axis corresponds again to the positive real E-axis, and the upper half of the k plane has filled the E plane.

But we did not assume that $z(k) = z(-k)$, so in general $Z(k^2/2m) \neq Z((-k)^2/2m)$; that is, Z(E) will be a double-valued function. To visualize Z(E) we imagine that the E-plane is a two-sheeted structure, as shown in Fig.(11-1). As k descends onto the negative real axis from the second quadrant, the corresponding point in the E-plane moves down through a "cut" onto a second "Riemann sheet" rather than returning to the starting point. Then the values of Z(E) on the upper sheet will be uniquely identified with the values of z(k) found in the upper half of the k-plane, while those on the second sheet correspond to the lower half-plane. As $\kappa \to 2\pi$, we are returning to the positive real k-axis; this means that as we approach the cut from below on the second sheet, we re-emerge onto the top sheet. Thus we represent the double-valuedness of Z(E) by defining it as a function in a "cut E-plane with two Riemann sheets".

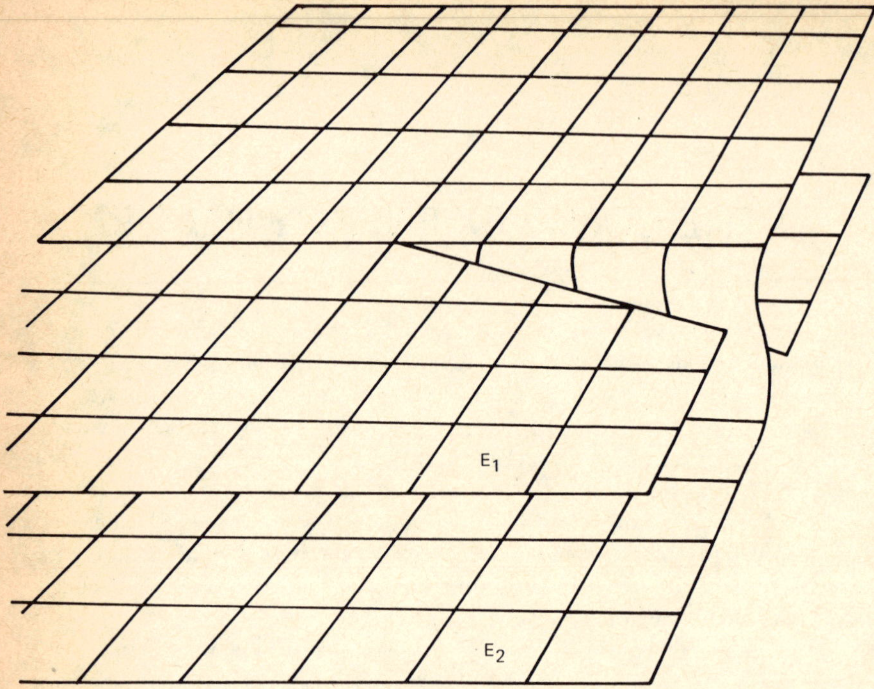

Fig.(11-1) A cut E-plane with two Riemann sheets.

Analyticity of $A_\ell(E)$

What does $a_\ell(k)$ look like in this cut E-plane? The first Riemann sheet contains the bound state poles, which are mapped onto the negative real E-axis along with the upper potential cut. Furthermore, the real positive values of k, which we would call "physical", are on the upper edge of the cut along the real E-axis. Thus the physically attainable energy values can be located in this top sheet, so we call it "the physical sheet". The lower half of the k-plane, with its resonances, antibound states, and the lower potential cut, is on the second or "unphysical" sheet. That it is "unphysical" is emphasized by the antibound states; although they corresponded to real negative values of $E = k^2/2m$, they did not produce physically satisfactory wave functions.

Thus we can define a unique mapping of an analytic function of k into the E-plane provided we keep in mind its two-sheeted structure. In particular,

$A_\ell(E)$ will have singularities in each E-plane only as required by the corresponding singularities of $a_\ell(k)$ in the k-plane. On the physical sheet, therefore, we will find a "left-hand cut", due to the potential cut in the upper half of the k-plane, running along the negative real axis from $E = -\frac{1}{4}\mu^2$ to $-\infty$, as well as bound-state poles on the negative real axis. In the unphysical sheet there is the lower potential cut plus the resonance and antibound state poles. Each sheet also has the "right-hand cut" through which the other sheet is reached, located on the positive real axis. This cut is present only as the junction at which physical and unphysical sheets merge. It should be noted that the left-hand cuts are entirely different; they may (as for the Yukawa potential) contain an infinite number of branch points, each one with its own attached cut. Moving through those cuts will not necessarily lead to the unphysical sheet. Indeed, the branch point in (10.35b) is logarithmic, and requires the presence of its own *infinite* set of Riemann sheets. Thus a *terra incognita* lurks beneath the left-hand cut, into which we shall not move.

Therefore the analytic structure of $A_\ell(E)$ is that shown in Fig.(11-2). Now let us write a Cauchy integral for $A_\ell(E)$, with E on the physical sheet,

$$A_\ell(E) = \frac{1}{2\pi i} \oint_c \frac{A_\ell(E')}{E' - E}\, dE', \tag{11.1}$$

where c is a closed contour surrounding the point E. The contour can be deformed to that labeled c′ in Fig.(11-3). As $|E'| \to \infty$, we know from (10.30b) that $|A_\ell(E)| \to 0$, so the contribution of the large circular parts of the contour vanishes, and we find

$$A_\ell(E) = \sum_B \frac{r_B}{E - E_B} + \frac{1}{2\pi i}\int_0^\infty dE' \frac{A_\ell(E' + i\epsilon)}{E' - E} + \frac{1}{2\pi i}\int_\infty^0 dE' \frac{A_\ell(E' - i\epsilon)}{E' - E}$$

$$+ \frac{1}{2\pi i}\int_{-\infty}^{-1/4\mu^2} dE' \frac{A_\ell(E' + i\epsilon)}{E' - E} + \frac{1}{2\pi i}\int_{-1/4\mu^2}^{-\infty} dE' \frac{A_\ell(E' - i\epsilon)}{E' - E}, \tag{11.2}$$

where E_B and r_B are the bound state energies and their residues and ϵ is infinitesimal. The integrals above and below the cuts can be simplified by defining the *discontinuities*

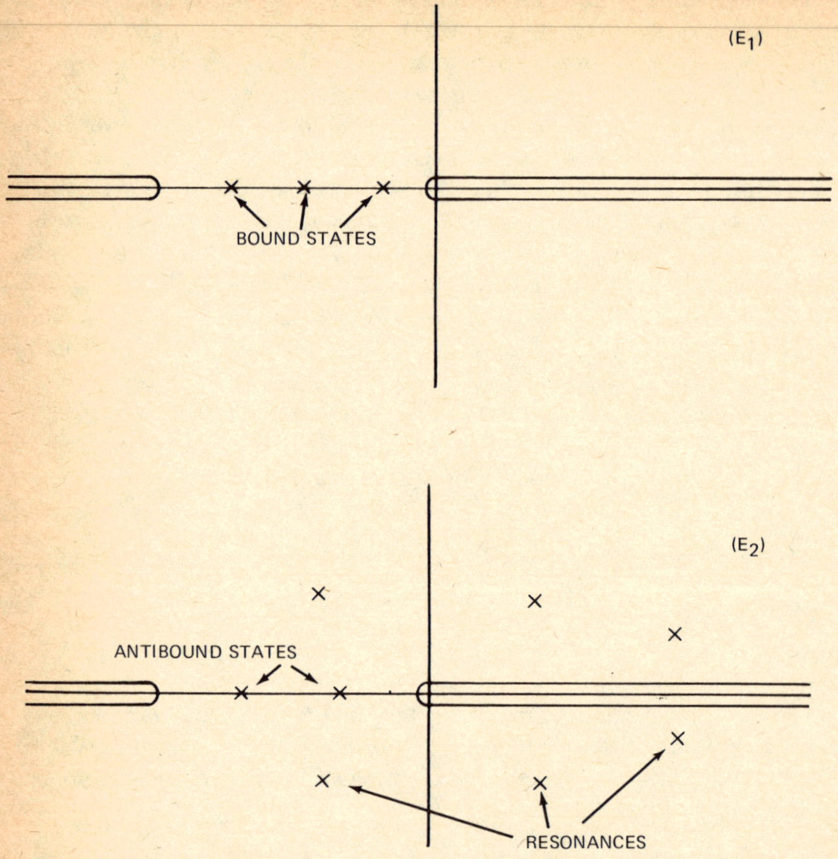

Fig.(11-2) Structure of $A_\ell(E)$ in the cut E plane.

$$\Delta_\ell^{(\ell)}(E') = \frac{1}{2i}[A_\ell(E' + i\epsilon) - A_\ell(E' - i\epsilon)], \quad E' < -\tfrac{1}{4}\mu^2$$

$$\Delta_\ell^{(r)}(E') = \frac{1}{2i}[A_\ell(E' + i\epsilon) - A_\ell(E' - i\epsilon)], \quad E' > 0, \tag{11.3}$$

yielding then

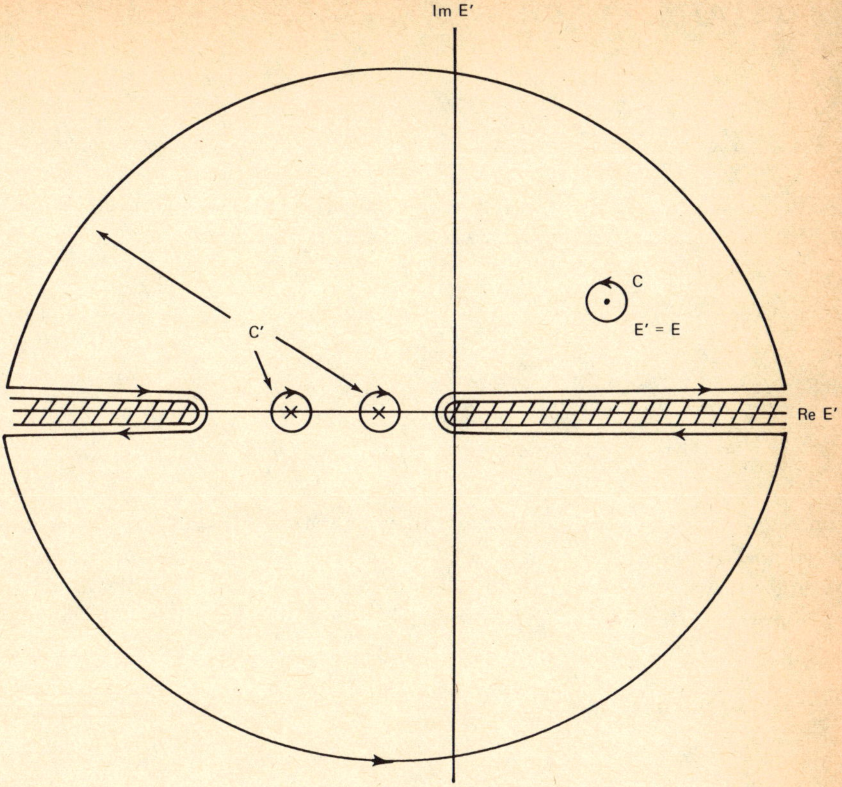

Fig.(11-3) Contours involved in writing a dispersion integral for $A_\ell(E)$.

$$A_\ell(E) = \sum_B \frac{r_B}{E - E_B} + \frac{1}{\pi}\int_0^\infty dE' \frac{\Delta_\ell^{(r)}(E')}{E' - E} + \frac{1}{\pi}\int_{-\infty}^{-1/4\mu^2} dE' \frac{\Delta_\ell^{(\ell)}(E')}{E' - E}. \tag{11.4}$$

The discontinuities (11.3) may be identified by making use of the symmetries (10.13). Since

$$S_\ell(k) = S_\ell^*(-k^*)$$

we have

$$1 + 2ika_\ell(k) = [1 + 2i(-k^*)a_\ell(-k^*)]^*$$

i.e.

$$a_\ell(k) = a_\ell^*(-k^*). \tag{11.5}$$

The points k and $-k^*$ have the same imaginary part, so they map onto energies E and E^* on the same sheet. It follows that on either sheet

$$A_\ell(E) = A_\ell^*(E^*). \tag{11.6}$$

In particular, for real E this becomes

$$A_\ell(E) = A_\ell^*(E),$$

implying that $A_\ell(E)$ is real. Functions satisfying the condition (11.6) are known as *real analytic functions* for this reason. Because of the cuts, however, E cannot be on the real axis unless $-\frac{1}{4}\mu^2 < E < 0$; otherwise, (11.6) relates values appearing above and below the cut. Specifically, for E' real we have

$$A_\ell(E' - i\epsilon) = A_\ell^*(E' + i\epsilon),$$

so

$$\Delta_\ell^{(\ell,r)}(E') = \frac{1}{2i}[A_\ell(E' + i\epsilon) - A_\ell^*(E' + i\epsilon)]$$

$$= \mathrm{Im}A_\ell(E' + i\epsilon). \tag{11.7}$$

Using (11.7), the discontinuity across the right-hand cut is particularly easy to evaluate, since there $E' + i\epsilon$ corresponds, as $\epsilon \to 0$, to the physical values of positive energy, and by the optical theorem (9.25),

$$\mathrm{Im}A_\ell(E + i\epsilon) = \frac{(2\ell + 1)}{4\pi}\sqrt{2mE}\,\sigma_\ell(E), \tag{11.8}$$

$\sigma_\ell(E)$ being the ℓ-wave contribution to the total cross section. Thus we find

$$A_\ell(E) = \sum_B \frac{r_B}{E - E_B} + \frac{(2\ell+1)}{4\pi^2} \int_0^\infty dE' \frac{\sqrt{2mE'}\,\sigma_\ell(E')}{E' - E} + \frac{1}{\pi} \int_{-\infty}^{-1/4\mu^2} dE' \frac{\operatorname{Im} A_\ell(E' + i\epsilon)}{E' - E}. \tag{11.9}$$

The resonance poles, being entirely on the unphysical sheet, do not make any direct contribution to (11.9). Their presence will nonetheless be felt, however, through (11.8), since a resonance produces a bump in the total cross section. It is instructive to see how this comes about in more detail.

Suppose that there is a resonance pole at $k = K$, with $\operatorname{Re} K > 0$ and $\operatorname{Im} K < 0$. Then the symmetries (10.13a) and (10.13b) imply that there is also a resonance pole at $k = -k^*$ and that there are *zeros* of $S_\ell(k)$ at $k = -K$ and $k = K^*$. Near the pole, $S_\ell(k)$ will be dominated by its contribution, and we may write

$$S_\ell(k) \approx \frac{(k+K)(k-K^*)}{(k-K)(k+K^*)}, \tag{11.10}$$

including all of the known symmetries. It follows that

$$a_\ell(k) \approx \frac{\operatorname{Im} K}{(k-K)(k+K^*)} = \frac{i}{2}\left[\frac{1}{(k-K)} - \frac{1}{(k+K^*)}\right], \tag{11.11}$$

so $a_\ell(k)$ shows only the poles. When $a_\ell(k)$ is mapped onto the E planes, both of these poles are on the second sheet, their positions being complex conjugates of each other, as shown in Fig.(11-4).

On the integration contour *above* the right-hand cut, the pole at $k = -K^*$ is nearby; it can be reached by merely slipping through the cut. The pole at $k = K$, however, is far away, since it can only be reached by a path that goes around the origin, as indicated. Therefore in this region we have

$$A_\ell(E' + i\epsilon) \approx -\frac{i}{2}\left(\frac{1}{k+K^*}\right). \tag{11.12a}$$

The situation is exactly reversed below the right-hand cut; now it is the pole at $k = K$ which is nearby, and

$$A_\ell(E' - i\epsilon) \approx \frac{i}{2}\left(\frac{1}{k-K}\right). \tag{11.12b}$$

Thus in each case the partial wave amplitude is dominated by the resonant poles,

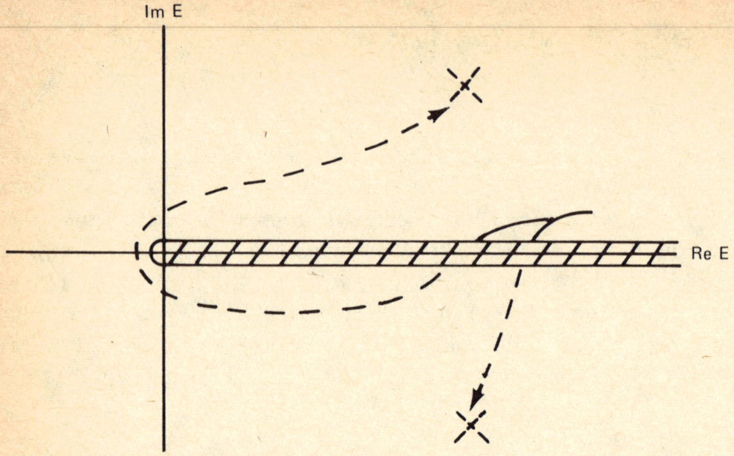

Fig.(11-4) Complex conjugate pairs of poles giving rise to resonances, and how they can be reached from above the cut on the physical sheet. Dashed lines indicate the segments of the contours on the unphysical sheet.

and therefore becomes large, in that region; but it is easy to see that the contributions have opposite signs in (11.12a) and ((11.12b), so the difference is large if the poles are near.

The Forward Scattering Amplitude

Knowing the singularities of $A_\ell(E)$ and the partial-wave series

$$A(E, \theta) = \sum_{\ell} (2\ell + 1) A_\ell(E) P_\ell(\cos\theta) \tag{11.13}$$

is not enough to enable us to deduce the singularities of $A(E, \theta)$ as a function of E. In the first place, the analytic properties of $A(E, \theta)$ in the complex E-plane will depend on the convergence of the series. If (11.13) is convergent in a given domain of E, then the singularities of $A(E, \theta)$ will be the sum of the singularities of $A_\ell(E)$. But the domain of convergence will, in general, depend on $\cos\theta$. Furthermore, it is possible that certain singularities of $A_\ell(E)$ will, when summed over ℓ, cancel each other out, leaving $A(E, \theta)$ analytic where $A_\ell(E)$ was not. In fact, this is exactly what happens for potential scattering.

The proof of the analytic structure of $A(E, \theta)$ is much more difficult than that of $A_\ell(E)$, and we shall not go into it here. The results can be seen by considering the forward scattering amplitude; since $P_\ell(1) = 1$ for all ℓ, we have

$$A(E, 0) = \sum_\ell (2\ell + 1)A_\ell(E). \tag{11.14}$$

In this case the series is convergent, so that the singularities of $A(E, 0)$ are at most those of $A_\ell(E)$. Inserting (11.9) into (11.14), we find

$$A(E, 0) = {\sum_B}' \frac{r_B}{E - E_B} + \frac{1}{4\pi^2}\int_0^\infty dE' \frac{\sqrt{2mE'}\,\sigma_T(E')}{E' - E} + \frac{1}{\pi}\int_{-\infty}^{-1/4\mu^2} dE' \frac{\operatorname{Im}A(E' + i\epsilon, 0)}{E' - E}, \tag{11.15}$$

where the sum over bound states now includes all different angular momenta. The right-hand cut is now an integral over the total cross section, and will converge only if $\sqrt{E'}\,\sigma_T(E') \to 0$ as $E' \to \infty$. (We shall deal presently with the possibility that this requirement is not fulfilled.)

The discontinuity across the left-hand cut is the sum of the discontinuities of the partial-wave amplitudes. Since these were determined by the detailed form of the potential, we expect the same to be true for $A(E, 0)$. Furthermore, the singularities of $A_\ell(E)$ were found by considering the Born series, and the singularities of $A(E, 0)$ can be exposed in a similar way.

Let us return to the Yukawa potential (10.31) to see how this works. The first Born approximation for this potential is

$$A^B(E, \theta) = -\frac{m}{2\pi}\int d^3r e^{i(\mathbf{k}-\mathbf{k}')\cdot\mathbf{r}} V(r) = -\frac{2m\lambda}{4k^2 \sin^2 \tfrac{1}{2}\theta + \mu^2}. \tag{11.16}$$

For $\theta \neq 0$, it has a pole at

$$k^2 = -\frac{\mu^2}{4 \sin^2 \tfrac{1}{2}\theta},$$

but in the forward direction this singularity moves to infinity in the E-plane and does not contribute. Thus the first Born approximation to $A(E, 0)$ is simply a constant. The partial wave amplitudes obtained from it, however, are not so simple in the E-plane; we have

$$a_\ell^B(k) = -2m\lambda \int_{-1}^{1} d\cos\theta \, \frac{P_\ell(\cos\theta)}{4k^2\sin^2 \tfrac{1}{2}\theta + \mu^2}. \tag{11.17}$$

The integrand of (11.17) becomes singular when k^2 takes on negative values with $0 \leqslant -\mu^2/4k^2 \leqslant 1$, implying that $a_\ell^B(k)$ has a corresponding cut in the k-plane, running from $-\tfrac{1}{4}\mu^2$ to $-\infty$. The presence of this singularity can be verified by performing the integration. The result is

$$a_\ell^B(k) = -\frac{m\lambda}{2k^2} Q_\ell\left(1 + \frac{\mu^2}{2k^2}\right),$$

where $Q_\ell(z)$ is the Legendre function of the second kind, which has a cut from $z = -1$ to $z = +1$.

Thus in the first Born approximation the partial-wave amplitude has exactly the singularity structure of (11.9), but $A(E, 0)$ is analytic where $A_\ell(E)$ has the left-hand cut. In the succeeding Born approximations for the Yukawa potential, the same thing happens. The nth Born approximation to $A(E, 0)$ is analytic on the negative real E-axis, but projecting out the partial wave amplitude introduces a singularity in $A_\ell(E)$.

Consequently the last term in (11.15) vanishes for a Yukawa potential, and in fact for any superposition of Yukawa potentials such as (10.39). The scattering amplitude has only the bound state poles and the right-hand cut. Since $A^B(E, 0)$ is a real constant, however, we will not have $|A(E, 0)| \to 0$ as $|E| \to \infty$, as required in order to write the dispersion relation (11.15). As earlier in connection with the square well potential, simply subtracting away the Born approximation makes everything manageable again; we can write the dispersion relation for $A(E, 0) - A^B(E, 0)$. It then follows that

$$A(E, 0) = A^B(E, 0) + \sum_B{}' \frac{r_B}{E - E_B} + \frac{1}{4\pi^2}\int_0^\infty dE' \, \frac{\sqrt{2mE'}\,\sigma_T(E')}{E' - E}. \tag{11.18}$$

If we know the energies and residues of the bound states and the total cross section, we can immediately write down forward scattering amplitude!

Thus we have, at least in the forward direction, accomplished our goal of bypassing the potential. There is no longer any direct need of the Schrödinger equation in obtaining $A(E, 0)$. To describe the scattering completely, however, we must know $A(E, \theta)$ for all θ, and for $\theta \neq 0$, the simplicity of (11.18) disappears. Singularities dependent on θ come back into the E-plane, as in (11.16); furthermore, the discontinuity of $A(E, \theta)$ on the right-hand cut is no longer related to the total cross section. The first problem can be eliminated by noting that $\Delta^2 = 4k^2 \sin^2 \frac{1}{2}\theta$ is the square of the momentum transferred to the particle during the scattering process. If we write the scattering amplitude as a function of Δ^2 rather than of θ, then the Born approximations will be independent of E, depending only on Δ^2. With $A(E, \theta) = F(E, \Delta^2)$, the dispersion relation will become

$$F(E, \Delta^2) = F^B(\Delta^2) + \sum_B{}' \frac{r_B(\Delta^2)}{E - E_B} + \frac{1}{\pi}\int_0^\infty dE' \frac{\operatorname{Im} F(E', \Delta^2)}{E' - E}, \tag{11.19}$$

which will be valid for all values of Δ^2 for which $F(E, \Delta^2)$ is well defined. In this case it becomes necessary to study the properties of $F(E, \Delta^2)$ as an analytic function of the momentum transfer variable Δ^2, which we shall discuss in Chapter 13.

In potential scattering, therefore, the scattering amplitude satisfies a dispersion relation in the energy which has only a right-hand cut in addition to the bound state pole terms. The left-hand cuts in the partial wave amplitudes are introduced by the integration over $\cos\theta$ in projecting them out. Their disappearance in the full amplitude $F(E, \Delta^2)$, however, is a result that is crucially dependent on the use of potential theory. In relativistic scattering theory, the left-hand cuts do not vanish. Potential theory is not adequate to describe relativistic scattering, but it can be used to get some hints at the answers which are obtained in a quantized field theory. We shall therefore overlook its inadequacies in order to learn something further about the left-hand cuts.

Relativistic Scattering and the Left-Hand Cut

The left-hand cut occurs for negative real energies. In non-relativistic scattering, there is no way to explore this region except via bound states. Relativistically, however, negative energy is associated with antiparticle states, and we might therefore guess that the left-hand cut is somehow associated with the scattering of antiparticles. The full statement of this idea, known as crossing symmetry, will be studied in some detail in Chapter 13. For now, however, we shall introduce it in a brief consideration of the relativistic formulation of potential scattering.

Suppose that we are describing spin–zero particles, which in their free state obey the Klein–Gordon equation

$$(-\nabla^2 + m^2)\Psi(\mathbf{r}, t) = -\frac{\partial^2}{\partial t^2}\Psi(\mathbf{r}, t). \tag{11.20}$$

If we represent $\Psi(\mathbf{r}, t)$ generally as a three-dimensional Fourier transform

$$\Psi(\mathbf{r}, t) = \int d^3k\Phi(\mathbf{k})e^{i(\mathbf{k}\cdot\mathbf{r}-Et)},$$

then (11.20) becomes

$$\int d^3k\Phi(k)(k^2 + m^2 - E^2)e^{i(\mathbf{k}\cdot\mathbf{r}-Et)} = 0$$

implying that

$$E = \pm\sqrt{k^2 + m^2}. \tag{11.21}$$

The double-valuedness of E in (11.21), of course, is what leads to the existence of negative energy states. Suppose we now specify $\Psi(r, 0)$. Then it is easily shown that a general solution of (11.20) is

$$\Psi(\mathbf{r}, t) = \Psi_+(\mathbf{r}, t) + \Psi_-(r, t),$$

$$\Psi_\pm(\mathbf{r}, t) = \int d^3k\Phi_\pm(k)e^{i(\mathbf{k}\cdot\mathbf{r}\pm\sqrt{k^2+m^2}\,t)}, \tag{11.22}$$

with $\Phi_\pm(k)$ constrained only by

$$\Phi_+(\mathbf{k}) + \Phi_-(\mathbf{k}) = \frac{1}{(2\pi)^3}\int d^3r\Psi(\mathbf{r}, 0)e^{-i\mathbf{k}\cdot\mathbf{r}} \tag{11.23}$$

In (11.22), we see that $\Phi_+(k)$ governs the time development of the particle states in $\Psi(r, t)$, while $\Phi_-(k)$ governs that of the antiparticles.

A particularly valuable way of understanding (11.22), first suggested by Feynman, involves noticing that the only real difference between the two parts of $\Psi(r, t)$ is in the sign of the time. That is, the antiparticle wave function $\Phi_-(r, t)$ looks very much like that of a particle propagating backwards in time. To show this clearly, let us take the non-relativistic limit of (11.22) by assuming that $\Phi_\pm(k)$ is appreciable only for $|k| \ll m$. Then the particle part

$$\Psi_+(\mathbf{r}, t) \approx e^{-imt} \int d^3k\Phi_+(k)e^{i(\mathbf{k}\cdot\mathbf{r}-(k^2/2m)t} \tag{11.24a}$$

satisfies the Schrödinger equation

$$-\frac{1}{2m}\nabla^2\Psi'_+(\mathbf{r}, t) = i\frac{\partial}{\partial t}\Psi'_+(\mathbf{r}, t), \tag{11.25a}$$

with $\Psi'_+(r, t) = \Psi_+(r, t)e^{imt}$. The same treatment of

$$\Psi_-(\mathbf{r}, t) \approx e^{imt} \int d^3k\Phi_-(k)e^{i(\mathbf{k}\cdot\mathbf{r}+(k^2/2m)t)} \tag{11.24b}$$

leads, with $\Psi'_-(r, t) = \Psi_-(r, t)e^{-imt}$, to

$$-\frac{1}{2m}\nabla^2\Psi'_-(\mathbf{r}, t) = -i\frac{\partial}{\partial t}\Psi'_-(\mathbf{r}, t), \tag{11.25b}$$

that is,

$$-\frac{1}{2m}\nabla^2\Psi'_-(\mathbf{r}, -t) = i\frac{\partial}{\partial t}\Psi'_-(\mathbf{r}, -t). \tag{11.25c}$$

Thus $\Psi'_-(r, -t)$ propagates like a particle; in other words, an antiparticle can be considered a particle propagating backwards in time.

Now let us describe the scattering of an incident beam by adding a potential in (11.20). We shall assume a particularly simple case which demonstrates well the points we wish to emphasize. Let the potential be the zero-component of a four-vector, the other three components vanishing in the chosen frame, and be spherically symmetric. Then the appropriate generalization of (11.20) is

$$(-\nabla^2 + m^2)\Psi(\mathbf{r}, t) = \left(i\frac{\partial}{\partial t} - V(r)\right)^2\Psi(\mathbf{r}, t). \tag{11.26}$$

If we write $\Psi(r, t)$ as

$$\Psi(r, t) = \int_{-\infty}^{\infty} dE\phi(\mathbf{r}, E)e^{-iEt}, \tag{11.27}$$

it follows that

$$[\nabla^2 - m^2 + (E - V(r))^2]\,\phi(\mathbf{r}, E) = 0. \tag{11.28}$$

With $k^2 = E^2 - m^2$, (11.21) becomes

$$[\nabla^2 + k^2 - 2\sqrt{k^2 + m^2}V(r) + (V(r))^2]\,\psi(\mathbf{r}, k) = 0, \tag{11.29}$$

where $\psi(\mathbf{r}, k) \equiv \phi(\mathbf{r}, \sqrt{k^2 + m^2})$. Comparing (11.29) and (9.1), we see that except for a redefinition of the potential, the two are identical. The solutions $\psi(\mathbf{r}, k)$ corresponding to scattering an incident beam are clearly still defined by

$$\psi(\mathbf{r}, k) \sim e^{i\mathbf{k}\cdot\mathbf{r}} + f(k, \theta)\frac{e^{ikr}}{r}. \tag{11.30}$$

An expansion of $\psi(\mathbf{r}, k)$ in spherical waves leads to

$$u_\ell''(k, r) + \left(k^2 - \left[\sqrt{k^2 + m^2}V(r) - V^2(r)\right] - \frac{\ell(\ell + 1)}{r^2}\right)u_\ell(k, r) = 0,$$

in agreement with (10.3), and the partial wave expansion of the scattering amplitude is

$$f(k, \theta) = \frac{1}{k}\sum_\ell (2\ell + 1)e^{i\delta_\ell}\sin\delta_\ell P_\ell(\cos\theta).$$

The boundary conditions used in Chapter 10 can also be applied to (11.30, and the same techniques used there can be adapted to study the analyticity of its solutions as a function of k. For the regular solution $\phi_\ell(k, r)$, defined by the boundary condition that it vanish as $r \to 0$, Poincaré's theorem can be applied to infer that $u_\ell(k, r)$ is analytic in the k-plane except for branch points at $k = \pm im$ introduced by the k-dependence of the potential term

$$U(r) = \sqrt{k^2 + m^2}V(r) - V^2(r). \tag{11.31}$$

The scattering matrix $S_\ell(k) = e^{2i\delta_\ell(k)}$ is defined exactly as in (10.6), by

$$\phi_\ell(k, r) \sim e^{-i(kr - \frac{1}{2}(\ell+1)\pi)} + S_\ell(k)e^{i(kr - \frac{1}{2}(\ell+1)\pi)},$$

and it is easily shown that $S_\ell(k)$ still has the symmetries (10.13a) and (10.13b). By using (11.31), the Jost function analysis can be carried over directly. Defining $f_\ell(k, r)$ as the solution of (11.30) which satisfies (10.8), we find again that $S_\ell(k) = f_\ell(k, 0)/f_\ell(-k, 0)$ is analytic in a cut plane which, if $m \geqslant \frac{1}{2}\mu$, is identical with that shown in Fig.(10-4). The bound states and resonances are similarly identified, and thus the singularity structure of the scattering amplitude is unchanged.

The dispersion relation (11.9) will therefore still be obtained if we map the analytic structure of $A_\ell(E)$ onto the k^2 plane. But since $k^2 = E^2 - m^2$, both positive and negative real energy values correspond to positive and real k^2; the left-hand cut still remains a mystery. This result is not surprising, since the left-hand cut is the manifestation of the detailed form of the potential, which we still have not specified.

Let us suppose that the potential in (11.26) represents the interaction of the particle being described with a target particle, which for simplicity we also take to be of mass m, in the center-of-momentum system of the two. Then we may write a wave function

$$\Psi(\mathbf{r}, t) = \psi(r, k)e^{-iEt} \tag{11.32a}$$

which has the time propagation factor characterizing a state containing only particles. That is, assuming that $\psi(\mathbf{r}, k)$ satisfies the boundary condition (11.30), we have

$$\Psi(\mathbf{r}, t) \sim e^{i(\mathbf{k}\cdot\mathbf{r}-Et)} + f(k, \theta)\frac{e^{i(kr-Et)}}{r}, \tag{11.32b}$$

which could be interpreted as the sum of an incident beam of particles with momentum $\mathbf{k}$ and a diverging spherical wave of scattered particles. We have seen, however, that antiparticles may be interpreted as particles moving backwards in time. In that case, (11.32) must also represent a process moving backwards in time, in which a converging spherical wave of incident antiparticles scatters coherently to produce a beam of antiparticles moving away from the target. Writing $e^{-iEt} = e^{-i(-E)(-t)}$ shows that the antiparticles have energy $-E$, and since they are moving away from the target the momentum of the outgoing beam is $-\mathbf{k}$.

Let us look at the kinematic description of these two processes. In the center-of-momentum system, the total energy W is defined by

$$(W, 0) = (\sqrt{k^2 + m^2}, \mathbf{k}) + (\sqrt{k^2 + m^2}, -\mathbf{k}),$$

i.e.

$$W^2 = 4(k^2 + m^2). \tag{11.33}$$

A more convenient frame in which to picture the two processes is the rest frame of the target particle, i.e. the laboratory frame. In this frame the incident particle has momentum **p** and energy $\omega = \sqrt{\mathbf{p}^2 + m^2}$ and it follows that

$$W^2 = ((\omega, p) + (m, 0))^2 = 2m^2 + 2m\omega. \tag{11.34a}$$

The alternative interpretation of (11.32) is that it represents an outgoing anti-particle with energy $-E$ and momentum $-\mathbf{p}$. Thus the corresponding quantity for antiparticle scattering is

$$W^2 = ((-\omega, -\mathbf{p}) + (m, 0))^2 = 2m^2 - 2m\omega. \tag{11.34b}$$

Comparing (11.34a) and (11.34b) shows immediately that the difference between the two cases is manifested in the sign of ω; if ω describes incoming particles, then $-\omega$ describes incident antiparticles. From (11.26) is follows that

$$k^2 = \tfrac{1}{2}m(\omega - m) = \tfrac{1}{2}m\nu \tag{11.35}$$

where ν is the kinetic energy of the incident particle in the lab system. The description of an incident antiparticle then corresponds to $\omega < -m$, that is, to $k^2 < -m^2$.

Now we can consider the k^2 plane as in the non-relativistic case. The physical region for particle–particle scattering, corresponding to real positive values of k, is infinitesimally above the right-hand cut, as before, i.e. $k^2 = (K^2 + i\epsilon)$ for real positive K^2. The physical region for antiparticle–particle scattering is then $k^2 = -(K^2 + i\epsilon)$ for real positive K^2, so it is infinitesimally *below* the left-hand cut. If $A(\omega, 0)$ is the relativistic version of the scattering amplitude describing the physical forward scattering process $AA \to AA$ with the incident particle having total energy ω in the laboratory frame, then the antiparticle scattering $\bar{A}A \to \bar{A}A$, with the same total energy, is described by $A(-\omega, 0)$ – the *same* function, but with a negative argument. As in (11.8), the discontinuities across the two cuts may therefore be expressed by means of the optical theorem,

$$\operatorname{Im} A(\omega + i\epsilon, 0) = \frac{\sqrt{\omega^2 - m^2}}{4\pi}\,\sigma(\omega), \quad \omega > m$$

$$\operatorname{Im} A(-\omega + i\epsilon, 0) = \operatorname{Im} A^*(-\omega - i\epsilon, 0)$$

$$= \frac{\sqrt{\omega^2 - m^2}}{4\pi} \bar{\sigma}(-\omega), \quad \omega < -m, \tag{11.36}$$

where $\sigma(\omega)$ and $\bar{\sigma}(\omega)$ are the total cross sections for particle–particle and anti-particle–particle scattering.

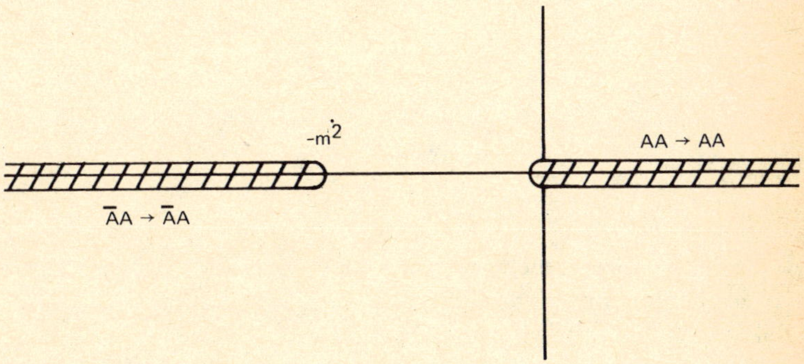

Fig.(11-5) Physical regions in the k^2 plane for particle–particle and antiparticle–particle scattering.

It follows that the relativistic version of (11.5), written as a function of ω, is

$$A(\omega, 0) = A^B(\omega, 0) + \sum_B \frac{r_B}{\omega - \omega_B} + \frac{1}{4\pi^2} \int_m^\infty d\omega' \frac{\sqrt{\omega'^2 - m^2}\, \sigma(\omega')}{\omega' - \omega}$$

$$+ \frac{1}{4\pi^2} \int_{-m}^{-\infty} d\omega' \frac{\sqrt{\omega'^2 - m^2}\, \bar{\sigma}(-\omega')}{\omega' - \omega} + \int_{-m}^{(2m^2 - \mu^2)/2m} d\omega' \frac{\operatorname{Im} A(\omega' + i\epsilon, 0)}{\omega' - \omega}, \tag{11.37}$$

where ω_B is the value of ω corresponding to a bound state B. If the mass of this bound state is M, then

$$\omega_B = \frac{M^2 - 2m^2}{2m}$$

Taking advantage of the symmetry between ω' and $-\omega'$ in the first two integrals, we may write (11.37) as

$$A(\omega, 0) = A^B(\omega, 0) + \sum_B \frac{r_B}{\omega - \omega_B} + \int_{-m}^{(2m^2-\mu^2)/2m} \frac{\operatorname{Im} A(\omega' + i\epsilon, 0)}{\omega' - \omega}$$

$$+ \frac{1}{4\pi^2} \int_m^\infty d\omega' \frac{\sqrt{\omega'^2 - m^2}}{\omega'^2 - \omega^2} \left\{\omega'[\sigma(\omega') + \bar{\sigma}(\omega')] + \omega[\sigma(\omega') - \bar{\sigma}(\omega')]\right\} . \tag{11.38}$$

Thus by using crossing symmetry we can express the discontinuity across most of the left-hand cut in terms of the antiparticle total cross section. In non-relativistic potential theory this term vanished, since there was no possibility of antiparticles. In a relativistic theory it is inevitably present, but as (11.38) shows, it is still possible to obtain an equation analogous to (11.18). The only unknown remaining is the discontinuity across the "unphysical" part of the left-hand cut, from $-m$ to $(2m^2 - \mu^2)/2m$. This cut was present in the partial wave amplitudes; whether it is present in a relativistic $A(\omega, 0)$ will be discussed in the next chapter. It may be entirely absent, or it may be present for entirely different reasons.

If (11.38) is to be valid, the integral involving the total cross sections must converge. As $\omega' \to \infty$, then, we must have

$$\omega' \frac{\sqrt{\omega'^2 - m^2}}{\omega'^2 - \omega^2} \left\{\omega'[\sigma(\omega') + \bar{\sigma}(\omega')] + \omega[\sigma(\omega') - \bar{\sigma}(\omega')]\right\} \to 0,$$

implying that

$$\omega'[\sigma(\omega') + \bar{\sigma}(\omega')] \to 0 \tag{11.39a}$$

and

$$[\sigma(\omega') - \bar{\sigma}(\omega')] \to 0. \tag{11.39b}$$

Since both σ and $\bar{\sigma}$ are positive quantities, (11.39a) requires that both must vanish faster than $1/\omega'$ as $\omega' \to \infty$, which guarantees that (11.39b) will be satisfied.

Unfortunately, however, the measured values of total cross sections do not appear to satisfy (11.39a). We shall discuss their actual behavior in some detail in Chapter 18; for the present, we shall assume that σ and $\bar{\sigma}$ become constant asymptotically. Then the integral in (11.38) certainly diverges.

The divergent term may be isolated by considering the symmetric and antisymmetric combinations

$$A^{\pm}(\omega, 0) = \tfrac{1}{2}[A(\omega, 0) \pm A(-\omega, 0)] \tag{11.40}$$

instead of $A(\omega, 0)$. Then the total cross sections appearing are

$$\sigma^{\pm}(\omega') = \tfrac{1}{2}[\sigma(\omega') \pm \bar{\sigma}(\omega')] = \frac{\sqrt{\omega'^2 - m^2}}{4\pi} \operatorname{Im} A^{\pm}(\omega', 0), \tag{11.41}$$

and (11.38) becomes the two equations

$$A^{-}(\omega, 0) = \sum_B{}' \frac{r_B \omega}{\omega^2 - \omega_B^2} + \frac{\omega}{2\pi^2} \int_m^\infty d\omega' \frac{\sqrt{\omega'^2 - m^2}}{\omega'^2 - \omega^2} \sigma^{-}(\omega') + U^{-}(\omega), \tag{11.42a}$$

$$A^{+}(\omega, 0) = A^B(0) - \sum_B \frac{r_B \omega_B}{\omega^2 - \omega_B^2} + \frac{1}{2\pi^2} \int_m^\infty d\omega' \frac{\sqrt{\omega'^2 - m^2}}{\omega'^2 - \omega^2} \omega' \sigma^{+}(\omega') + U^{+}(\omega'), \tag{11.42b}$$

where $U^{\pm}(\omega)$ denotes the contributions of the unphysical part of the cut,

$$U^{\pm}(\omega) = \frac{1}{2\pi} \int_m^{(2m^2-\mu^2)/2m} d\omega' \operatorname{Im} A(\omega', 0) \left(\frac{1}{\omega' - \omega} \pm \frac{1}{\omega' + \omega} \right). \tag{11.43}$$

Now there is no problem with the convergence of the odd part $A^{-}(\omega, 0)$ of the amplitude in (11.42a) so long as $\sigma^{-}(\omega)$ vanishes asymptotically, that is, provided particle and antiparticle total cross sections become equal. This equality was proved on fairly general grounds by Pomeranchuk, and will be discussed in Chapter 16.

Nothing is gained in (11.42b), however, since it depends on the sum of the cross sections, rather than the difference. In order to alleviate this failure, we attempt to subtract away the infinity by assuming that we know $A(\omega_0, 0)$ for some ω_0. Then we may use (11.42b) to write a dispersion relation for $A^+(\omega, 0) - A^+(\omega_0, 0)$, obtaining after some algebra

$$A^+(\omega, 0) = A^+(\omega_0, 0) - \sum_B r_B \omega_B \frac{(\omega_0^2 - \omega^2)}{(\omega^2 - \omega_B^2)(\omega_0^2 - \omega_B^2)} + U^+(\omega) - U^+(\omega_0)$$

$$+ \frac{\omega^2 - \omega_0^2}{2\pi^2} \int_m^\infty d\omega' \, \frac{\sqrt{\omega'^2 - m^2}\,\sigma^+(\omega')}{(\omega'^2 - \omega^2)(\omega'^2 - \omega_0^2)}. \tag{11.44}$$

By subtracting off $A^+(\omega_0, 0)$, we have gained a convergence factor $(\omega'^2 - \omega_0^2)^{-1}$ in the integrand. Thus (11.44) will be convergent if $\sigma^+(\omega')$ is asymptotically constant, or even if it diverges less rapidly than ω', as $\omega' \to \infty$.

Equation (11.44) is called a *subtracted* dispersion relation, because the subtraction of $A^+(\omega_0, 0)$ was a crucial step in making it convergent. Clearly, if (11.44) were still to diverge, we could subtract again at some point ω_1, obtaining thereby another factor of $(\omega'^2 - \omega_1^2)$ in the denominator. By continuing this procedure, we can always obtain a convergent integral if $\sigma^+(\omega')$ increases less rapidly than some power of ω', at the expense of needing to know the value of $A(\omega', 0)$ at a correspondingly large number of points.

An alternative way of expressing the idea behind subtraction is to note that the divergence of the integral in (11.39b) implies that $A(\omega, 0)$ does not vanish as $|\omega| \to \infty$. Therefore the circular part of the contour at infinity, which we neglected in writing (11.2), will not be negligible. It may be, however, that $|A^+(\omega, 0)/\omega^2|$ will be vanishing in this case. Then we may write a dispersion relation for the function

$$f(\omega) = \frac{A^+(\omega, 0)}{\omega^2 - \omega_0^2}$$

for any value of ω_0. The singularities of $f(\omega)$ will be identical with those of $A^+(\omega, 0)$ except for additional poles at $\omega = \pm\omega_0$, leading to the $A^+(\omega_0, 0)$ which appears in (11.41).

Thus the dispersion relation (11.38) for the forward scattering amplitude can be cured of its divergence problems by rather simple means. The subtraction method leads to satisfactory results in all cases for which the scattering amplitude can be bounded by a polynomial in the energy. The only failure corresponds to essential singularities at infinity in the energy plane. (Such a singularity did arise in the case of the square well potential, but subtracting away the first

Born term removed it.) Consequently it seems reasonable to expect that, with enough subtractions, a dispersion relation can be used to express the scattering amplitude directly in terms of physically measurable quantities.

To be specific, let us consider the simpler case (11.42a). The energy values corresponding to physical scattering situations are obtained by letting ω descend onto the right-hand cut; if we write

$$\omega = \omega_r + i\epsilon$$

this corresponds to $\omega_r > m$, with ϵ positive and becoming infinitesimal. Then

$$\frac{1}{\omega' - \omega} = \frac{\omega' - \omega_r}{(\omega' - \omega_r)^2 + \epsilon^2} - i \frac{\epsilon}{(\omega' - \omega_r)^2 + \epsilon^2} \tag{11.45}$$

The imaginary part of (11.45) corresponds to one of the standard forms used to represent the delta-function,

$$\lim_{\epsilon \to 0} \frac{\epsilon}{\chi^2 + \epsilon^2} = \pi\delta(\chi) \tag{11.46a}$$

The real part is used to define the principal part of a divergent integral,

$$P\int \frac{f(\chi)}{\chi} d\chi = \lim_{\epsilon \to 0} \int \frac{\chi f(\chi)}{\chi^2 + \epsilon^2} d\chi. \tag{11.46b}$$

Thus for physical scattering we may take the real part of (11.42a), with ω_r real and greater than m, to obtain

$$\text{Re}\, A^-(\omega_r, 0) = \sum_B{}' \frac{r_B \omega_r}{\omega_r^2 - \omega_B^2} + \frac{\omega_r}{2\pi^2} P \int_m^\infty d\omega' \frac{\sqrt{\omega'^2 - m^2}}{\omega'^2 - \omega_r^2} \sigma(\omega')$$

$$+ \text{Re}\, U^-(\omega_r, 0). \tag{11.47}$$

(There are no problems in letting $\epsilon \to 0$ in the bound state poles or in U^-, since $\omega_r > m$.)

The real part of the forward scattering amplitude can be measured by comparing the forward differential cross section, which measures $|A(\omega, 0)|^2$, with the value of $\text{Im}A(\omega, 0)$ obtained from the total cross section; it can also be obtained from interference effects between coulomb and strong scattering. The positions of the bound-state poles are known, and their residues are simply related to the compling constants. The total cross sections are, of course, easily measurable. Consequently the dispersion relation (11.47) can be compared directly with experimental data, except for the contribution $U^-(\omega_r, 0)$ of the unphysical part of the left-hand cut. That this last unknown either vanishes entirely or has a simple, although less easily measured, physical significance will be shown in the next chapter.

Chapter 12

UNITARITY AND ANALYTICITY

Two principal advantages were gained in the last chapter by transforming to the energy plane. First, we were able to eliminate the direct contributions of the resonance poles, leaving only the bound states and two cuts. More importantly, the partial-wave amplitude became a real analytic function in this plane, and the discontinuities across the cuts could be obtained, by means of the optical theorem, by measuring total cross sections.

The essential ingredient in the latter step was the unitarity of the scattering matrix. It was the statement that probability must be conserved which led, in Chapter 9, to the reality of the phase shifts. The same idea can be applied formally, in S-matrix theory, to learn the structure of the discontinuities without any reliance on the existence or detailed form of a potential. For this purpose it is necessary here to begin by reviewing briefly the formal statement of S-matrix theory and unitarity.

The S-Matrix

Let us consider a scattering process in which the system is initially in one of a complete set of states $|i, \alpha\rangle$, where α denotes all of the quantum numbers necessary to specify the state, including the number of particles present as well as their energies, momenta, and the discrete quantum numbers studied in Part I. These states may be thought of as the eigenstates of an initial Hamiltonian in which the particles involved are non-interacting.

The scattering takes place via a subsequent interaction, following which the system is left in one of a set of non-interacting final states denoted similarly by $|f, \beta\rangle$. Now the probability amplitude that a system initially specified by $|i, \alpha\rangle$ will scatter into the final state $|f, \beta\rangle$ is given by the overlap of the two states,

$$S_{\beta\alpha} = \langle f, \beta | i, \alpha\rangle. \tag{12.1}$$

If we define S as the operator transforming $|f, \alpha\rangle$ into $|i, \alpha\rangle$,

$$|i, a\rangle = S|f, a\rangle \tag{12.2}$$

then $S_{\beta a}$ is simply its matrix element

$$S_{\beta a} = \langle f, \beta|S|f, a\rangle. \tag{12.3}$$

We assume that the final states and the initial states span the same Hilbert space. Then (12.2) implies that S is a transformation within this space. Furthermore, the initial states may be assumed orthonormal,

$$\langle i, a|i, \beta\rangle = \delta_{a\beta}, \tag{12.4a}$$

and likewise the final states,

$$\langle f, a|f, \beta\rangle = \delta_{a\beta}. \tag{12.4b}$$

It follows then from (12.2) that

$$\langle i, a|i, \beta\rangle = \langle f, a|\, S^\dagger S|f, \beta\rangle = \langle f, a|f, \beta\rangle$$

for arbitrary a and β, and consequently that

$$S^\dagger S = 1. \tag{12.5a}$$

A similar proof shows that

$$SS^\dagger = 1. \tag{12.5b}$$

Therefore S is a unitary operator.

The unitarity of S guarantees the conservation of overall probability, since the probability that a system in the initial state

$$|\psi_i\rangle = \sum_a c_a|i, a\rangle, \text{ with } \sum_a |c_a|^2 = 1,$$

will be found in *any* final state is

$$\sum_\beta |\langle\psi_i|f, \beta\rangle|^2 = \sum_{aa'\beta} c_a^* c_{a'} \langle i, a|S^\dagger|i, \beta\rangle\langle i, \beta|S|i, a'\rangle$$

$$= \sum_{aa'} c_a^* c_{a'} \langle i, a|S^\dagger S|i, a'\rangle$$

$$= \sum_a |c_a|^2 = 1. \tag{12.6}$$

In order to account for the possibility that there is no scattering, which is included in S, we extract a unit operator and write

$$S = 1 + iT \tag{12.7}$$

to define the transition operator T. Then the unitarity conditions (12.5) become

$$\frac{1}{i}[T - T^\dagger] = TT^\dagger = T^\dagger T. \tag{12.8}$$

The matrix elements of T describe the probability amplitudes that some interaction has taken place. Taking the matrix elements of (12.8) yields, on inserting a complete set of states,

$$\frac{1}{i}\langle f, \beta|T - T^\dagger|f, a\rangle = \sum_\gamma \langle f, \beta|T|f, \gamma\rangle\langle f, \gamma|T^\dagger|f, a\rangle,$$

that is,

$$\frac{1}{i}[T_{\beta a} - T^*_{a\beta}] = \sum_\gamma T_{\beta\gamma}T^*_{a\gamma}. \tag{12.9}$$

Since energy must be conserved, the matrix elements of T between states with different total energy will vanish. To represent this condition it is appropriate to extract a delta function from $T_{\beta a}$ and write

$$T_{\beta a} = 2\pi\delta(E_\beta - E_a)\,\mathscr{T}_{\beta a}. \tag{12.10}$$

The transition matrix $\mathscr{T}_{\beta a}$ defined in (12.10) is then identical with that in (1.1). Inserting (12.10) into (12.9) yields

$$\frac{1}{2i}[\mathscr{T}_{\beta a} - \mathscr{T}^*_{a\beta}]\,\delta(E_\beta - E_a) = \pi\sum_\gamma \delta(E_\beta - E_\gamma)\delta(E_\beta - E_a)\,\mathscr{T}_{\beta\gamma}\mathscr{T}^*_{a\gamma}$$

$$= \pi\delta(E_\beta - E_a)\sum_\gamma \delta(E_\beta - E_\gamma)\,\mathscr{T}_{\beta\gamma}\mathscr{T}^*_{a\gamma} \tag{12.11}$$

since $\delta(E_\beta - E_\gamma)\delta(E_\gamma - E_a) = \delta(E_\beta - E_a)\delta(E_\gamma - E_\beta)$. If we now restrict our attention to energy-conserving transitions, the delta function may be cancelled to yield

$$\frac{1}{2i}[\mathscr{T}_{\beta a} - \mathscr{T}^*_{a\beta}] = \pi \sum_\gamma \delta(E_\beta - E_\gamma)\, \mathscr{T}_{\beta\gamma} \mathscr{T}^*_{a\gamma}. \tag{12.12}$$

(Note, however, that if $E_\beta \neq E_a$ both sides of (12.11) vanish, and there is no guarantee that (12.12) is valid.)

Equation (12.12) is the generalized form of the unitarity relation for $\mathscr{T}$. The case of particular interest is that with $a = \beta$, which yields

$$\text{Im}\, \mathscr{T}_{aa} = \pi \sum_\gamma \delta(E_a - E_\gamma) |\mathscr{T}_{a\gamma}|^2, \tag{12.13}$$

the generalized optical theorem. This relation thus is more general than the earlier derivation in potential theory. It holds in the form (12.13) for any choice of the state a. It need not refer to a single particle scattering from a potential; it could be two particles scattering from each other, or a many-particle interaction in which a refers to a state containing an arbitrary number of particles. The only crucial point is that the right-hand side in (12.13) is summed over *all* states γ which conserve energy.

Applications of Unitarity

The simplest example is the situation already encountered in potential scattering. The initial and final systems are taken to be non-relativistic single particle states, conveniently represented by plane wave states $|\mathbf{k}_a\rangle$. Using the usual normalization

$$\langle \mathbf{k}_a | \mathbf{k}_\beta \rangle = (2\pi)^3 \delta^3(\mathbf{k}_\beta - \mathbf{k}_a) \tag{12.14}$$

we have for the summation over states

$$\sum_\gamma |f, \gamma\rangle\langle f, \gamma| = \frac{1}{(2\pi)^3} \int d^3k |\mathbf{k}_\gamma\rangle\langle \mathbf{k}_\gamma|, \tag{12.15}$$

so that (12.9) becomes

$$\frac{1}{2i}\langle k_\beta|(T - T^\dagger)|k_\alpha\rangle = \frac{1}{(2\pi)^3}\int d^3k_\gamma \langle k_\beta|T|k_\gamma\rangle\langle k_\gamma|T^\dagger|k_\alpha\rangle. \tag{12.16}$$

Extracting the energy delta-function then yields

$$\frac{1}{i}[\mathscr{T}_{\beta\alpha} - \mathscr{T}^*_{\alpha\beta}] = \frac{1}{(2\pi)^2}\int k_\gamma^2 dk_\gamma d\Omega_\gamma \delta\left(\frac{1}{2m}(k_\beta^2 - k_\gamma^2)\right)\mathscr{T}_{\beta\gamma}\mathscr{T}^*_{\alpha\gamma}. \tag{12.17}$$

If the potential is spherically symmetric, then the transition operator must also be; consequently the matrix element $\mathscr{T}_{\beta\alpha}$ can depend only on the magnitude $|\mathbf{k}_\alpha| = |\mathbf{k}_\beta|$, i.e., on the energy E_β, and on the angle $\theta_{\beta\alpha}$ between the two momenta, so we write

$$\mathscr{T}_{\beta\alpha} = \mathscr{T}(E_\beta, \theta_{\beta\alpha}) = \mathscr{T}_{\alpha\beta}. \tag{12.18}$$

Then (12.17) becomes

$$\operatorname{Im}\mathscr{T}(E_\beta, \theta_{\beta\alpha}) = \frac{mk_\beta}{2(2\pi)^2}\int d\Omega_\gamma \mathscr{T}(E_\beta, \theta_{\beta\gamma})\mathscr{T}(E_\beta, \theta_{\gamma\alpha}), \tag{12.19}$$

and if we expand $\mathscr{T}(E, \theta)$ in Legendre polynomials

$$\mathscr{T}(E, \theta) = \sum_\ell (2\ell + 1)t_\ell(E)P_\ell(\cos\theta) \tag{12.20}$$

it follows from (12.19) that

$$\operatorname{Im}t_\ell(E_\beta) = \frac{m}{2\pi}k_\beta|t_\ell(E_\beta)|^2. \tag{12.21}$$

Comparing (12.21) with (9.23a) verifies that

$$a_\ell(k) = \frac{m}{2\pi}t_\ell(E) \tag{12.22}$$

and therefore, as in (1.8),

$$\frac{m}{2\pi}\mathscr{T}(E, \theta) = f(k, \theta) = A(E, \theta). \tag{12.23}$$

Thus we have rederived the results of potential theory by assuming that there are no inelastic processes. In the presence of inelastic states, however, the summation over states must include them, and the results will be altered accordingly. We can include this possibility in the simple framework used above by assuming that there are two distinct types of particle with equal mass, and that the potential can transform one into the other. (For example, the particles could be K^+ and K° mesons, and the potential could absorb or emit charge to interchange them.) The basis plane wave states will then be represented by $|k_\alpha, m_\alpha\rangle$, with $m_\alpha = 1, 2$ distinguishing the two particles. The normalization condition becomes

$$\langle k_\alpha, m_\alpha | k_\beta, m_\beta\rangle = (2\pi)^3 \delta^3(k_\beta - k_\alpha)\delta_{m_\alpha m_\beta} \tag{12.24}$$

and the summation over states is

$$\sum_\ell |f, \gamma\rangle\langle f, \gamma| = \sum_{m_\gamma = 1,2} \int \frac{d^3k_\gamma}{(2\pi)^3} |k_\gamma, m_\gamma\rangle\langle k_\gamma, m_\gamma| . \tag{12.25}$$

If the potential is spherically symmetric, (12.18) will be replaced by

$$\mathscr{T}_{\beta\alpha} = \langle k_\beta, m_\beta | \mathscr{T} | k_\alpha, m_\alpha\rangle = \mathscr{T}^{m_\beta m_\alpha}(E_\beta, \cos\theta_{\beta\alpha}), \tag{12.26}$$

and (12.19) becomes

$$\frac{1}{2i}[\mathscr{T}^{m_\beta m_\alpha}(E_\beta, \cos\theta_{\beta\alpha}) - \mathscr{T}^{m_\alpha m_\beta *}(E_\beta, \cos\theta_{\beta\alpha})] =$$

$$\frac{mk_\beta}{2(2\pi)^2} \int d\Omega_\gamma \sum_{m_\gamma} \mathscr{T}^{m_\beta m_\gamma}(E_\beta, \theta_{\beta\gamma}) \mathscr{T}^{m_\alpha m_\gamma *}(E_\beta, \theta_{\gamma\alpha}). \tag{12.27}$$

There are four different amplitudes defined in (12.26), namely $\mathscr{T}^{11}$, $\mathscr{T}^{12}$, $\mathscr{T}^{21}$, and $\mathscr{T}^{22}$. (It follows from time reversal invariance, however, that $\mathscr{T}^{12} = \mathscr{T}^{21}$.) If we expand $\mathscr{T}^{m_\beta m_\alpha}$ as in (12.20),

$$\mathscr{T}^{m_\beta m_\alpha}(E_\beta, \cos\theta_{\beta\alpha}) = \sum_\ell (2\ell + 1) t_\ell^{m_\beta m_\alpha}(E_\beta) P_\ell(\cos\theta_{\beta\alpha}),$$

then we obtain instead of (12.21)

$$\operatorname{Im} t_\ell^{11}(E_\beta) = \frac{m}{2\pi} k_\beta(|t_\ell^{11}(E_\beta)|^2 + |t_\ell^{12}(E_\beta)|^2)$$

$$\text{Im}\ t_\ell^{22}(E_\beta) = \frac{m}{2\pi}\,k_\beta(|t_\ell^{21}(E_\beta)|^2 + |t_\ell^{22}(E_\beta)|^2). \tag{12.28}$$

Equivalently, we may consider the forward elastic scattering amplitudes and get

$$\text{Im}\ \mathscr{T}^{11}(E_\beta, 0) = \frac{mk_\beta}{2(2\pi)^2}\int d\Omega_\gamma(|\mathscr{T}^{11}(E_\beta, \theta_{\gamma\beta})|^2 + |\mathscr{T}^{12}(E_\beta, \theta_{\gamma\beta})|^2). \tag{12.29}$$

Since the differential cross section for $m_\alpha \to m_\beta$ is

$$\frac{d\sigma^{m_\alpha \to m_\beta}}{d\Omega}(E_\beta, \theta_{\beta\alpha}) = \left(\frac{m}{2\pi}\right)^2 |\mathscr{T}^{m_\alpha m_\beta}(E_\beta, \theta_{\beta\alpha})|^2, \tag{12.30}$$

this yields simply

$$\text{Im}\ \mathscr{T}^{11}(E_\beta, 0) = \frac{k_\beta}{2m}[\sigma^{(1\to 1)} + \sigma^{(1\to 2)}]$$

$$= \frac{k_\beta}{2m}\sigma_T^{(1)}, \tag{12.31a}$$

where $\sigma_T^{(1)}$ is the total cross section for scattering of particle 1. In a similar way it follows that

$$\text{Im}\ \mathscr{T}^{22}(E_\beta, 0) = \frac{k_\beta}{2m}\sigma_T^{(2)}. \tag{12.31b}$$

Analogous results can be obtained in a more general situation where arbitrary inelastic processes are possible. Let the states α and β of $\mathscr{T}_{\beta\alpha}$ be single particle states as in (12.18). The summation over states can be written

$$\sum_\gamma |f, \gamma\rangle\langle f, \gamma| = \frac{1}{(2\pi)^3}\int d^3k_\gamma |\mathbf{k}_\gamma\rangle\langle\mathbf{k}_\gamma| + \sum_\gamma{}' |f', \gamma\rangle\langle f', \gamma| \tag{12.32}$$

where the first term identifies the elastic states (that is, the states which have the same discrete quantum numbers as $|f, \alpha\rangle$ and $|f, \beta\rangle$), and $|f', \gamma\rangle$ are the inelastic states. They may be single-particle states with different quantum numbers, as above, or many-particle states; in each case they must be identified by the appropriate set of quantum numbers. As one of these, we choose $\mathbf{K}_\gamma$, the total momentum of all particles in the inelastic state. Then $\mathscr{T}_{\beta\gamma} = \langle \mathbf{k}_\beta | \mathscr{T} | f', \gamma \rangle$ will be a function of the angle $\theta_{\beta\gamma}$ between $\mathbf{k}_\beta$ and $\mathbf{K}_\gamma$, as well as the total energy E_β and the appropriate remaining variables. Denoting the latter by X_γ, we write

$$\langle \mathbf{k}_\beta | \mathscr{T} | f', \gamma \rangle = \mathscr{T}_{\text{inel}}(E_\beta, \theta_{\beta\gamma}, X_\gamma) \tag{12.33}$$

and expand the resulting function in Legendre polynomials,

$$\mathscr{T}_{\text{inel}}(E_\beta, \theta_{\beta\gamma}, X_\beta) = \sum_\ell (2\ell + 1) t_\ell^{\text{inel}}(E_\beta, X_\gamma) P_\ell(\cos\theta_{\beta\gamma}) \tag{12.34}$$

with

$$t_\ell^{\text{inel}}(E_\beta, X_\gamma) = \int_{-1}^{1} d\cos\theta_{\beta\gamma} P_\ell(\cos\theta_{\beta\gamma})\, \mathscr{T}_{\text{inel}}(E_\beta, \theta_{\beta\gamma}, X_\gamma).$$

The contribution of the inelastic final states to (12.12) will be

$$I(E_\beta, \theta_{\beta\gamma}) = \sum_\gamma \langle \mathbf{k}_\beta | \mathscr{T} | f', \gamma \rangle \langle \mathbf{k}_a | \mathscr{T} | f', \gamma \rangle^* \delta(E_\beta - E_\gamma)$$

$$= \int d\Omega_{\beta\gamma} \int dX_\gamma \rho(E_\beta, X_\gamma)\, \mathscr{T}_{\text{inel}}(E_\beta, \theta_{\beta\gamma}, X_\gamma),\, \mathscr{T}^*_{\text{inel}}(E_\beta, \theta_{\gamma\alpha}, X_\gamma),$$

where $\rho(E_\beta, X_\gamma)$ is the density of final inelastic states; I is a function of E_β and $\theta_{\beta\gamma}$ only, since all dependence on other quantum numbers is summed or integrated away. Using (12.34), we then find that

$$I(E_\beta, \theta_{\beta a}) = 4\pi \sum_\ell (2\ell + 1) \dot{P}_\ell(\cos\theta_{\beta a}) \int dX_\gamma \rho(E_\beta, X_\gamma) | t_\ell^{\text{inel}}(E_\beta, X_\gamma)|^2. \tag{12.36}$$

If the elastic amplitude is expanded as in (12.20), it follows that

$$\text{Im } t_\ell(E_\beta) = \frac{mk_\beta}{2\pi} |t_\ell(E_\beta)|^2 + 4\pi^2 \int dX_\gamma \rho(E_\beta, X_\gamma) | t_\ell^{\text{inel}}(E_\beta, X_\gamma)|^2. \tag{12.37a}$$

In deriving (12.37a) we have assumed that E_β is real. Having obtained it, however, we may ask whether it could also be valid somehow for complex E_β. Since $E_\beta = E_\beta^*$ on the real axis, we may write it in the alternative form

$$\frac{1}{2i}[t_\ell(E_\beta) - t_\ell^*(E_\beta^*)] - \frac{\sqrt{2m^3}}{2\pi} E_\beta^{1/2} t_\ell(E_\beta) t_\ell^*(E_\beta^*)$$

$$- 4\pi^2 \int dX_\gamma \rho(E_\beta, X_\gamma) t_\ell^{inel}(E_\beta, X_\gamma)(t_\ell^{inel}(E_\beta^*, X_\gamma))^* = 0, \qquad (12.37b)$$

As pointed out earlier, $t_\ell^*(E_\beta^*)$ is a function of E_β, not E_β^*; thus (12.37b) contains only functions of E_β. That is, all of the terms of the equation can be considered analytic functions of the complex variable E_β. Thus the left-hand side of (12.37b) defines a function of E_β which vanishes along the entire real axis. It must therefore also vanish in every domain of the E_β plane which can be reached by analytic continuation from the real axis. Thus (12.37b) can be used for complex E_β, and the imaginary part can be identified with the discontinuity as in (11.7).

With

$$I_\ell = 2\pi m k_a \int dX_\gamma \rho(E_\beta, X_\gamma) |t_\ell^{inel}(E_\beta, X_\gamma)|^2, \qquad (12.38)$$

equation (12.37a) becomes equivalent to

$$\text{Im}\, f_\ell(k) = |f_\ell(k)|^2 + I_\ell. \qquad (9.23b)$$

It is clear that I_ℓ is real and positive from (12.38); furthermore, if $x = \text{Re} f_\ell(k)$ and $y = \text{Im} f_\ell(k)$, we have

$$I_\ell = y - (x^2 + y^2),$$

which has a maximum value of ¼ occurring when $x = 0$, $y = ½$. Thus

$$0 \leqslant I_\ell \leqslant ¼. \qquad (12.39)$$

In the presence of inelasticity it is often convenient to define the elasticity coefficient η_ℓ by writing

$$f_\ell = (\eta_\ell e^{2i\delta_\ell} - 1)/2i$$

with δ_ℓ real; then we have

$$I_\ell = ¼(1 - \eta_\ell^2) \qquad (12.40)$$

so $0 \leqslant \eta_\ell \leqslant 1$, and $\eta_\ell = 1$ for purely elastic scattering.

Unitarity and Relativistic Scattering

Although these examples are couched in the language of a single particle being scattered by a potential, they are valid independently of potential theory. We can use them equally well to describe the collision of two (or more) particles at relativistic energies, by defining the complete set of states appropriately. For simplicity let us first consider a situation in which the universe contains only a single type of particle, a scalar meson with mass m. An asymptotically free single-particle state can then be denoted by $|k\rangle$, where $k = (k_0, \mathbf{k})$ is the particle's four-momentum. We now take the invariant normalization

$$\langle k|k'\rangle = 2k_0(2\pi)^3\delta^3(\mathbf{k} - \mathbf{k}');$$

then the sum over a complete set of single-particle states is given covariantly by

$$P(1) = \frac{1}{(2\pi)^3} \int d^4p\,\theta(p_0 - m)\delta(p^2 - m^2)|p\rangle\langle p|, \tag{12.41}$$

where the step function and the delta function are inserted to restrict P(1) to particles on the mass shell. Thus on any physical single-particle state $|\psi_1\rangle$, we have

$$P(1)|\psi_1\rangle = |\psi_1\rangle,$$

so P(1) is the projection operator for one-particle states. States containing more than one particle are easily represented as direct products of single-particle states, and the sum P(N) over a complete set of N-particle states is exactly analogous to P(1), for example

$$P(2) = \frac{1}{(2\pi)^6} \int |p\rangle|p'\rangle d^4p d^4p'\theta(p_0 - m)\theta(p_0' - m)\delta(p^2 - m^2)\delta(p'^2 - m^2)\langle p|\langle p'|. \tag{12.42}$$

The sum in (12.9) must include all possible numbers of particles, since relativistically we know that particles can be created or destroyed; thus

$$\sum_\gamma |f, \gamma\rangle\langle f, \gamma| = \sum_{N=1}^{\infty} P(N). \tag{12.43}$$

In writing (12.10) we acknowledged that energy was conserved in the reaction; there we were picturing a potential of some kind which could absorb or give up momentum in the scattering process. The interactions we are considering now, however, must conserve total momentum as well as total energy. To take account of this conservation, we redefine the T-matrix by

$$T_{\beta a} = (2\pi)^4 \delta^4(k(\beta) - k(a)) \mathscr{T}_{\beta a} \tag{12.44}$$

where $k(a)$ is the total four-momentum of the state a. Then (12.12) is replaced by

$$\frac{1}{i} [\mathscr{T}_{\beta a} - \mathscr{T}^*_{a\beta}] = (2\pi)^4 \sum_\gamma \delta^4(k(\beta) - k(\gamma)) \mathscr{T}_{\beta\gamma} \mathscr{T}^*_{a\gamma}. \tag{12.45}$$

Let us take a and β in (12.45) to be two-particle states. Everything is Lorentz-invariant, so we can choose the center-of-momentum system and write

$$\mathscr{T}_{\beta a} = \langle k_0, \mathbf{k}'| \langle k_0, -\mathbf{k}'| \mathscr{T} |k_0, \mathbf{k}\rangle |k_0, -\mathbf{k}\rangle = \mathscr{T}(k_0, \theta_{\beta a})$$

where $k_0 = \sqrt{k^2 + m^2}$ and $\theta_{\beta a}$ is the scattering angle, $\mathbf{k} \cdot \mathbf{k}' = k^2 \cos\theta_{\beta a}$. In terms of this amplitude, the differential cross section is given by

$$\frac{d\sigma}{d\Omega_{CM}} = \frac{1}{(16\pi k_0)^2} |\mathscr{T}(k_0, \theta_{\beta a})|^2. \tag{12.46}$$

The generalized unitarity relation is now of the form

$$\mathrm{Im}\, \mathscr{T}(k_0, \theta_{\beta a}) = \tfrac{1}{2}(2\pi)^4 \sum_\gamma \delta^4(k(\beta) - k(\gamma)) \mathscr{T}_{\beta\gamma} \mathscr{T}^*_{a\gamma}, \tag{12.47}$$

the sum over γ including all states as in (12.43), and with $a = \beta$ the optical theorem becomes

$$\mathrm{Im}\, \mathscr{T}(k_0, 0) = 4kk_0\sigma(k_0). \tag{12.48}$$

We denote the contribution of N-particle states to (12.47) by $\mathscr{T}_N(k_0, \theta_{\beta a})$. Then for N = 2, we have

$$\mathscr{T}_2(k_0, \theta_{\beta a}) = \frac{1}{8\pi^2} \int d^4p \int d^4p' \delta^4(K - p - p') \langle k_0, \mathbf{k}'| \langle k_0, -\mathbf{k}'| \mathscr{T} |p\rangle |p'\rangle$$

$$\theta(p_0 - m)\theta(p_0' - m)\delta(p^2 - m^2)\delta(p'^2 - m^2)\langle p|\langle p'|\mathscr{T}|k_0, \mathbf{k}\rangle|k_0, -\mathbf{k}\rangle \quad (12.49a)$$

where $K = (2k_0, 0)$ is the total four-momentum of the initial state a. This expression is an eight-dimensional integration with six dimensions' worth of delta functions; as expected from (12.19), it yields an integral over solid angle. Specifically, after integrating over d^4p' and over dp_0, we have

$$\mathscr{T}_2 = \frac{1}{8\pi^2}\int \frac{d^3p}{2p_0}\langle k_0, \mathbf{k}'|\langle k_0, -\mathbf{k}'|\mathscr{T}|p_0, \mathbf{p}\rangle|p_0, -\mathbf{p}\rangle\theta(2k_0 - p_0 - m)$$

$$\frac{1}{4k_0}\delta(k_0 - p_0)\langle p_0, \mathbf{p}|\langle p_0, -\mathbf{p}|\mathscr{T}|k_0, \mathbf{k}\rangle|k_0, -\mathbf{k}\rangle \quad (12.49b)$$

with $p_0' = \sqrt{\mathbf{p}^2 + m^2}$. Integrating over the magnitude of $\mathbf{p}'$ then yields, with $\theta_{\gamma a}$ ($\theta_{\beta\gamma}$) denoting the angle between $\mathbf{p}$ and $\mathbf{k}(\mathbf{k}')$,

$$\mathscr{T}_2(k_0, \theta_{\beta a}) = \frac{1}{(8\pi)^2}\frac{k}{k_0}\theta(k_0 - m)\int d\Omega_p\,\mathscr{T}(k_0, \theta_{\beta\gamma})\mathscr{T}^*(k_0, \theta_{\gamma a}) \quad (12.49c)$$

which is essentially equivalent to (12.19). The step function $\theta(k_0 - m)$ indicates explicitly, however, that the two-particle intermediate states contribute only in physical situations with $k_0 = \sqrt{k^2 + m^2} \geqslant m$.

A similar treatment of the contribution $\mathscr{T}_3$ of three-particle states is possible. We need not go into the details here; it suffices to note that the creation of an additional particle will require the presence of more energy in the initial state. This requirement will be manifested in the appearance, as in (12.49c), of a step function causing $\mathscr{T}_3$ to vanish unless the total energy $W = 2k_0$ of the two-particle state a is larger than 3m, i.e. $k_0 \geqslant 3m/2$. Likewise, in $\mathscr{T}_4$ there will be a step function $\theta(k_0 - 2m)$, and in $\mathscr{T}_N$, $\theta(k_0 - \frac{1}{2}Nm)$. Only the single-particle intermediate states differ from this general form; for them, we have

$$\mathscr{T}_1 = \pi\int d^4p\,\delta^4(K - p)\langle k_0, \mathbf{k}'|\langle k_0, -\mathbf{k}'|\mathscr{T}|p\rangle\theta(p_0 - m)\delta(p^2 - m^2)$$

$$\times\langle p|\mathscr{T}|k_0, \mathbf{k}\rangle|k_0, -\mathbf{k}\rangle$$

$$= 8\pi m\delta(k_0 - \tfrac{1}{2}m)|\langle k_0, \mathbf{k}|\langle k_0, -\mathbf{k}|\mathscr{T}|m, 0\rangle|^2.$$

The delta function shows that in physical scattering, with $k_0 > m$, the formation of the bound state is forbidden by energy-momentum conservation; the matrix element is the coupling constant g of the three-particle vertex, which is independent of the direction of **k**.

If we now write $\mathscr{T}_N(k_0, 0) = \theta(k_0 - \frac{1}{2}Nm)t_N(k_0)$ for $N > 1$ and $\mathscr{T}_1 = 8\pi m\delta(k_0 - \frac{1}{2}m)g^2$, (12.47) is of the form

$$\mathrm{Im}\ \mathscr{T}(k_0, 0) = 8\pi m\delta(k_0 - \tfrac{1}{2}m)g^2 + \sum_{N=2}^{\infty} \theta(k_0 - \tfrac{1}{2}Nm)t_N(k_0). \qquad (12.50a)$$

Equivalently, using (12.48) and assuming that $k_0 \geqslant m$, we have

$$\sigma(k_0) = \frac{1}{4kk_0} \sum_{N=2}^{\infty} \theta(k_0 - \tfrac{1}{2}Nm)t_N(k_0), \qquad (12.50b)$$

showing clearly that N-particle processes contribute to the total cross section only when there is sufficient energy. For example, when $m < k_0 < 3m/2$, only elastic scattering is allowed; when k_0 is increased through $3m/2$, however, $t_3(k_0)$ suddenly begins to contribute, since three-particle final states are then possible. When this happens, we say that the three-particle channel has opened, with $2k_0 = 3m$ giving its *threshold* energy. Similarly, the four-particle threshold is at $2k_0 = 4m$, etc.

In addition to appearing in $\mathrm{Im}\ \mathscr{T}(k_0, 0)$ and $\sigma(k_0)$, threshold effects can also be seen in the partial wave amplitudes. If we expand

$$\mathscr{T}(k_0, \theta) = 8\pi \sum_{\ell} (2\ell + 1)a_\ell(k_0)P_\ell(\cos\theta) \qquad (12.51a)$$

then if $m < k_0 < 3m/2$ it follows from (12.49c) that

$$\mathrm{Im}\, a_\ell(k_0) = \frac{k}{k_0} |a_\ell(k_0)|^2. \qquad (12.51b)$$

Above the inelastic thresholds, however, the partial waves of higher t_N will contribute as in (12.37a).

We have seen in Chapter 11, however, that for dispersion relations a more convenient variable is $\omega = (2k_0^2 - m^2)/m$. Therefore let us define

$$A(\omega, \theta) = \frac{1}{8\pi m} \mathscr{T}(k_0, \theta)$$

$$A_N(\omega) = \frac{1}{8\pi m} t_N(k_0).$$

to obtain for $\omega > 0$

$$\mathrm{Im}\, A(\omega, 0) = \delta(\omega + \tfrac{1}{2}m)g^2 + \sum_{N=2}^{\infty} \theta(\omega - \tfrac{1}{2}m(N^2 - 2))A_N(\omega). \qquad (12.52a)$$

(With this definition, $A(\omega, 0)$ is directly analogous with the function appearing in (11.37).) Now let us assume $A(\omega, 0)$ can be continued into the complex ω plane, as in Chapter 11. Here again arguments such as those leading to (12.37b) can be made to identify $A(\omega, 0)$ as a real analytic function of ω, with its discontinuity across the positive real axis given by (12.52a). The intermediate single-particle state, which yields the delta function, can be recognized as a pole contribution by comparison with (11.46a). The N-particle intermediate states yield the appropriate discontinuity beginning at $\omega = \tfrac{1}{2}m(N^2 - 2)$.

In particular, however, let us notice the effect of A_N as ω approaches its threshold. The discontinuity makes a sudden change as ω passes through $\tfrac{1}{2}m(N^2 - 2)$, showing us that there is a branch point there. As each new channel opens, a corresponding branch point occurs in the ω plane. Thus if $A(\omega, 0)$ is a real analytic function of ω, its singularities in the half-plane $\mathrm{Re}\,\omega > 0$ are the direct results of the thresholds of the possible intermediate states. There can be no other branch points on the positive real axis; if there were, their effects would be shown in (12.52a).

The same procedure used to obtain (12.52a) can also be used for $\theta \neq 0$. The threshold factors arise in the same way, and if the angular dependence is expressed as a function of the momentum transfer $\Delta^2 = 4k^2 \sin^2 \tfrac{1}{2}\theta$ the singularities of $F(\omega, \Delta^2) = A(\omega, \theta)$ in the ω plane are independent of Δ^2. The resulting generalization of (12.52a) is

$$\mathrm{Im}\, F(\omega, \Delta^2) = \delta(\omega + \tfrac{1}{2}m)g^2 + \sum_{N=2}^{\infty} \theta(\omega - \tfrac{1}{2}m(N^2 - 2))F_N(\omega, \Delta^2), \qquad (12.52b)$$

with g^2 independent of Δ^2 because we assume the bound state to be spinless. For the present, however, we shall confine our attention to the forward direction, since it is here that the optical theorem is applicable.

The analytic structure for Re $\omega < 0$ follows in a similar way if we assume crossing symmetry. Then real negative values of ω correspond to antiparticle–particle scattering. Consequently the singularities in the left half-plane may be put in one-to-one correspondence with the thresholds of intermediate states in antiparticle–particle scattering. In the simple case we have been considering particle and antiparticle must be identical (otherwise the single-particle state could not be a bound state of the two-particle system) so for $\omega < 0$ we have the same result as (12.52a) with the appropriate change of sign,

$$\text{Im } A(\omega, 0) = \delta(\omega - \tfrac{1}{2}m)g^2 + \sum_{N=2}^{\infty} \theta(-\omega - \tfrac{1}{2}m(N^2 - 2))A_N(-\omega, 0). \quad (12.53)$$

That is, the antiparticle scattering is identical to that of particles, so $A(\omega, 0) = A(-\omega, 0)$.

If we now assume that the only singularities of $A(\omega, 0)$ are those given in (12.52a) and (12.53), as indicated by potential theory, then its analytic structure is shown in Fig.(12-1). There are poles at $\omega = \pm\frac{1}{2}m$, corresponding to the single-particle states. In addition there are branch points at $\omega = \pm\frac{1}{2}m(N^2 - 2)$. All of these can be grouped into two cuts running along the real axis from $\pm m$ to $\pm\infty$. If we neglect the Born term and the possible necessity of subtractions, then, $A(\omega, 0)$ satisfies the dispersion relation

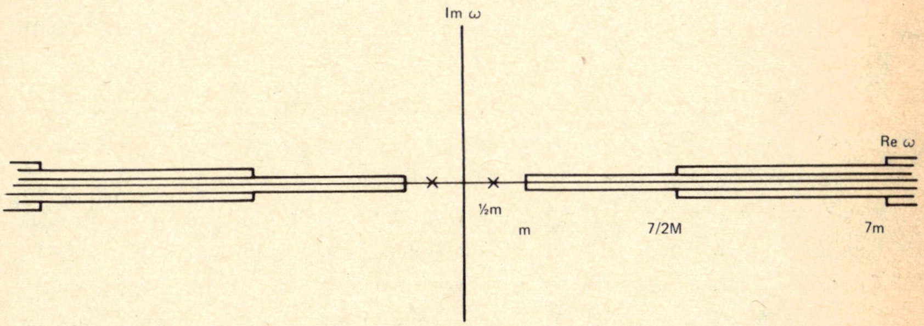

Fig.(12-1) Threshold branch points of the equal-mass scattering amplitude in the ω-plane.

$$A(\omega, 0) = g^2\left(\frac{1}{\omega - \frac{1}{2}m} - \frac{1}{\omega + \frac{1}{2}m}\right) + \frac{1}{4\pi^2}\int_m^\infty d\omega' \frac{\sqrt{\omega'^2 - m^2}\,\sigma(\omega')}{\omega' - \omega}$$

$$+ \frac{1}{4\pi^2}\int_{-m}^{-\infty} d\omega' \frac{\sqrt{\omega'^2 - m^2}\,\bar{\sigma}(-\omega')}{\omega' - \omega} \tag{12.54}$$

which agrees exactly with (11.37) except that the unphysical cut term is absent. Since $\sigma(\omega) = \bar{\sigma}(\omega)$ in this case, we can proceed in the same way to obtain a dispersion relation

$$A(\omega, 0) = \frac{mg^2}{\omega^2 - \frac{1}{4}m^2} + \frac{1}{2\pi^2}\int_m^\infty d\omega' \frac{\sqrt{\omega'^2 - m^2}}{\omega'^2 - \omega^2}\,\omega'\sigma(\omega') \tag{12.55}$$

as in (11.42b).

What has happened to the unphysical cut term? It corresponded to the part of the cut in the partial wave amplitude which began at $k = \frac{1}{2}i\mu$, where μ was the mass of the exchanged particle. In the system we are studying, μ must be identified with m; the scattering of two of these particles is then pictured as proceeding via exchange of the same particle. The unphysical cut begins at $\omega = \frac{1}{2}m$, corresponding with the pole of the antiparticle scattering. It is easy to see that anything that can be exchanged in particle–particle scattering must be a bound state for antiparticle–particle interactions. But these are two different situations, and should not be confused. What has really happened is that the unphysical part of the partial-wave amplitude's cut has disappeared entirely, as in (11.18), and the pole has appeared in an unrelated way through unitarity.

πN and NN Dispersion Relations

The real world is much more complex than the one-particle version we used to derive (12.55), but the principle remains the same. In every case, we shall now assume that the singularities of $A(\omega, 0)$ are those implied by unitarity in particle–particle and antiparticle–particle scattering.

The best example of the results of this assumption is in pion–nucleon scattering. Because the masses of the two particles are different in this case, the kinematics is considerably more complicated. Denoting the nucleon mass by m and the pion mass by μ, we have for the energy of the incident pion in the laboratory frame

$$\omega = \frac{1}{2m}(W^2 - m^2 - \mu^2),$$

where W is the total energy in the center-of-momentum frame. The momentum in that frame is given by

$$k^2 = \frac{m^2(\omega^2 - \mu^2)}{2m\omega + m^2 + \mu^2} = \frac{m^2(\omega^2 - \mu^2)}{W^2}.$$

To be rigorous here we should take account of the spin of the nucleon; in fact, however, its existence does not alter the forward dispersion relations, and we shall therefore ignore it. The appropriate generalizations of (12.46) and (12.48) then are

$$\frac{d\sigma}{d\Omega_{CM}}(\pi^{\pm}p) = \frac{m^2}{W^2}|A_{\pm}(\omega, \theta)|^2$$

$$\sigma(\pi^{\pm}p) = \frac{4\pi m}{kW} \operatorname{Im} A_{\pm}(\omega, 0) = \frac{4\pi}{\sqrt{\omega^2 - \mu^2}} \operatorname{Im} A_{\pm}(\omega, 0), \tag{12.56}$$

where $A_{\pm}(\omega, \theta)$ is the amplitude describing $\pi^{\pm}p$ scattering.

The singularities of these amplitudes follow from a consideration of the intermediate states which can contribute to the unitarity relation. For $\pi^{+}p$, there are no bound states; for $\pi^{-}p$, the neutron has the correct quantum numbers to contribute as a single-particle state in (12.52). The resulting bound-state pole term can be shown to have the form

$$A_{-}^{pole}(\omega, 0) = -\frac{g^2}{\pi}\left(\frac{\mu}{2m}\right)^2 \frac{\omega}{\omega^2 - (\mu^2/2m)^2}, \tag{12.57}$$

where g^2 is identical with the pion–nucleon coupling constant defined in Chapter 1. There are no other bound states in the pion–nucleon system; the next singularity encountered as ω increases is the elastic scattering threshold, $\omega = \mu$. For larger values of ω, there will occur branch points at the threshold of each new channel that opens. In contrast to the simplicity of (12.52), however, there is now a veritable forest of thresholds. Some of these are shown in Fig.(12-2). In addition to the branch points produced by creating extra pions, at

$$\omega_N = N\mu + (N^2 - 1)\mu^2/2m,$$

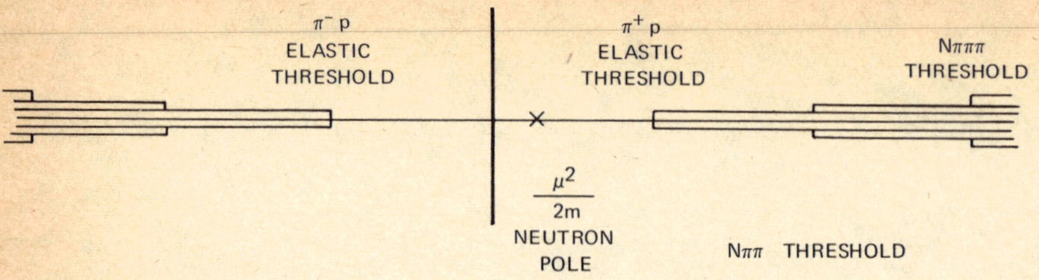

Fig.(12-2) Analytic structure of the $\pi^{\pm}$p scattering amplitude resulting from threshold singularities. (Not to scale.)

there are thresholds for intermediate states such as (KΛ) and (KΣ), as well as for each of these with N additional pions; for particle–antiparticle production, such as (K$\bar{\text{K}}$N) and (N$\bar{\text{N}}$N) final states; and so on in profusion.

The presence of all of these multitudinous branch points is, however, of little importance to the dispersion relations, since their effects will be manifested only in the total cross sections. They are all accounted for in the two cuts. Defining

$$A^{\pm}(\omega, 0) = \tfrac{1}{2}[A_+(\omega, 0) \pm A_-(\omega, 0)]$$

as in (11.40), we find easily a pair of dispersion relations analogous to (11.42). A subtraction is necessary in $A^+(\omega, 0)$, and if it is made at $\omega_0 = \mu$, the results are

$$A^+(\omega, 0) = A^+(\mu, 0) + \frac{g^2\mu^2}{2\pi m(4m^2 - \mu^2)} \frac{(\omega^2 - \mu^2)}{(\omega^2 - (\mu^2/2m)^2)} + \frac{\omega^2 - \mu^2}{2\pi^2} \int_{\mu}^{\infty} d\omega' \frac{\omega'\sigma^+(\omega')}{\sqrt{\omega'^2 - \mu^2}(\omega'^2 - \omega^2)} \tag{12.58a}$$

$$A^-(\omega, 0) = \frac{g^2}{\pi} \left(\frac{\mu}{2m}\right)^2 \frac{\omega}{\omega^2 - (\mu^2/2m)^2} + \frac{\omega}{2\pi^2} \cdot \int_{\mu}^{\infty} d\omega' \frac{\sqrt{\omega'^2 - \mu^2}\,\sigma^-(\omega')}{\omega'^2 - \omega^2}, \tag{12.58b}$$

where

$$\sigma^{\pm} = \tfrac{1}{2}[\sigma(\pi^+p) \pm \sigma(\pi^-p)],$$

as in (11.41).

A comparison of (12.58) with experiment may be attempted by taking the real part of each equation, as indicated in (11.47), obtaining for example

$$\operatorname{Re} A^{+}(\omega, 0) = \operatorname{Re} A^{+}(\mu, 0) + \frac{f^2(\omega^2 - \mu^2)}{m(\omega^2 - (\mu^2/2m)^2)}$$

$$+ \frac{\omega^2 - \mu^2}{2\pi^2} P \int_{\mu}^{\infty} d\omega' \quad \frac{\omega'\sigma^{+}(\omega')}{\sqrt{\omega'^2 - \mu^2}(\omega'^2 - \omega^2)} \tag{12.59}$$

for $\omega \geqslant \mu$. Since the pion–nucleon coupling constant is known and the total cross sections are measured up to fairly high values of ω, (12.59) can be used to calculate the real part of the amplitude. It is only necessary to assume some simple dependence for $\sigma^{+}(\omega)$ as $\omega \to \infty$ to describe the upper end of the integral, such as

$$\sigma^{+}(\omega) \approx \sigma(\infty) + a\omega^{-1/2}$$

for ω larger than the measured energies. A similar treatment of (12.58b), with simple parametrization of $\sigma^{-}(\omega) \approx b\omega^{-1/2}$, allows us to obtain numerical predictions for $\operatorname{Re} A^{-}(\omega, 0)$. The comparison of these predictions with experiment is shown in Fig.(12-3). It should be noted that, while the results are relatively insensitive to the constants $\sigma(\infty)$, a, and b, the basic form of $\sigma^{\pm}$ is still a rather crucial assumption. If the results shown in Fig.(12-3) disagreed with experiment, we would probably discard the assumed simple parametrization of $\sigma^{\pm}$ rather than the dispersion relations. The good agreement obtained should not, therefore, be considered a real verification of equations (12.58), so much as a vindication of our belief in them.

As a second example let us discuss briefly the nucleon–nucleon scattering amplitude, neglecting still the existence of the nucleon's spin. There is a single bound state of the two-nucleon system, namely the deuteron; the right-hand cut begins at $\omega = m$, and all of the many inelastic thresholds are contained in it. Therefore this part of the dispersion relation is no more complicated than in the earlier cases we have studied. It is the crossed channel processes which are significantly different. Here we must consider the intermediate states accessible in nucleon–antinucleon scattering. As we saw in Chapter 3, the pion is a bound state of nucleon and antinucleon, and thus (12.53) will lead to a pion pole. The

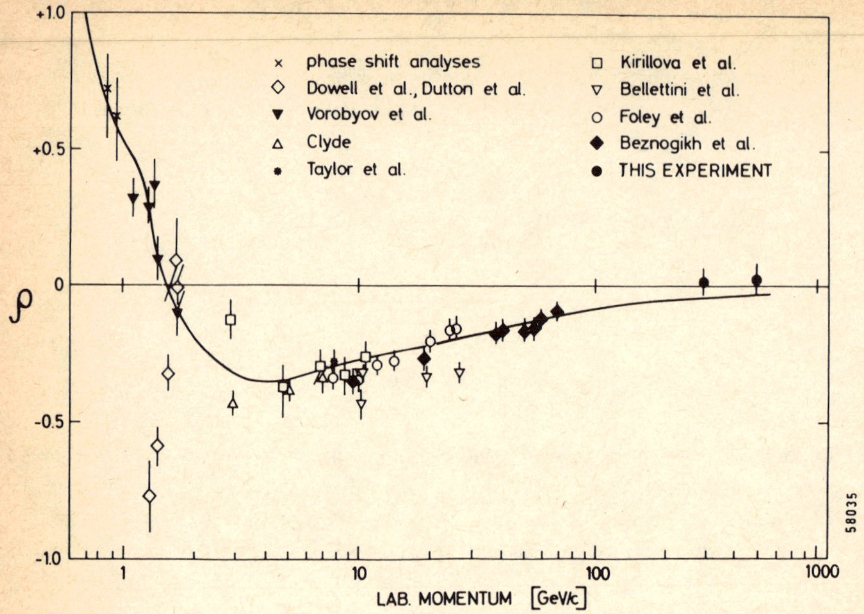

Fig.(12-3) Comparison of the experimentally measured ratio of the real part of the forward scattering amplitude to the imaginary part with the predictions of the pion–nucleon dispersion relations. [From U. Amaldi *et al., Phys. Lett.* **43B**, 231 (1973).]

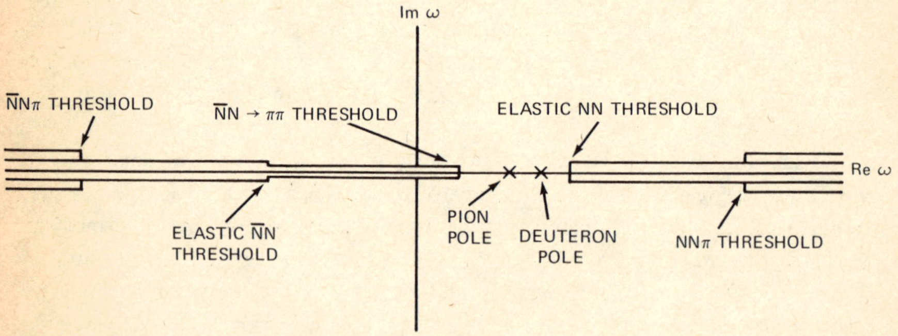

Fig.(12-4) Threshold branch points of the nucleon–nucleon scattering amplitude.

two-particle intermediate state with the lowest threshold is not that giving elastic scattering, however, but the two-*pion* state. In fact, there are a number of thresholds in this amplitude which lie *below* the elastic threshold. The total cross section, however, is defined only above the elastic threshold. Consequently a part of the left-hand cut must remain as an "unphysical" contribution, as in (11.37). The (unsubtracted) dispersion relation for proton–proton scattering is therefore of the form

$$A(\omega, 0) = (\text{pole terms}) + \frac{1}{4\pi^2}\int_m^\infty d\omega' \frac{\sqrt{\omega'^2 - m^2}\,\sigma_{pp}(\omega')}{\omega' - \omega}$$

$$+ \frac{1}{4\pi^2}\int_{-m}^{-\infty} d\omega' \frac{\sqrt{\omega'^2 - m^2}\,\sigma_{\bar{p}p}(-\omega')}{\omega' - \omega} + \frac{1}{\pi}\int_{-(m^2-\mu^2)/2m}^{-m} d\omega' \frac{\mathrm{Im}A(\omega' + i\epsilon)}{\omega' - \omega} \qquad (12.60)$$

The unphysical term in this equation looks very similar to that in (11.37), but we reiterate that the two are conceptually unrelated.

Unitarity is therefore the key to understanding the singularities of the scattering amplitude. It allows us to establish a direct connection between branch points and thresholds, as well as reaffirming the correspondence established in potential theory between poles and bound states. We have neglected a number of details in this chapter in order to show the fundamental idea as clearly as possible. In particular, the existence of spin necessitates a more careful treatment of the pion–nucleon and nucleon–nucleon amplitudes to include these degrees of freedom. The results remain essentially unchanged, however, as given in (12.58) and (12.60).

If dispersion relations are valid, they provide an elegant encapsulation of our belief in the mathematical simplicity of nature. In quantum field theory, they can be proved; but their existence is probably due to something much more basic. Three ideas are essential – unitarity, crossing symmetry, and analyticity. All of these are present in field theory, but they may exist independently of it. For this reason, they have become articles of faith in the strong interactions.

Chapter 13

RELATIVISTIC SCATTERING THEORY

Crossing symmetry is based on the observation that an incoming particle looks very much like an outgoing antiparticle. This is clearly a relativistic idea, and we used it in a relativistic way in the last chapter to obtain the dispersion relations for the forward scattering amplitude. But the full glory of crossing symmetry is not revealed in so simple an application; to exploit the notion to its fullest, we must learn to describe scattering processes more generally, in a relativistically invariant notation.

The amplitude describing an arbitrary reaction is a matrix element of the transition operator, as we have shown in Chapter 12, taken between the initial and final states. Suppose that the initial state contains N_i particles labeled by their four-momenta $p_1, \ldots p_{N_i}$, and the final state $N_f = N - N_i$, labeled analogously by momenta $p_{N_i+1}, \ldots p_N$. Then the amplitude will depend on these N four-momenta, as well as on any discrete quantum numbers involved. If the transition matrix is defined as in (12.44), however, everything is relativistically invariant, including the amplitude obtained from it. Consequently the amplitude must be a scalar function depending only on the relativistically invariant quantities which can be constructed from the p_i.

The number of such invariants is easy to calculate. The N four-momenta yield 4N variables; each of these, however, is constrained by the mass equation $(p_i)^2 = m_i^2$, so only 3N are free. Furthermore, the requirement that four-momentum is conserved yields four constraints, one for each component of

$$\sum_{i=1}^{N_i} p_i = \sum_{i=N_i+1}^{N} p_i.$$

Finally, Lorentz invariance implies a further set of six constraints, equivalent to the invariance of the entire N-particle system under translations and rotations. Thus the amplitude can depend only on 3N – 10 relativistically invariant variables.

For elastic scattering of two particles, $N_i = N_f = 2$, so N = 4, and the amplitude depends, as expected, on two variables. The result depends only on N, however, and not on N_i or N_f. Thus the amplitude for the decay of one particle into three – beta decay, for example, or $\eta^\circ \to 3\pi$ – is also a function of two variables. When a third particle is created in a two-particle collision, we have N = 5, and correspondingly five invariant variables; the same is true for a four-particle decay. In general, each additional particle means three extra variables. In this chapter, we shall study the description of N = 4 processes, which are the only ones sufficiently simple to have allowed much progress.

Relativistic Description of Two-Particle Scattering

The simplest kinematical situation is that described in Chapter 12, involving elastic scattering of two particles of equal mass m. Let us describe this reaction by labeling the incoming particles A and B, and the outgoing ones C and D. The process AB → CD may then be denoted by

$$p_1 + p_2 \to p_3 + p_4, \tag{13.1}$$

where p_1 and p_2 are the four-momenta of the incoming particles, and p_3 and p_4 those of the outgoing ones. Conservation of four-momentum requires that

$$p_1 + p_2 = p_3 + p_4 \tag{13.2}$$

in any frame.

To form invariants, we must use the scalar products of these four-vectors. For example, we may define

$$s = (p_1 + p_2)^2 = p_1^2 + p_2^2 + 2p_1 \cdot p_2. \tag{13.3a}$$

In the center-of-momentum (CM) system, s may be evaluated as in (11.33), with $p_1 = (k_0, \mathbf{k})$ and $p_2 = (k_0, -\mathbf{k})$, to yield

$$s = 4k_0^2 = 4(k^2 + m^2) = W_s^2, \tag{13.3b}$$

the square of the total energy. We may also combine p_1 with p_3 to form an invariant,

$$t = (p_1 - p_3)^2. \tag{13.4a}$$

If $p_3 = (k_0, \mathbf{k}')$ and $p_4 = (k_0, -\mathbf{k}')$, then the CM value of t is

$$t = ((k_0, \mathbf{k}) - (k_0, \mathbf{k}'))^2 = (0, \mathbf{k} - \mathbf{k}')^2 = -\Delta^2 \tag{13.4b}$$

$$= -2k^2(1 - \cos\theta_s)$$

where θ_s is the CM scattering angle. Note that $-t$ is the square of the momentum transfer, which appeared in (11.19), and that $t = 0$ for forward scattering.

Clearly s and t provide two independent kinematical variables which can serve as the relativistic replacement of ω and θ, since

$$\omega = (s - 2m^2)/2m, \tag{13.5a}$$

$$\cos\theta_s = 1 + \frac{2t}{s - 4m^2} \tag{13.5b}$$

If we write the amplitude as a function of s and t,

$$F(s, t) = mA(\omega, \theta_s) \tag{13.6}$$

(extracting a factor m for convenience), then it will be relativistically invariant. That is, it will be a function of s and t, which are invariants, rather than of any quantity which depends explicitly on a given frame. Since the total cross section is a measured number and hence an invariant quantity, the optical theorem can be written

$$\sigma(s) = \frac{8\pi}{\sqrt{s(s - 4m^2)}} \operatorname{Im}F(s, 0). \tag{13.7}$$

The differential cross section, however, is obviously dependent, through $d\Omega$, on the frame in which it is measured. Instead of measuring the scattering intensity at a given angle, we could do things invariantly by measuring a differential cross section with respect to t. Since the scattering is independent of the azimuthal angle ϕ, we have

$$\frac{d\sigma}{d\Omega_{CM}} = \frac{1}{2\pi}\frac{d\sigma}{d(\cos\theta_s)} = \frac{s - 4m^2}{4\pi}\frac{d\sigma}{dt}. \tag{13.8}$$

Thus $d\sigma/dt$, the differential cross section with respect to momentum transfer, is given by

$$\frac{d\sigma}{dt} = \frac{4\pi}{s(s - 4m^2)} |F(s, t)|^2 \tag{13.9}$$

and is an invariant quantity.

In defining t, we rather arbitrarily chose to combine p_1 with p_3. A similar definition using p_4 leads to a third variable

$$u = (p_1 - p_4)^2 = (0, \mathbf{k} + \mathbf{k}')^2 \tag{13.10}$$

$$= -2k^2(1 + \cos\theta_s)$$

which is as good a choice as t, since

$$\cos\theta_s = 1 + \frac{2u}{s - 4m^2}. \tag{13.11}$$

The difference between t and u is subjective; t = 0 defines "forward scattering" in the sense that particle C travels forward, (i.e. in the same direction that A was travelling before the collision), while u = 0 implies that particle D travels forward, i.e. particle C travels backward. If particles C and D are identical, these situations will be indistinguishable, and the amplitude must therefore be unchanged if t is replaced by u. The two variables are not independent, since it is easily seen that

$$s + t + u = 4m^2. \tag{13.12}$$

Crossing and Physical Regions

The values of s, t, and u which describe the physical scattering process AB → CD are limited by kinematical constraints. It follows from (13.3) that

$$s = 4(k^2 + m^2) \geqslant 4m^2; \tag{13.13a}$$

likewise, (13.4) and (13.10) imply that

$$0 \geqslant t \geqslant 4m^2 - s \tag{13.13b}$$

$$0 \geqslant u \geqslant 4m^2 - s. \tag{13.13c}$$

If we draw an s – t plane, the "physical region" may be shown as in Fig.(13-1a).

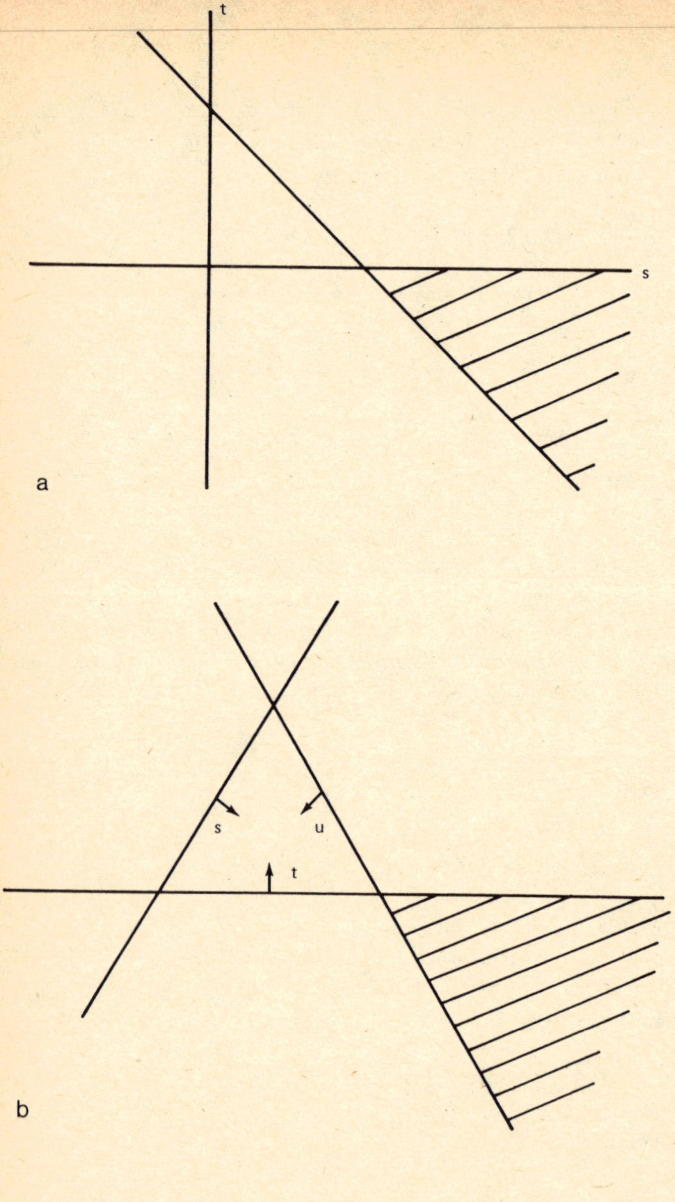

Fig.(13-1) The physical region for the process $AB \to CD$, plotted using (a) rectangular s–t axes and (b) triangular stu axes.

The similarity between t and u disappears in this plot, however, since t is chosen as an axis. In order to maintain the conceptual symmetry between them, it is convenient to use instead the variables $x = \frac{1}{2}(s - u)$ and $y = \sqrt{t}/2$, as shown in Fig.(13-1b). Then the lines $s = 0$, $t = 0$, $u = 0$ form an equilateral triangle. These lines are the s, t, and u axes, and the values of these variables are the distances from them. The physical region is still defined by $s \geq 4m^2$, $t \geq 0$, $u \geq 0$, i.e. by the shaded region.

The advantage of choosing the axes in this way becomes apparent when we consider crossing symmetry. We have seen that an incoming particle is like an outgoing antiparticle, and vice versa, with the sign of the four-momentum reversed. Thus if p_3 describes the outgoing particle C, for example, an incoming antiparticle $\overline{C}$ is described by $-p_3$. Likewise, $-p_2$ may refer to an outgoing antiparticle $\overline{B}$ rather than the incoming particle B. Consequently the process $A\overline{C} \to \overline{B}D$ is described analogously to (13.1) by

$$p_1 + (-p_3) \to (-p_2) + p_4. \tag{13.14}$$

The conservation of four-momentum requires that $p_1 - p_3 = p_4 - p_2$, reproducing (13.2).

In the CM frame of $A\overline{C} \to \overline{B}D$, (13.14) will have

$$\begin{aligned} p_1 &= (q_0, \mathbf{q}) \qquad & -p_2 &= (q_0, \mathbf{q}') \\ -p_3 &= (q_0, -\mathbf{q}) \qquad & p_4 &= (q_0, -\mathbf{q}') \end{aligned} \tag{13.15}$$

with $q_0 = \sqrt{q^2 + m^2}$, and therefore

$$\begin{aligned} s &= (p_1 + p_2)^2 = (0, \mathbf{q} - \mathbf{q}')^2 \\ &= -2q^2(1 - \cos\theta_t) \\ t &= (p_1 - p_3)^2 = (2q_0, 0)^2 = 4q_0^2 = W_t \\ u &= (p_1 - p_4)^2 = -2q^2(1 + \cos\theta_t), \end{aligned} \tag{13.16}$$

where θ_t is the CM scattering angle for (13.14). It follows from (13.16) that the physical region for this process is defined by

$$0 \geq s \geq 4m^2 - t$$

$$t \geq 4m^2$$

$$0 \geq u \geq 4m^2 - t. \tag{13.17}$$

In the crossed channel $A\bar{C} \to \bar{B}D$, therefore, the physical significances of t and s have been interchanged; t is now the square of the total energy, while $-s$ is the square of the momentum transfer. The change in meaning is reflected in the relocation of the physical region, which is now the area above the triangle, as shown in Fig.(13-2a).

It is obvious that the same procedure could be carried through for the other crossed channel $A\bar{D} \to C\bar{B}$, yielding

$$s = -2q_u^2(1 + \cos\theta_u)$$

$$t = -2q_u^2(1 - \cos\theta_u)$$

$$u = 4(q_u^2 + m^2) = W_u^2, \tag{13.18}$$

with q_u the CM momentum and θ_u the scattering angle. The corresponding physical region is the third wedge, as shown in Fig.(13-2b).

Defining the three variables s, t, and u, and drawing the plane defined by them in such a way that the three are symmetrically treated, thus leads to a simple picture incorporating all three processes as shown in Fig.(13-3). The same set of variables can be used to describe all three processes, distinguishing them by the different physical regions. To specify one of the three reactions, it is common

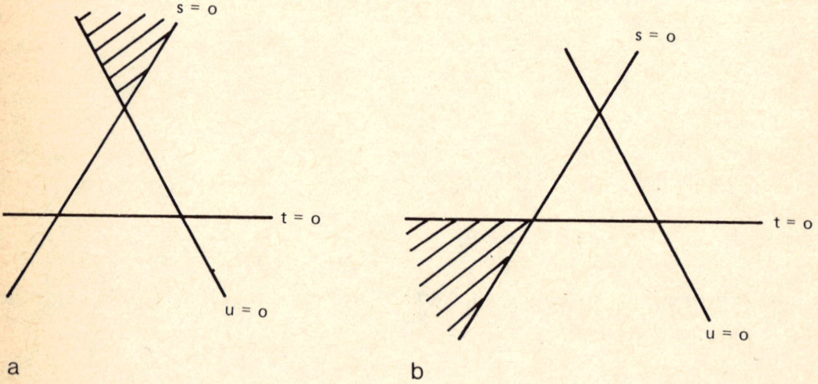

Fig.(13-2) Physical regions for the processes (a) $A\bar{C} \to \bar{B}D$ and (b) $A\bar{D} \to C\bar{B}$, plotted as in Fig.(13-1b).

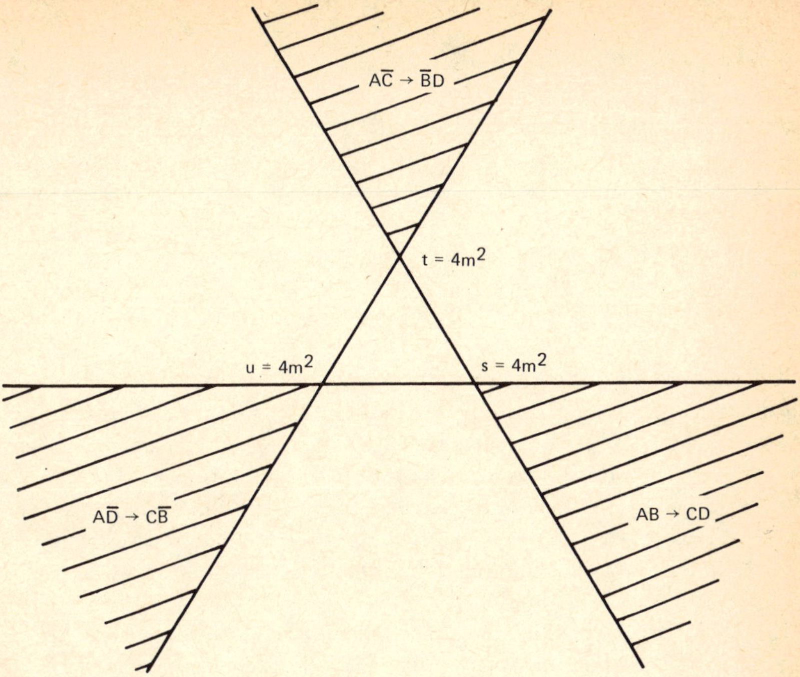

Fig.(13-3) Physical regions for the three processes related by crossing symmetry, plotted simultaneously.

to refer to the energy variable of the channel; thus $AB \to CD$ is the s-channel reaction, $A\bar{C} \to \bar{B}D$ the t-channel, and $A\bar{D} \to C\bar{B}$ the u-channel.

The general idea of crossing symmetry now becomes more accessible. Let us first assume that all four particles are the same, and that they are identical with their antiparticle. Then all three channels describe the same reaction $AA \to AA$. It follows that the amplitude must not depend on how we label the channels; it must be symmetric under any permutation of s, t, and u. To be precise, let us assume that $F(W^2, -\Delta^2)$ describes the scattering $AA \to AA$, with W^2 and Δ^2 defined physically rather than with reference to any specific channel. That is, W^2 is the square of the CM energy, and $-\Delta^2$ the square of the momentum transfer. If we call the process the s-channel, then

$$F(W^2, -\Delta^2) = F(s, t), \quad s \geqslant 4m^2, \;\; 0 \geqslant t \geqslant 4m^2 - s \qquad (13.19a)$$

If instead we called it the t-channel, we would have

$$F(W^2, -\Delta^2) = F(t, s), \quad t \geqslant 4m^2, \quad 0 \geqslant s \geqslant 4m^2 - t, \tag{13.19b}$$

and similarly for the u-channel,

$$F(W^2, -\Delta^2) = F(u, t), u \geqslant 4m^2, \quad 0 \geqslant t \geqslant 4m^2 - u. \tag{13.19c}$$

The definitions (13.19) have *not* implied anything about the behavior of $F(W^2, -\Delta^2)$ outside the physical region $W^2 \geqslant 4m^2$ and $0 \leqslant \Delta^2 \leqslant W^2 - 4m^2$, however. In the stu plane, $F(W^2, -\Delta^2)$ could have *any* value outisde the physical regions.

It is here that the assumption of crossing symmetry is invoked. We assume that the function F(s, t) defined by (13.19a) can be continued *analytically* from the s-channel physical region, where it was defined, into the t-channel, and that the resulting function F(s, t) is identical with F(t, s). In this region F(s, t) has $W^2 \leqslant 0$, $-\Delta^2 \geqslant 4m^2$, so that it is meaningless as a description of the s-channel process. Crossing symmetry says that even so, it is the function describing the t channel. In the same way, continuing F(s, t) into the u-channel physical region yields F(u, t).

Thus relativistic kinematics plus the equivalence of incoming particles and outgoing antiparticles allows us to define three separate physical regions of the stu plane; crossing symmetry goes much farther, implying that a single analytic function of these variables describes all three processes. When the processes are identical, it follows that the amplitude must be symmetric under any permutation of stu. In order to show this symmetry explicitly, it is helpful to write the amplitude as a function of all three variables,

$$F(s, t) = F(s, t, u). \tag{13.20}$$

(It should be realized, of course, that the three variables are not independent, and that (13.20) is a mathematically incorrect convenience.) Crossing symmetry then implies that

$$F(s, t, u) = F(t, u, s) = F(u, s, t) = F(s, u, t) = F(u, t, s) = F(t, s, u)$$

is a single symmetric analytic function, the singularities of which are specified by unitarity in the different channels as in Chapter 12.

More generally, the scattering of non-identical particles can also be assumed crossing symmetric. If the s-channel reaction AB → CD is described as in (13.19a) by $F_s(W^2, -\Delta^2)$, and the t-channel and u-channel reactions correspondingly by

$F_t(W^2, -\Delta^2)$ and $F_u(W^2, -\Delta^2)$, where F_s, F_t, and F_u are a priori unrelated functions, when crossing symmetry requires that there exist a single function F(s, t, u) for which

$$\begin{aligned} F(s, t, u) &= F_s(s, t) \text{ in the s channel,} \\ &= F_t(t, s) \text{ in the t channel,} \\ &= F_u(u, t) \text{ in the u channel,} \end{aligned} \tag{13.21}$$

and which may be continued analytically from one channel to another.

The emphasis on this last point, the possibility of continuing F(s, t, u) analytically between the physical regions; is crucial. Otherwise one could define F(s, t, u) with arbitrary natural boundaries. For example, it could satisfy (13.21) in all three physical regions, and vanish identically everywhere else. In such a situation crossing symmetry would be an empty idea, since the three processes would be totally unrelated. But we have seen in quantum mechanics that it does seem to be possible to treat dynamical quantities as complex variables and continue the amplitude as a function of them. Also, it has been possible to prove crossing symmetry within the framework of quantum field theory.

Unequal Masses

Nucleon–nucleon scattering involves only particles of equal mass, and so is described by the physical regions obtained from (13.13), (13.17) and (13.18). Generally, however, the particles involved in a scattering process will have different masses, and therefore we must consider this more complicated situation. In particular, pion–nucleon scattering involves two unequal masses, and we shall begin by obtaining the physical regions for $\pi N \to \pi N$ and its crossed channels.

Let us take p_1 and p_3 as the four-momenta of the pions, and p_2 and p_4 for the nucleons. Then in the s-channel CM frame we have

$$\begin{aligned} p_1 &= (\sqrt{k^2 + \mu^2}, \mathbf{k}) \qquad & p_2 &= (\sqrt{k^2 + m^2}, -\mathbf{k}) \\ p_3 &= (\sqrt{k^2 + \mu^2}, \mathbf{k}') \qquad & p_4 &= (\sqrt{k^2 + m^2}, -\mathbf{k}') \end{aligned} \tag{13.22}$$

(Conservation of energy guarantees that $k^2 = k'^2$.) Then

$$s = (p_1 + p_2)^2 = 2k^2 + m^2 + \mu^2 + 2\sqrt{(k^2 + \mu^2)(k^2 + m^2)}$$

$$t = (p_1 - p_3)^2 = -2k^2(1 - \cos\theta_s)$$

$$u = (p_1 - p_4)^2 = 2k^2 + m^2 + \mu^2 - 2\sqrt{(k^2 + \mu^2)(k^2 + m^2)} - 2k^2(1 + \cos\theta_s) \tag{13.23}$$

relate s, t, and u to the momentum and scattering angle. From (13.23) it is easily verified that

$$s + t + u = 2m^2 + 2\mu^2 \tag{13.24}$$

which reflects the more general relation that s + t + u is equal to the sum of the squares of the masses; this results from

$$s + t + u = 3p_1^2 + p_2^2 + p_3^2 + p_4^2 + 2(p_1 \cdot p_2 - p_1 \cdot p_3 - p_1 \cdot p_4)$$

$$= p_1^2 + p_2^2 + p_3^2 + p_4^2$$

since $p_1 + p_2 = p_3 + p_4$.

Inverting (13.23), we find that

$$k^2 = \frac{[s - (m + \mu)^2]\,[s - (m - \mu)^2]}{4s} \tag{13.25a}$$

$$\cos\theta_s = 1 + \frac{t}{2k^2}$$

$$= 1 + \frac{2ts}{[s - (m + \mu)^2]\,[s - (m - \mu)^2]}\,. \tag{13.25b}$$

The physical region of the s-channel is defined by $s \geqslant (m + \mu)^2$, $|\cos\theta_s| \leqslant 1$, so it is bounded by the curves $\cos\theta_s = \pm 1$. From (13.23), we see that $\cos\theta_s = +1$ when $t = 0$, while $\cos\theta_s = -1$ when

$$t = -\frac{[s - (m + \mu)^2]\,[s - (m - \mu)^2]}{s} \tag{13.26}$$

or, equivalently, when

$$su = (m^2 - \mu^2)^2. \tag{13.27}$$

The curve defined by (13.27) is shown in Fig.(13-4); it intersects the line $t = 0$ at $s = (m \pm \mu)^2$, and approaches $u = 0$ asymptotically as $s \to \infty$. Thus the physical region of the s-channel differs from that obtained in the equal mass case in that it overlaps the line $u = 0$, extending to the hyperbola (13.27).

The u-channel process in this case is $\pi\overline{N} \to \pi\overline{N}$, which, by charge conjugation, is identical to the s-channel process. Its physical variables are defined as in (13.23), with s and u interchanged, leading to $u \geqslant (m + \mu)^2$ and $|\cos\theta_u| \leqslant 1$. The boundaries are again given by $t = 0$ and the hyperbola (13.27), so the u-channel physical region is symmetric with that of the s-channel, as shown in Fig.(13-4).

The t-channel, however, is $\pi\pi \to N\overline{N}$, which is different from the other two. In its CM frame, we have

$$p_1 = (\sqrt{p^2 + \mu^2}, \mathbf{p}) \qquad -p_2 = (\sqrt{q^2 + m^2}, \mathbf{q})$$

$$-p_3 = (\sqrt{p^2 + \mu^2}, -\mathbf{p}) \qquad p_4 = (\sqrt{q^2 + m^2}, -\mathbf{q}).$$

with

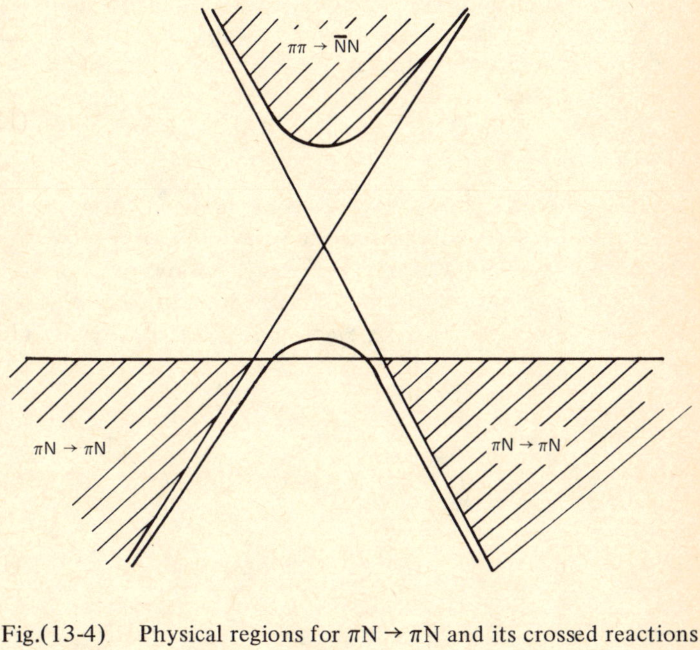

Fig.(13-4) Physical regions for $\pi N \to \pi N$ and its crossed reactions.

$$q^2 = p^2 + \mu^2 - m^2.$$

Then

$$s = -p^2 - q^2 + 2pq\cos\theta_t$$

$$t = 4(p^2 + \mu^2) = 4(q^2 + m^2)$$

$$u = -p^2 - q^2 - 2\,pq\cos\theta_t, \tag{13.28}$$

and the physical region is given by $t \geqslant 4m^2$, $|\cos\theta_t| \leqslant 1$. The latter criterion leads to a boundary equation

$$su = (m^2 - \mu^2)^2,$$

identical with (13.27); in fact, the resulting curve is just the other branch of the hyperbola, which approaches $s = 0$ and $u = 0$ asymptotically but lies inside the wedge. Thus the physical region of the t-channel will be slightly smaller than in the equal mass case, as shown.

A particularly interesting example of unequal mass scattering kinematics is also provided by the reaction $\eta\pi \rightarrow \pi\pi$. If p_1 is chosen as the four-momentum of the η, then all three channels are equivalent. The boundaries of the physical regions are defined by a cubic equation symmetric in s, t, and u,

$$stu = \mu^2(M^2 - \mu^2)^2, \tag{13.29}$$

where M denotes the mass of the η, which leads to Fig.(13-5). In this case, however, the fact that the mass of the η exceeds the sum of the three pion masses implies that the process $\eta \rightarrow 3\pi$ can also take place. (It is forbidden by G-parity in the strong interactions, of course, but we are only concerned here with the kinematics.) That is, we may cross only *one* pion (p_2) and consider the process as

$$p_1 \rightarrow -p_2 + p_3 + p_4.$$

In the rest frame of the η, we may write

$$p_1 = (\mu, \mathbf{0}) \qquad -p_2 = (\sqrt{p^2 + \mu^2}, \mathbf{p})$$

$$p_3 = (\sqrt{q^2 + \mu^2}, \mathbf{q}) \qquad p_4 = (\sqrt{(\mathbf{p} + \mathbf{q})^2 + \mu^2}, -(\mathbf{p} + \mathbf{q})).$$

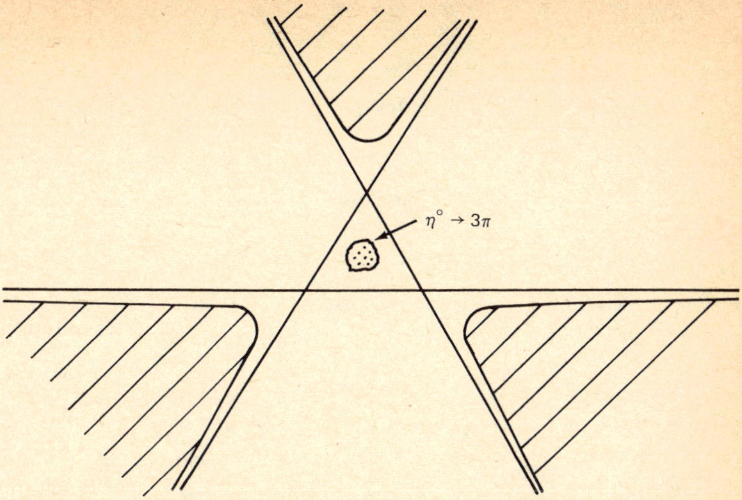

Fig.(13-5) Physical regions for the reactions $\eta^\circ\pi \to \pi\pi$ and the decay $\eta^\circ \to 3\pi$.

Then

$$s = M^2 + \mu^2 - 2M\sqrt{p^2 + \mu^2}$$

$$t = M^2 + \mu^2 - 2M\sqrt{q^2 + \mu^2}$$

$$u = M^2 + 3\mu^2 - s - t \qquad (13.30)$$

leads to a physical region in which all three variables are positive, i.e., in the center of the triangle, as shown.

A similar result will be obtained whenever one of the masses is large enough to allow a decay into the other three. If the three final-state particles have different masses, the symmetry of the decay region will be lost; but it will nonetheless occur as a "blob" inside the triangle. Each possible kinematic configuration for a three-particle decay corresponds to a point inside this region. By looking at the distribution of a large number of events for such a decay, we can learn the angular momentum and parity of the parent particle. This technique was first developed by Dalitz in studying the decay $K \to 3\pi$, and a plot of the distribution of events within the blob is often called a Dalitz plot.

Thus, in general, the physical regions for arbitrary masses consist of the three two-particle channels, which will be approximately triangular regions shifted from those of Fig.(13-3) by the mass differences, plus a central Dalitz region if one mass is larger than the sum of the other three. Crossing symmetry is then the statement that a single function F(s, t, u) with only physically implied singularities describes all of these processes.

Singularities in the stu Plane

We have seen in Chapter 12 that the singularities of the scattering amplitude in each channel can be obtained using unitarity. For the case considered earlier with only one kind of particle involved, the amplitude was analytic in the complex ω-plane except for singularities introduced by unitarity in the direct channel, in (12.52), and in the crossed channel, in (12.53). Specifically, choosing the s-channel as the direct channel, we can translate the singularities of (12.52) into the complex $s = 2m(\omega + m)$ plane. The pole at $\omega = -\frac{1}{2}m$ then occurs at $s = m^2$, and the thresholds are at $s = (Nm)^2$. It is helpful to visualize the fact that these thresholds are due to N-particle intermediate states by drawing the appropriate process as shown in Fig.(13-6).

Considering the u-channel as the crossed channel, we have in the same way singularities arising from the thresholds occurring in the u-channel. Since these correspond to having the total energy sufficiently large to create an N-particle state in the u-channel, they appear at $u = (Nm)^2$, i.e. at $s = 4m^2 - t - (Nm)^2$. When $t = 0$, this corresponds to $\omega = -\frac{1}{2}(N^2 - 2)m$, in agreement with (12.53). Thus in the complex s-plane, F(s, 0, u) satisfies the dispersion relation

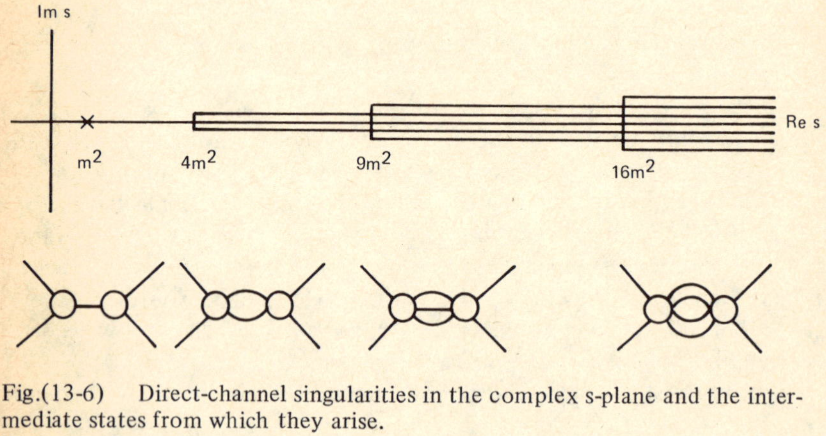

Fig.(13-6) Direct-channel singularities in the complex s-plane and the intermediate states from which they arise.

$$F(s, 0, u) = F^B + \frac{g^2}{s - m^2} + \frac{g^2}{3m^2 - s} + \frac{1}{\pi}\int_{4m^2}^{\infty} ds' \frac{\mathrm{Im}F(s', 0, 4m^2 - s')}{s' - s}$$

$$+ \frac{1}{\pi}\int_{0}^{-\infty} ds' \frac{\mathrm{Im}F(s', 0, 4m^2 - s')}{s' - s} \tag{13.31a}$$

$$= F^B + \frac{g^2}{s - m^2} + \frac{g^2}{u - m^2} + \frac{1}{\pi}\int_{4m^2}^{\infty} ds' \frac{\mathrm{Im}F(s', 0, u')}{s' - s}$$

$$+ \frac{1}{\pi}\int_{4m^2}^{\infty} du' \frac{\mathrm{Im}F(s', 0, u')}{u' - u}, \tag{13.31b}$$

where F^B denotes the Born term and we have written the left-hand cut term explicitly as a function of u. Using the optical theorem (13.7) in each channel leads directly from (13.31) to the earlier result (12.55).

These results can now be immediately extended to $t < 0$, however, since the threshold singularities are unchanged, to yield

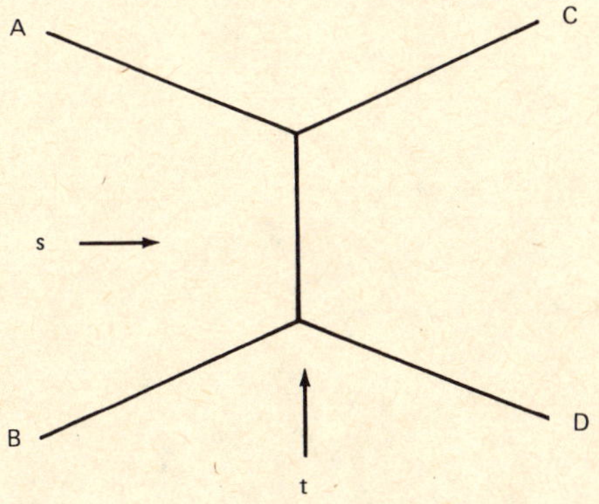

Fig.(13-7) The s-channel Born term diagram.

$$F(s, t, u) = F^B(t) + \frac{g^2}{s - m^2} + \frac{g^2}{u - m^2} + \frac{1}{\pi}\int_{4m^2}^{\infty} ds' \frac{\mathrm{Im}F(s', t, 4m^2 - s' - t)}{s' - s}$$

$$+ \frac{1}{\pi}\int_{4m^2}^{\infty} du' \frac{\mathrm{Im}F(4m^2 - u' - t, t, u')}{u' - u} \tag{13.32a}$$

i.e., writing $u' = 4m^2 - s' - t$,

$$F(s, t, u) = F^B(t) + \frac{g^2}{s - m^2} + \frac{g^2}{u - m^2} + \frac{1}{\pi}\int_{4m^2}^{\infty} ds' \frac{\mathrm{Im}F(s', t, u')}{s' - s}$$

$$+ \frac{1}{\pi}\int_{-t}^{-\infty} ds' \frac{\mathrm{Im}F(s', t, u')}{s' - s}. \tag{13.32b}$$

The left-hand singularity now begins at $s' = -t$ rather than at the origin. The residues of the pole terms are independent of t if the bound state is spinless, as we are assuming here; otherwise the residue contains $P_\ell(\cos\theta)$, as shown in Chapter 9. The Born term $F^B(t)$ has already been evaluated in (11.6) under the assumption that the potential corresponds to the exchange of a particle of mass μ. Here $m = \mu$, and the graph corresponding to this term is that shown in Fig. (13-7). Viewed in the t-channel, this graph corresponds simply to a bound-state pole, consistent with the form of the result

$$F^B(t) = \frac{g^2}{t - m^2} \tag{13.33}$$

which follows from (11.16). Therefore

$$F(s, t, u) = \frac{g^2}{s - m^2} + \frac{g^2}{t - m^2} + \frac{g^2}{u - m^2}$$

$$+ \frac{1}{\pi}\int_{4m^2}^{\infty} ds' \frac{\mathrm{Im}F(s', t, u')}{s' - s} + \frac{1}{\pi}\int_{4m^2}^{\infty} du' \frac{\mathrm{Im}F(s', t, u')}{u' - u}. \tag{13.32c}$$

The Born term in the s-channel is thus identical with the pole term we would obtain from unitarity in the t-channel. In fact, using diagrams to represent multiple particle exchange as in Fig.(13-8), we see that the Nth Born approximation, corresponding to exchanging N particles in the s-channel, is equivalent to the N-particle intermediate state contribution to the unitarity equation in the t-channel. We would therefore expect it to produce a threshold singularity at $t = (Nm)^2$. It can be verified from an explicit calculation of the Born series that exactly those singularities are indeed obtained. Thus, except for the bound state pole at $t = m^2$, the amplitude is analytic as a function of t for $t < 4m^2$, and consequently (13.32c) will be valid for all t within that region.

The assumption that the strong interactions are propagated via the exchange of all possible particles therefore enables us to unify the singularity structures of the various channels. In (13.32), we have found singularities in the energy plane from unitarity in the s and u channels, with the singularities in the momentum transfer being given by unitarity in the t channel. Crossing symmetry tells us that interchanging the variables s, t, and u should affect only their physical interpretation, but not the function itself. In other words, we are led to the very simple hypothesis, originally proposed by Mandelstam, that the scattering amplitude F(s, t, u) is an analytic function of two complex variables having *only* those singularities required by unitarity in each of the three reaction channels.

Since the singularities following from unitarity are all on the real axis of the appropriate energy plane, we may indicate them in the stu plane. We have a four-dimensional complex space, labeled by the real and imaginary parts of s and t (or, alternatively, s and u, or t and u). The stu plane is the two-dimensional subspace containing the two real axes, and the singularities lie entirely in this plane, as shown in Fig.(13-9).

The dispersion relation (13.32c) is written for fixed t; it involves a contour lying on (or infinitesimally above and below) the real s axis, as shown, yielding the appropriate singularities. Because F(s, t, u) has singularities on the real t axis,

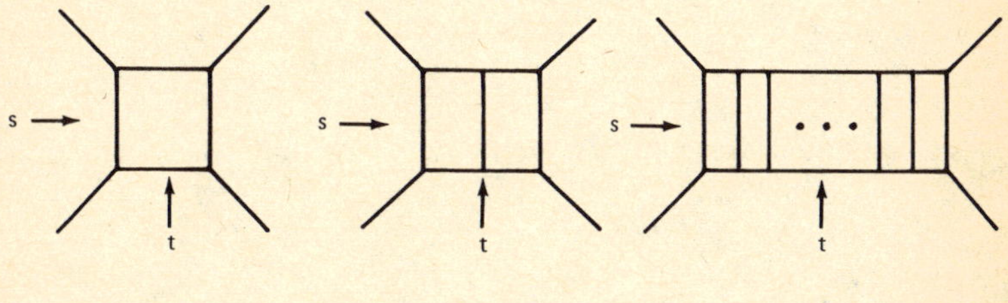

Fig.(13-8) Higher order Born terms in the s channel, which are equivalent to multiparticle intermediate states in the t channel.

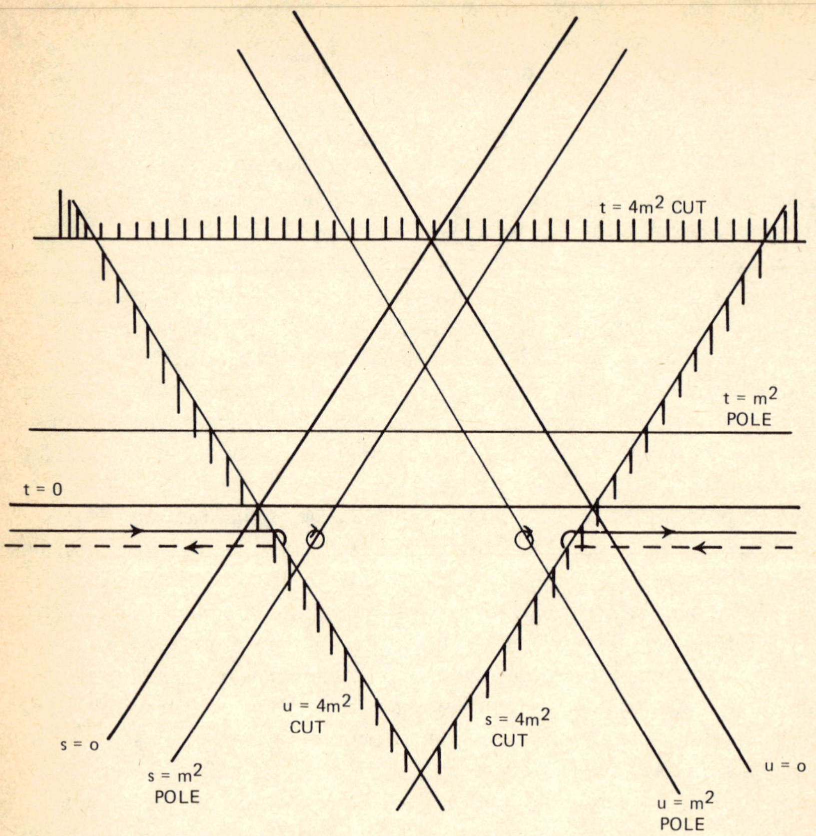

Fig.(13-9) Locations of singularities plotted in the real stu plane, for the equal mass case. An integration contour at fixed negative t is shown.

this contour cannot be drawn for Re$t \geqslant 4m^2$; it would require moving into the complex t-plane in that case.

Alternatively, we could write a dispersion relation at fixed u,

$$F(s, t, u) = \frac{g^2}{s - m^2} + \frac{g^2}{t - m^2} + \frac{g^2}{u - m^2}$$

$$+\frac{1}{\pi}\int_{4m^2}^{\infty} ds' \frac{\mathrm{Im}F(s', t', u)}{s' - s} + \frac{1}{\pi}\int_{4m^2}^{\infty} dt' \frac{\mathrm{Im}F(s', t', u)}{t' - t} \tag{13.34}$$

which will be valid where F(s, t, u) is analytic for real u, i.e. for $u < 4m^2$. Likewise, for $s < 4m^2$ we may write

$$F(s, t, u) = \frac{g^2}{s - m^2} + \frac{g^2}{t - m^2} + \frac{g^2}{u - m^2}$$

$$+\frac{1}{\pi}\int_{4m^2}^{\infty} dt' \frac{\mathrm{Im}F(s, t', u')}{t' - t} + \frac{1}{\pi}\int_{4m^2}^{\infty} du' \frac{\mathrm{Im}F(s, t', u')}{u' - u}. \tag{13.35}$$

Equations (13.32), (13.34), and (13.35) are known as *single* dispersion relations, because we fix one of the variables s, t, and u at a value for which F(s, t, u) is analytic. To complete the cycle, however, we should write the amplitude as a *double* dispersion relation, making use of analyticity in two variables; doing this yields the Mandelstam representation, which we shall discuss presently.

To conclude our relativistic formulation of single dispersion relations, we consider again the pion–nucleon and the nucleon–nucleon amplitudes. For the former, the analytic structure in the stu plane is shown in Fig.(13-10), with the first few thresholds indicated explicitly in each channel. At fixed t, the first threshold branch points in both the s and u channels correspond to elastic scattering and therefore pass through the edge of the physical region at $s = (m + \mu)^2$ or $u = (m + \mu)^2$ and $t = 0$. The dispersion relation in this case is of the form (neglecting spin and subtractions)

$$F_{\pi p}(s, t, u) = (\text{pole terms}) + \frac{1}{\pi}\int_{(m+\mu)^2}^{\infty} ds' \frac{\mathrm{Im}F_{\pi p}(s', t, u')}{s' - s}$$

$$+\frac{1}{\pi}\int_{(m+\mu)^2}^{\infty} du' \frac{\mathrm{Im}F_{\pi p}(s', t, u')}{u' - u}, \tag{13.36}$$

which, for $t = 0$, is equivalent to (12.58). For $t < 0$, however, (13.36) is still

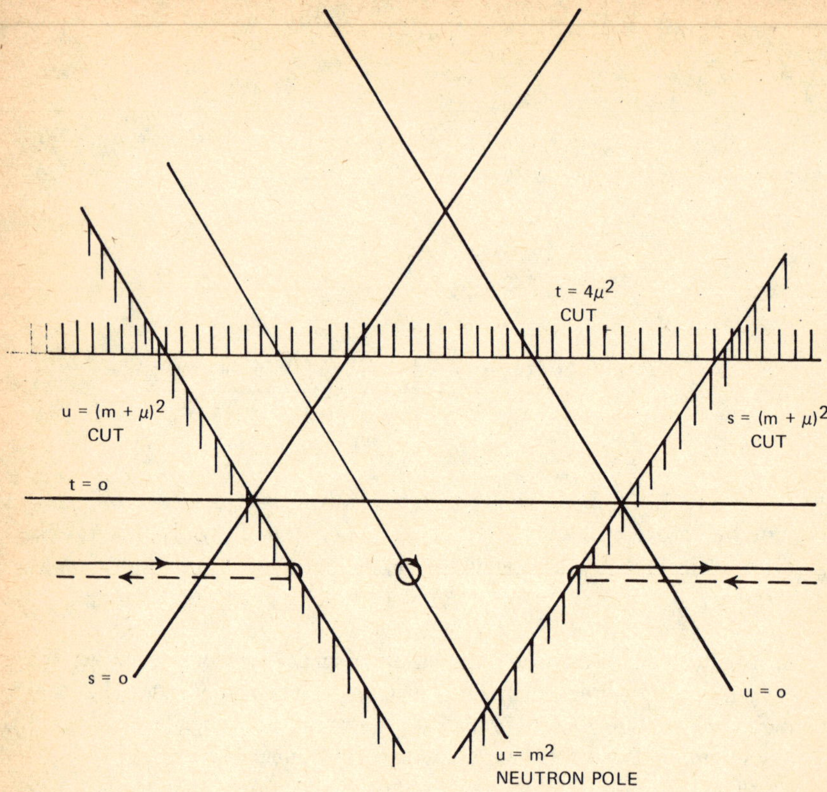

Fig.(13-10) The same as Fig.(13-9), for the pion–nucleon amplitude of (13.36). (Not to scale.)

valid, although we do not have the optical theorem as a physical source of the value of the discontinuity. Also, note that for fixed $t < 0$ the thresholds may lie *outside* the physical region. The reason for this is that unitarity governs only the allowed energies, and not the allowed momentum transfers; intermediate states which have unphysical values of momentum transfer may enter into the unitarity summation for non-forward scattering.

At fixed u, however, the lowest threshold lies below the elastic scattering threshold, corresponding to the contribution of $\pi\pi$ intermediate states in $\pi\pi \to p\bar{p}$. Hence for $u < 4m^2$ we have

$$F_{\pi p}(s, t, u) = (\text{pole terms}) + \frac{1}{\pi}\int_{(m+\mu)^2}^{\infty} ds' \frac{\text{Im}F(s', t', u)}{s' - s}$$

$$+ \frac{1}{\pi}\int_{4\mu^2}^{\infty} dt' \frac{\text{Im}F(s', t', u)}{t' - t}, \qquad (13.37)$$

invariably involving an integration over some part of an unphysical region. The situation at fixed s is exactly analogous.

The stu diagram and the associated thresholds for the process pn → pn are shown in Fig.(13-11). In this case the masses are all equal, but both the t and the

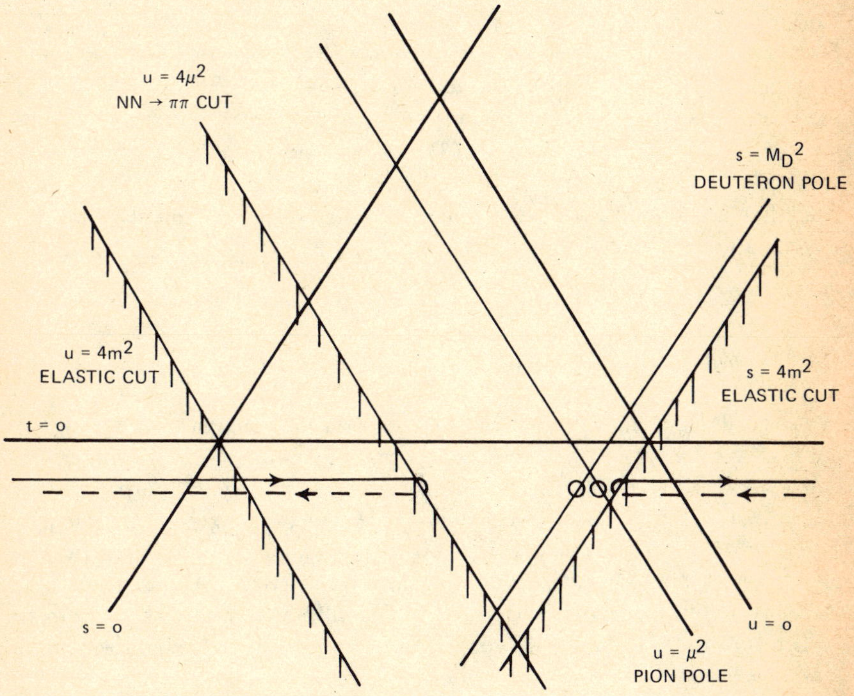

Fig.(13-11) The same as Fig.(13-9), for the nucleon–nucleon amplitude. Only s and u channel thresholds are shown. (Not to scale.)

u channels have the $\pi\pi$ intermediate state as the lowest threshold. The single dispersion relations are correspondingly more complicated by unphysical regions.

In formulating single dispersion relations, we have completely neglected the presence of spin for the particles involved. For pion–nucleon scattering, for example, there are two independent scattering amplitudes, representing scattering processes in which the spin state of the nucleon is or is not changed. Reactions involving more spin degrees of freedom will have a correspondingly larger number of amplitudes. It has been shown, however, that the basic ideas of crossing symmetry remain valid in these more complicated cases. The amplitude describing a particular spin process in one channel may, when continued into a crossed channel, contribute to several different processes; in general, there will be a "crossing matrix" describing the interplay of the spin amplitudes.

A very similar relationship also exists among the different isospin amplitudes which may be involved in a reaction. If the s channel describes elastic π^+p scattering, which is purely $I = 3/2$, the u channel describing elastic π^-p scattering will be a mixture of $I = 3/2$ and $I = 1/2$ amplitudes, while the t channel $\pi^-\pi^+ \to \bar{p}p$ will contain $I = 0$ and $I = 1$. Likewise, if the s-channel reaction is chosen to be purely $I = 1/2$, the other two channels will again contain mixtures of their two isospin amplitudes. How these amplitudes contribute to each other under crossing is most easily described by means of an isospin crossing matrix. Crossing matrices for spin, for isospin, and even for SU(3) have been studied at some length. Although they are necessary for detailed calculations, there is no fundamental alteration in the results we have obtained. We shall therefore omit any further consideration of them.

The Mandelstam Representation

The single despersion relations, being necessarily at fixed momentum transfer, do not manifest the symmetry between all three channels that appears so elegantly in the physical region plot. In order to regain this symmetry, we must treat all three variables s, t, and u equally, and not single one of them out to be held at a fixed real value. In other words, we must write a *double* dispersion relation, expressing the scattering amplitude as an analytic function of two complex variables chosen appropriately from s, t, and u with the restriction $s + t + u = 4m^2$.

Let us consider again the dispersion relation (13.32c) obtained in the equal mass case,

$$F(s, t, u) = \frac{g^2}{s - m^2} + \frac{g^2}{t - m^2} + \frac{g^2}{u - m^2}$$

$$+\frac{1}{\pi}\int_{4m^2}^{\infty} ds' \frac{\mathrm{Im}F(s', t, u')}{s' - s} + \frac{1}{\pi}\int_{4m^2}^{\infty} du' \frac{\mathrm{Im}F(s', t, u')}{u' - u}. \tag{13.32c}$$

As a function of t, F(s, t, u) is analytic in a complex t-plane with a pole at $t = m^2$ and a cut beginning at $t = 4m^2$, plus crossed channel effects. The pole is the Born term, which is indicated explicitly. The branch point at $t = 4m^2$, however, must clearly arise in the integrals on the right-hand side of (13.32c). Since F(s, t, u) is a real analytic function of t, each of the integrals must also be. Let us rewrite them in the form

$$F_s(s, t, u) = \frac{1}{\pi}\int_{4m^2}^{\infty} ds' \frac{\mathrm{Im}F(s', t, u')}{s' - s} = \frac{1}{\pi}\int_{4m^2}^{\infty} ds' \frac{\Delta_s(s', t, u')}{s' - s} \tag{13.38a}$$

$$F_u(s, t, u) = \frac{1}{\pi}\int_{4m^2}^{\infty} du' \frac{\mathrm{Im}F(s', t, u')}{u' - u} = \frac{1}{\pi}\int_{4m^2}^{\infty} du' \frac{\Delta_u(s', t, u')}{u' - u}. \tag{13.38b}$$

For fixed real t, we have seen that the discontinuities Δ_s and Δ_u of F(s, t, u) in the complex s plane can be identified with the imaginary part of the function there. Thus Δ_s and Δ_u are indeed real functions for real t. For complex t, however, this identification is no longer possible. Instead we assume that Δ_s and Δ_u are real analytic functions of t. Then $\Delta_s(s', t, 4m^2 - s' - t)$, for example, must satisfy

$$\Delta_s^*(s', t^*, 4m^2 - s' - t^*) = \Delta_s(s', t, 4m^2 - s' - t) \tag{13.39}$$

in the complex t plane, and if there are no convergence problems at infinity we can write a dispersion relation for $\Delta_s(s', t, u')$.

If we assume that the locations of the singularities in the complex t-plane are given, independently by s′, by the unitarity thresholds, then for fixed s′ we obtain

$$\Delta_s(s', t, u') = \frac{1}{\pi}\int_{4m^2}^{\infty} dt' \frac{\rho_{st}(s', t', 4m^2 - s' - t')}{t' - t} + \frac{1}{\pi}\int_{-s'}^{-\infty} dt' \frac{\rho_{su}(s', t', 4m^2 - s' - t')}{t' - t}, \tag{13.40}$$

where ρ_{st} and ρ_{su} are the discontinuities of Δ_s across the two cuts. Inserting (13.40) into (13.38a) yields, on writing $u' = 4m^2 - s' - t'$,

$$F_s(s, t, u) = \frac{1}{\pi^2} \int_{4m^2}^{\infty} ds' \int_{4m^2}^{\infty} dt' \frac{\rho_{st}(s', t', u')}{(s' - s)(t' - t)} + \frac{1}{\pi^2} \int_{4m^2}^{\infty} ds' \int_{4m^2}^{\infty} du' \frac{\rho_{su}(s', t', u')}{(s' - s)(t' - t)}. \tag{13.41a}$$

A similar treatment of $F_u(s, t, u)$ leads to

$$F_u(s, t, u) = \frac{1}{\pi^2} \int_{4m^2}^{\infty} du' \int_{4m^2}^{\infty} dt' \frac{\rho_{ut}(s', t', u)}{(u' - u)(t' - t)} + \frac{1}{\pi^2} \int_{4m^2}^{\infty} ds' \int_{4m^2}^{\infty} du' \frac{\rho_{us}(s', t', u')}{(s' - s)(u' - u)}. \tag{13.41b}$$

Furthermore, it is not difficult to see that $\rho_{su} = \rho_{us}$. Combining these two results in (13.32c) leads to the *Mandelstam representation*

$$F(s, t, u) = \frac{g^2}{s - m^2} + \frac{g^2}{t - m^2} + \frac{g^2}{u - m^2} + \frac{1}{\pi^2} \int_{4m^2}^{\infty} ds' \int_{4m^2}^{\infty} dt' \frac{\rho_{st}(s', t', u')}{(s' - s)(t' - t)}$$

$$+ \frac{1}{\pi^2} \int_{4m^2}^{\infty} ds' \int_{4m^2}^{\infty} du' \frac{\rho_{su}(s', t', u')}{(s' - s)(u' - u)} + \frac{1}{\pi^2} \int_{4m^2}^{\infty} dt' \int_{4m^2}^{\infty} du' \frac{\rho_{tu}(s', t', u')}{(t' - t)(u' - u)}. \tag{13.42}$$

This double dispersion relation shows elegantly the conceptual equivalence of the three physical regions. If F(s, t, u) describes a process that is completely crossing symmetric, then the appropriate symmetries hold among the discontinuities, namely

$$\rho_{st}(s', t', u') = \rho_{su}(u', s', t') = \rho_{tu}(t', u', s'). \tag{13.43}$$

Each of these functions is non-vanishing only in a limited region of the stu plane.

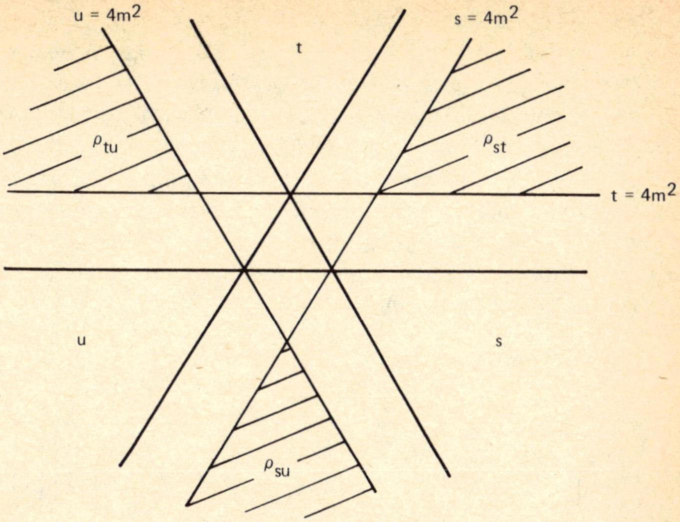

Fig.(13-12) Regions in which the double spectral functions may be non-vanishing.

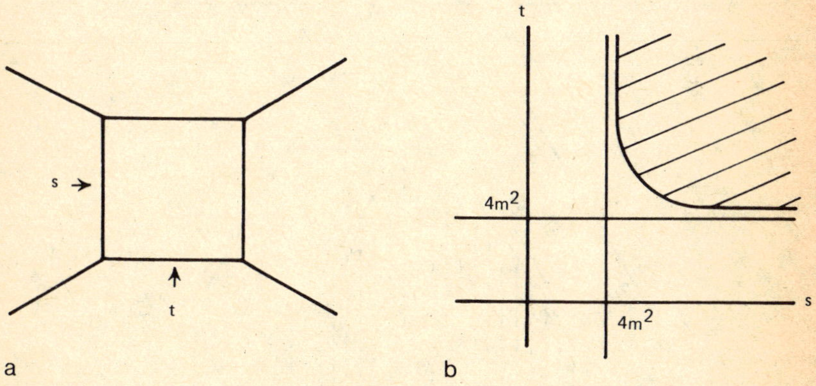

Fig.(13-13) (a) The second order Born term diagram and (b) the region of the s–t plane in which its double spectral function is non-vanishing.

$\Delta_s(s', t, 4m^2 - s' - t)$ vanishes for $s' \geqslant 4m^2$, and its discontinuity $\rho_{st}(s', t', u')$ is therefore zero except for $s' \geqslant 4m^2$ and $t' \geqslant 4m^2$. Likewise, $\rho_{su}(s', t', u') \equiv 0$ unless $s' \geqslant 4m^2$ and $u' \geqslant 4m^2$, while $\rho_{tu}(s', t', u') \equiv 0$ outside the region $t' \geqslant 4m^2$, $u' \geqslant 4m^2$. In other words, the *double spectral functions,* as the ρ's are usually called, are non-vanishing only where threshold singularities of two different channels overlap, as shown in Fig.(13-12).

In fact, the double spectral functions do not quite fill these triangular regions. Instead, they have boundaries which are smooth curves approaching the thresholds asymptotically. For example, the double spectral function ρ_{st} can be calculated for the second-order Born term represented by the graph shown in Fig. (13-13a). The mathematical techniques involved are rather esoteric; they can be

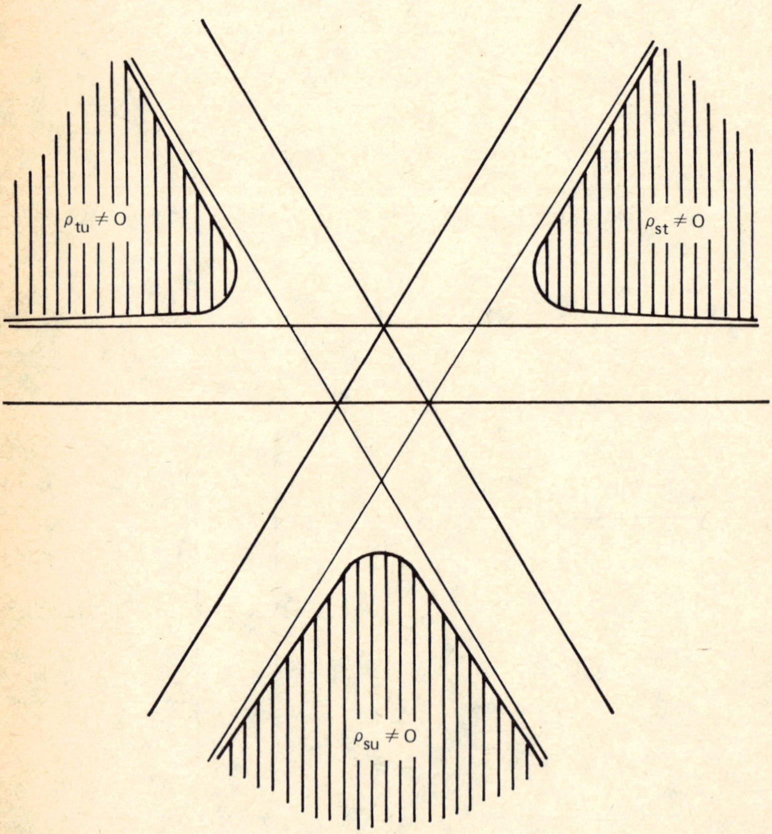

Fig.(13-14) Boundaries of the double spectral functions in the stu plane.

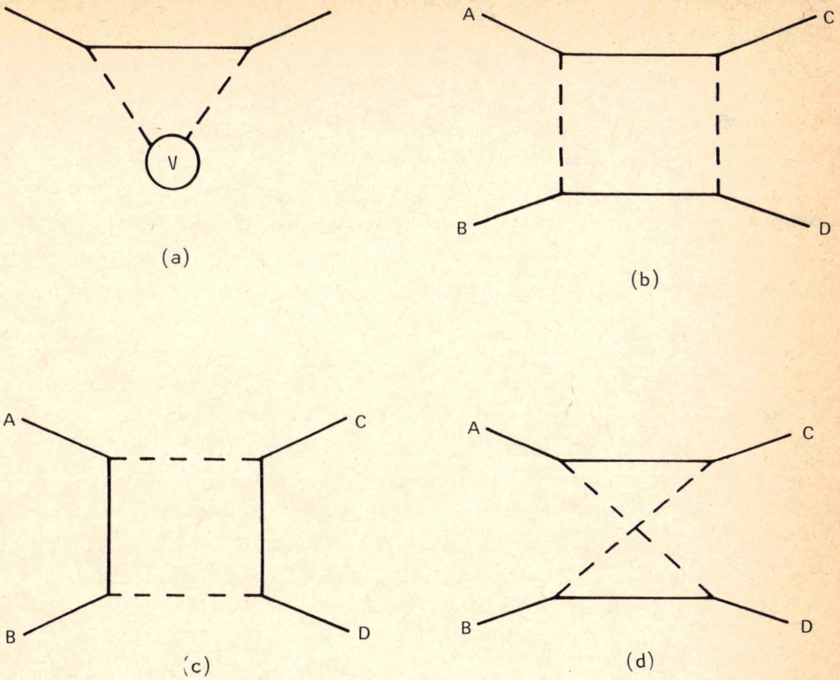

Fig.(13-15) The diagrams corresponding to the second Born term for interaction with (a) a potential V, (b) another particle in the s-channel, (c) another particle in the t-channel, (d) another particle in the u-channel.

found in the books by Eden indicated in the reading list. The result is that $\rho_{st}(s', t', u') \equiv 0$ outside the region bounded by the curve

$$(s - 4m^2)(t - 4m^2) = 4m^4, \tag{13.44}$$

shown in Fig.(13.13b). Note that (13.44) is symmetric in s and t, reflecting the fact that the graph pictured looks the same in the t channel as in the s channel. A similar analysis can be made for other graphs, leading to the boundaries of the double spectral functions as shown in Fig.(13-14).

The inadequacy of potential theory is easily visualized in the Mandelstam representation. The double spectral function arises from the second (and higher) Born terms, in which the potential interacts at least twice with the scattered particle, as shown in Fig.(13-15a). If a target particle is considered the source of the potential, the corresponding graph is Fig.(13-15b). This graph leads only to the double spectral function ρ_{st}, and consequently the dispersion relation (11.18), having only a right-hand cut, is obtained. Relativistically, however, two other graphs are possible, shown in Fig.(13-15c, d), which cannot be obtained in potential theory. These graphs are responsible for the appearance of the other two double spectral functions.

The Mandelstam representation (13.42) can be generalized to include more realistic scattering processes with spin, isospin, different masses, etc. Its primary virtue, however, does not lie in actual calculations, but in the elegance with which it expresses the scattering amplitude. The knowledge of hadron dynamics is localized in its double spectral functions, which play the role of a generalized "potential". If we knew how to construct these functions, it would be possible to write down the amplitude. But the only way to obtain them seems to be from perturbation theory, and therein arises again the familiar problem of the convergence of the strong interactions perturbative series. While the Mandelstam representation has been proved for every order of perturbation, we cannot yet prove that it holds for their sum. Thus the fundamental problem remains; we do not know how to calculate hadron dynamics. Analyticity, crossing symmetry, dispersion relations, and now the Mandelstam representation, provide insights into the structure of the strong interactions, but do not alleviate our inability to answer the basic questions.

Chapter 14

FURTHER TOPICS

In the preceding chapters we have seen how analyticity and crossing symmetry enable us to make a rather elegant hypothesis about the structure of the scattering amplitude without providing any detailed form for the dynamics which it describes. Now we shall attempt to add some simple physical assumptions to this framework in order to obtain more specific consequences. After rederiving the partial wave dispersion relations from the Mandelstam representation, we shall show a well-known method for obtaining approximate solutions to them. The "bootstrap" hypothesis will then be presented briefly, and we shall close this part of the book with a description of how analyticity and crossing symmetry determine the phase of the scattering amplitudes.

Partial Wave Dispersion Relations

We shall continue to work in the simplified universe containing only scalar, isoscalar mesons of mass m. Assuming that no subtractions are required, we have the Mandelstam representation

$$F(s, t, u) = \frac{g^2}{s - m^2} + \frac{g^2}{t - m^2} + \frac{g^2}{u - m^2} + \frac{1}{\pi^2} \iint ds' dt' \, \frac{\rho_{st}(s', t', u')}{(s' - s)(t' - t)}$$

$$+ \frac{1}{\pi^2} \iint ds' du' \, \frac{\rho_{su}(s', t', u')}{(s' - s)(u' - u)} + \frac{1}{\pi^2} \iint dt' du' \, \frac{\rho_{st}(s', t', u')}{(t' - t)(u' - u)} \tag{14.1}$$

To project out the s-channel partial wave amplitudes, the relations

$$\cos\theta_s = 1 - \frac{2t}{s - 4m^2} = 1 + \frac{2u}{s - 4m^2}$$

must be used. The s-channel pole term, being independent of $\cos\theta_s$, is virtually unaltered, yielding

$$\int_{-1}^{1} d\cos\theta_s P_\ell(\cos\theta_s)\frac{g^2}{s - m^2} = \frac{g^2}{s - m^2}\,\delta_{\ell 0}. \tag{14.2a}$$

For the t-channel pole, however, we find

$$\int_{-1}^{1} d\cos\theta_s P_\ell(\cos\theta_s)\frac{g^2}{t - m^2} = \frac{2g^2}{s - 4m^2}\int_{-1}^{1} d\cos\theta_s \frac{P_\ell(\cos\theta_s)}{\cos\theta_s - \left(1 + \dfrac{2m^2}{s - 4m^2}\right)}$$

$$= \frac{g^2}{s - 4m^2}\,Q_\ell\left(1 + \frac{2m^2}{s - 4m^2}\right) \tag{14.2b}$$

which has a cut from $s = +3m^2$ to $s = -\infty$. The u-channel pole yields similarly

$$\int_{-1}^{1} d\cos\theta_s P_\ell(\cos\theta_s)\frac{g^2}{u - m^2} = \frac{g^2}{s - 4m^2}\,Q_\ell\left(-1 - \frac{2m^2}{s - 4m^2}\right), \tag{14.2c}$$

also producing a cut from $s = 3m^2$ to $s = -\infty$.

Since the pole terms are explicitly known, however, it is more convenient to extract them from the amplitude and write a dispersion relation for the partial wave projection of the integrals in (14.1). This procedure is, of course, equivalent to removing the Born term. Thus we define

$$a_\ell(s) = \int_{-1}^{1} d\cos\theta_s \widetilde{F}(s, t, u) P_{\ell'}(\cos\theta_s)$$

$$\widetilde{F}(s,t,u) = \frac{1}{\pi^2}\iint ds'dt'\,\frac{\rho_{st}(s',t',u')}{(s'-s)(t'-t)} + \frac{1}{\pi^2}\iint ds'du'\,\frac{\rho_{su}(s',t',u')}{(s'-s)(u'-u)}$$

$$+\frac{1}{\pi^2}\iint dt'du'\,\frac{\rho_{tu}(s',t',u')}{(t'-t)(u'-u)}. \tag{14.3}$$

The first two terms are easily evaluated, yielding

$$\int_{-1}^{1} d\cos\theta_s P_\ell(\cos\theta_s)\iint ds'dt'\,\frac{\rho_{st}(s',t',u')}{(s'-s)(t'-t)}$$

$$= \frac{1}{s-4m^2}\iint ds'dt'\,\frac{\rho_{st}(s',t',u')}{(s'-s)}\,Q_\ell\left(1+\frac{2t'}{s-4m^2}\right) \tag{14.4a}$$

$$\int_{-1}^{1} d\cos\theta_s P_\ell(\cos\theta_s)\iint ds'du'\,\frac{\rho_{su}(s',t',u')}{(s'-s)(u'-u)}$$

$$= \frac{1}{s-4m^2}\iint ds'du'\,\frac{\rho_{su}(s',t',u')}{(s'-s)}\,Q_\ell\left(-1-\frac{2u'}{s-4m^2}\right). \tag{14.4b}$$

In both of these we see cuts in the s plane. The Legendre function

$$Q_\ell\left(1+\frac{2t'}{s-4m^2}\right)$$

is singular from $s = -\infty$ to $s = 4m^2 - t'$, and since t' takes on all values greater than $4m^2$ there results a cut running from $s = -\infty$ to $s = 0$ in the integral. Furthermore, there is an obvious cut, resulting from the factor $1/(s'-s)$, running from $4m^2$ to ∞. The same structure is present in (14.4b). For the third term of (14.3), we use partial fractioning to obtain

$$\int_{-1}^{1} d\cos\theta_s P_\ell(\cos\theta_s)\iint dt'du'\,\frac{\rho_{tu}(s',t',u')}{t'+u'-4m^2-s}\left[\frac{1}{t'-t}+\frac{1}{u'-u}\right]$$

$$= \frac{1}{s - 4m^2} \iint dt' du' \frac{\rho_{tu}(s', t', u')}{s' - s} \left[Q_\ell\left(1 + \frac{2t'}{s - 4m^2}\right) \right.$$

$$\left. + Q_\ell\left(-1 - \frac{2u'}{s - 4m^2}\right)\right]. \tag{14.4c}$$

Once again the Legendre functions produce a cut from $s = -\infty$ to $s = 0$, and since $\rho_{tu}(s', t', u')$ is non-vanishing only when $t' \geqslant 4m^2$, $u' \geqslant 4m^2$, the factor $1/(t' + u' - 4m^2 - s)$ leads to a cut from $s = 4m^2$ to $s = \infty$.

Consequently the partial wave amplitude $a_\ell(s)$ defined by (14.3) satisfies a dispersion relation of the form

$$a_\ell(s) = \frac{1}{\pi} \int_{4m^2}^{\infty} ds' \frac{\Delta_\ell^{(r)}(s')}{s' - s} + \frac{1}{\pi} \int_{0}^{-\infty} ds' \frac{\Delta_\ell^{(\ell)}(s')}{s' - s}, \tag{14.5}$$

where the discontinuities $\Delta_\ell^{(r)}$ and $\Delta_\ell^{(\ell)}$ can be calculated in terms of the double spectral functions from equations (14.4). This equation is particularly simple as a result of our choice of all masses equal. In more realistic situations a more complicated singularity structure arises from unequal-mass kinematics; we shall omit the details of those alterations, however.

The N/D Method

Without prior knowledge of the discontinuities, we cannot infer any further information about the partial wave amplitude from (14.5). Even knowing that the amplitude must satisfy unitarity is still not enough. It may happen, however, that we can construct an intuitive ansatz of some sort for the discontinuity of the left-hand cut, based on a potential or a particle exchange model. In this case, there is a procedure by which we can construct iteratively a function which has that discontinuity and satisfies elastic unitarity.

Let us neglect inelastic channels and assume that the elastic unitarity relation (12.51)

$$\text{Im}\, a_\ell(s) = \frac{2k}{W} |a_\ell(s)|^2$$

is valid for the entire right-hand cut. This equation can be rewritten conveniently as

$$\mathrm{Im}\left(\frac{1}{a_\ell(s)}\right) = -\frac{2k}{W}. \tag{14.6}$$

Now suppose that we may write

$$a_\ell(s) = \frac{N(s)}{D(s)}, \tag{14.7}$$

where the numerator N(s) is analytic except for the *left*-hand cut, while the denominator is analytic except for the *right*-hand cut. If such a partition of the discontinuities can be made, then dispersion relations can be written separately for N(s) and D(s). Their asymptotic behaviors must be assumed, of course; in so doing, the possibility of common factors in both N and D, canceling each other in (14.7), is eliminated. We assume that $N(s) \to 0$ and $D(s) \to 1$ as $|s| \to \infty$, and obtain the dispersion relations

$$N(s) = \frac{1}{\pi} \int_{-\infty}^{0} ds' \frac{\mathrm{Im}N(s')}{s' - s} \tag{14.8a}$$

$$D(s) = 1 + \frac{1}{\pi} \int_{4m^2}^{\infty} ds' \frac{\mathrm{Im}D(s')}{s' - s} \tag{14.8b}$$

From the unitarity relation in the form (14.6), however, it follows that on the right-hand cut

$$\mathrm{Im}\left(\frac{D(s)}{N(s)}\right) = -\frac{2k}{W} = -\left(\frac{s - 4m^2}{s}\right)^{1/2}$$

Since N(s) is analytic in this region, the discontinuity results entirely from D(s), and therefore for real $s \geqslant 4m^2$

$$\mathrm{Im}D(s) = -\frac{2k}{W} N(s). \tag{14.9}$$

Conversely, D(s) is analytic on the left-hand cut, so for real $s \leqslant 0$ we have

$$\mathrm{Im}N(s) = D(s)\mathrm{Im}a_\ell(s). \tag{14.10}$$

The dispersion relations (14.8) therefore become

$$N(s) = \frac{1}{\pi}\int_{-\infty}^{0} ds' \frac{D(s')\mathrm{Im}a_\ell(s')}{s' - s} \tag{14.11a}$$

$$D(s) = 1 - \frac{1}{\pi}\int_{4m^2}^{\infty} ds' \left(\frac{s' - 4m^2}{s'}\right)^{1/2} \frac{N(s')}{s' - s}. \tag{14.11b}$$

The N/D procedure thus leads to a system of two coupled integral equations expressing N(s) and D(s) in terms of each other and the discontinuity across the left-hand cut. If we assume that this discontinuity is known, say

$$\mathrm{Im}_L a_\ell(s) = A(s), \tag{14.12}$$

then a solution for $a_\ell(s)$ can be constructed by iteration. Taking first D(s) = 1, we find

$$N^{(1)}(s) = \frac{1}{\pi}\int_{-\infty}^{0} ds' \frac{A(s')}{s' - s}; \tag{14.13a}$$

the next step is

$$D^{(1)}(s) = \frac{1}{\pi}\int_{4m^2}^{\infty} ds' \left(\frac{s' - 4m^2}{s'}\right)^{1/2} \frac{N^{(1)}(s')}{s' - s}. \tag{14.13b}$$

Further iterations may be carried out in the same way. Whether the sequence will converge depends, of course, on A(s); but the calculation becomes rapidly more difficult, and when one considers that A(s) was an approximation in the first place, it seems hardly likely that the exact solution will be much better than the first iteration. For these reasons, N/D calculations usually stop with (14.13).

For example, let us consider the simple choice

$$A(s) = \frac{\alpha}{\sqrt{-s}} \tag{14.14}$$

for $s < 0$. Then

$$N^{(1)}(s) = \frac{\alpha}{\pi} \int_{-\infty}^{0} ds' \frac{1}{\sqrt{-s'}\,(s' - s)} = \frac{\alpha}{\sqrt{s}} \tag{14.15}$$

and therefore

$$D^{(1)}(s) = 1 - \frac{\alpha}{\pi} \int_{4m^2}^{\infty} ds' \frac{\sqrt{s' - 4m^2}}{s'(s' - s)}$$

$$= 1 - \frac{\alpha}{s}(\sqrt{4m^2 - s} - 2m). \tag{14.16}$$

Thus the first approximation to the amplitude is

$$a_\ell^{(1)}(s) = \frac{\alpha\sqrt{s'}}{s - \alpha(\sqrt{4m^2 - s} - 2m)}$$

$$= \frac{\alpha W}{W^2 - 2\alpha(ik - m)} \tag{14.17}$$

for which it is easily verified that (14.6) is satisfied.

Since N(s) is, by construction, analytic except for the left-hand cut, it can produce no other singularities in $a_\ell(s)$. Zeros of D(s), however, will lead to poles of $a_\ell(s)$. For example, (14.16) vanishes when $s = -\alpha(\alpha + 4m)$. Such poles must be interpreted as bound states of the potential producing A(s). The precise location of the pole will depend on the exact solution of the N/D equations; the first iteration (14.16) is valid only to order α, so the pole position may be inaccurate also beyond that order. If α is small, we may hope that the sequence will converge rapidly, so that (14.17) will be a good approximation to the true solution.

This example of the N/D procedure has little physical value, of course, since the A(s) in (14.14) was chosen for its solubility rather than because it represents a realistic potential. Furthermore, the elastic unitarity condition (14.6) is not correct above the first inelastic threshold, and the solution correspondingly lacks the higher branch points. More elaborate models of the N/D type can be constructed, and more meaningful forms of A(s) can be used, according to the particular situation one wishes to study.

Bootstraps

The basic hypothesis that all interactions take place via particle exchange can be used in combination with the N/D formalism to approximate a "self-consistent universe". The idea is that all particles are bound states (or resonances, "almost-bound states") of all particles, the forces responsible for the binding being generated by the exchange of those same particles. In terms of the single-particle universe we have been considering, we would say that the particle A with mass M is a bound state of two A's, with a binding force resulting from the exchange of an A. It turns out that this simplistic case cannot work, the potential being repulsive. For more realistic situations, however, it is indeed possible for a set of particles to pull themselves up into existence by their own bootstraps in this way, and the philosophy developed from these calculations is thus called "bootstrap theory".

The bootstrap of the ρ meson, originally calculated by Zachariasen, is still the best example of the theory. Let us consider the scattering of two pions, as shown in Fig.(14-1a). The ρ meson is observed as a J = 1 resonance in the I = 1 amplitude, and dominates the amplitude in the low-mass region. Since pions have negative G-parity, the exchange of a single pion is forbidden, and the lightest-mass exchange allowed is that of two pions. It is reasonable to assume that this t-channel two-pion state is also dominated by the ρ contribution. Thus the bootstrap philosophy states that the interaction responsible for the production of a ρ in the s channel is the exchange of a ρ in the t channel.

A quantitative treatment of this idea is given in the reference by Zachariasen. The Born approximation to the amplitude is calculated by considering the ρ simply as a particle of spin one, with mass m_ρ and a coupling constant $g_{\rho\pi\pi}$ describing the $\rho\pi\pi$ vertex. The corresponding graph, shown in Fig.(14-1b), leads to a first approximation to the amplitude, from which the J = 1 contribution can be projected out. This partial wave amplitude has a left-hand cut running from

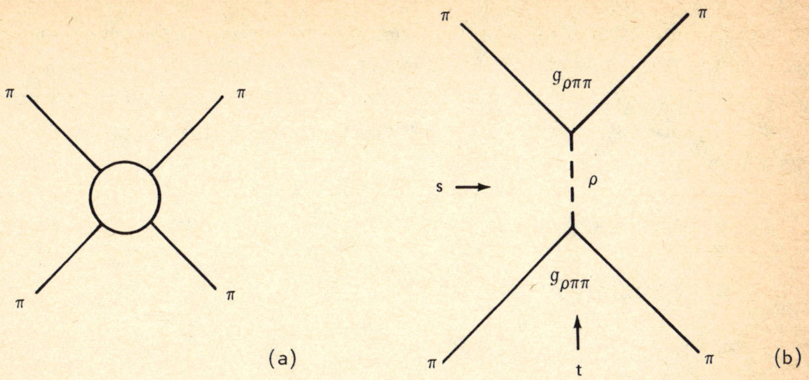

Fig.(14-1) Pion–pion scattering, shown (a) generally and (b) as propagated by ρ-exchange in the s-channel, i.e. ρ-formation in the t-channel.

$s = 4\mu^2 - m_\rho^2$ to $s = -\infty$, as expected from the results of Chapter 13. Using the discontinuity across this cut as the input A(s) in (14.12), we then calculate the partial wave amplitude via the N/D technique. As in (14.17), a bound-state pole will result. Now we require that this pole reproduce the ρ; then the pole must occur when $D(m_\rho^2) = 0$, and its residue must be appropriately related to $g_{\rho\pi\pi}$. Thus we obtain in principle two equations in two unknowns, m_ρ and $g_{\rho\pi\pi}$. In practice it is necessary to introduce a subtraction in D(s) in order to get a convergent integral for (14.16), and the results depend to some extent on the subtraction constant. The values obtained are $m_\rho \approx 350$ MeV and $g_{\rho\pi\pi}^2/4\pi \approx 2.4$, neither of which is in particularly good agreement with the measured values. The very existence of solutions, however, must be considered a surprisingly successful result, since there is no guarantee that the pole will occur for physical values of s, if at all. With the myriad approximations inherent in the N/D approach, the existence of other two-pion resonances which could also be exchanged, etc., the fact that the results obtained are even of the right order of magnitude may be taken as an indication that the bootstrap scheme is at least possible.

To extend the bootstrap idea to include more particles and avoid some of the problems resulting from their neglect is an appealing idea, and much effort has been expended in it. The complications involved with including many channels are considerable, however, and the full evaluation of physically realistic bootstraps has not yet been carried out.

Phase Considerations from Crossing Symmetry

Finally let us now mention a rather ingenious use of crossing symmetry plus analyticity through which the phase of a crossing-symmetric amplitude may be related to its energy dependence. Suppose that F(s, t, u) is the amplitude representing a process symmetric between the s and u channels, so that when t = 0 we may write

$$F(s, 0, u) = F(u, 0, s) = F_S(s).$$

The crossing condition implies that

$$F_S(s) = F_S(u) = F_S(4m^2 - s), \tag{14.18}$$

which is more clearly seen by using the variable $\omega = (s - 2m^2)/2m$, as in Chapter 11. With $f(\omega) = F_S(s)$, (14.18) becomes

$$f(\omega) = f(-\omega). \tag{14.19}$$

As we have seen, (14.19) is valid in a complex ω plane with cuts from $\omega = \pm m$ to $\omega = \pm\infty$, the s channel physical region lying infinitesimally above the right-hand cut, and the u channel correspondingly below the left-hand cut. Thus crossing symmetry states that for real ω, with ϵ infinitesimal, $f(\omega)$ satisfies

$$f(\omega + i\epsilon) = f(-\omega - i\epsilon). \tag{14.20}$$

Furthermore, since f is a real analytic function,

$$f(-\omega - i\epsilon) = f^*(-\omega + i\epsilon)$$

so in the upper half ω plane we can write, for real positive ω,

$$f(\omega) = f^*(\omega e^{i\pi}). \tag{14.21}$$

Now let us define $g(\omega) = f(\omega e^{i\pi/2})$. Then (14.21) becomes

$$g(\omega e^{-i\pi/2}) = g^*(\omega e^{i\pi/2}), \tag{14.22}$$

or, with $y = \omega e^{-i\pi/2}$ and therefore $y^* = \omega e^{i\pi/2}$,

$$g(y) = g^*(y^*). \tag{14.23}$$

This equation is identical with the definition (11.6) of a real analytic function. Thus it appears that if we write the amplitude as a function of the variable $y = \omega e^{-i\pi/2}$, the function must be real when y is real. By the arguments used earlier, however, (14.23) must also hold for all complex y to which g(y) and $g^*(y^*)$ can be continued analytically.

The reality condition (14.23) can be used in the following way. Suppose that we know the dependence of the total cross section, and therefore of the imaginary part of the amplitude, on ω. A crossing-symmetric amplitude can be obtained by simply replacing ω appropriately by $\omega e^{-i\pi/2}$. For example, if

$$\mathrm{Im}\, f(\omega) = c\omega^{a} \tag{14.24}$$

(with c real), then we may take

$$f_+(\omega) = -c(\omega e^{-i\pi/2})^{a}/\sin \tfrac{1}{2}\pi a. \tag{14.25}$$

Another functional form may simultaneously satisfy both (14.23) and (14.24), since $\mathrm{Re} f(\omega)$ is not determined, but (14.25) is the simplest possibility.

This technique can easily be extended to amplitudes describing processes which are not crossing symmetric. It is only necessary to separate them into symmetric and antisymmetric parts. For example, let us suppose that the s-channel amplitude describes $\pi^+ p$ scattering, and the u-channel $\pi^- p$. Then the crossing relations are the form

$$f_+(\omega) = f_-(-\omega), \tag{14.26}$$

with $f_\pm(\omega)$ denoting the $\pi^\pm p$ amplitude. The symmetric and antisymmetric combinations

$$\begin{aligned} f^\pm(\omega) &= \tfrac{1}{2}[f_+(\omega) \pm f_-(\omega)] \\ &= \tfrac{1}{2}[f_+(\omega) \pm f_+(-\omega)] \end{aligned} \tag{14.27}$$

then satisfy

$$f^\pm(-\omega) = \pm f^\pm(\omega). \tag{14.28}$$

The symmetric amplitude f^+ clearly can be shown, as above, to be a real analytic function of $\omega e^{-i\pi/2}$. The antisymmetric combination, however, leads to

$$f^-(\omega) = -[f^-(\omega e^{i\pi})]^*$$

instead of (14.21); but merely inserting a factor i here yields

$$if^{-}(\omega) = [if^{-}(\omega e^{i\pi})]^{*} \tag{14.29}$$

Therefore $if^{-}(\omega)$, like $f^{+}(\omega)$, must be a real analytic function of the variable $\omega e^{-i\pi/2}$. If, for example,

$$\operatorname{Im} f^{\pm}(\omega) = c_{\pm}\,\omega^{\alpha_{\pm}},$$

if follows as in (14.25) that we may take

$$f^{+}(\omega) = -\frac{c_{+}(\omega e^{-i\pi/2})^{\alpha_{+}}}{\sin \frac{1}{2}\pi\alpha_{+}}$$

$$f^{-}(\omega) = \frac{ic_{-}(\omega e^{-i\pi/2})^{\alpha_{-}}}{\cos \frac{1}{2}\pi\alpha_{-}} \tag{14.30}$$

The arguments leading to (14.23) can also be extended to the case $t \neq 0$, provided that the appropriate generalization of the crossing symmetric variable ω is chosen. Specifically, the choice

$$w = \frac{s-u}{4m} = \frac{2s+t-4m^2}{4m} \tag{14.31}$$

satisfies the requirement that crossing from the s to the u channel interchanges w and $-w$. It then follows that a crossing symmetric amplitude is a real analytic function of w.

Reading List for Analyticity

There are a number of books approaching the analyticity of the scattering amplitude from a variety of viewpoints. For introductory purposes, one of the best is

1. D. Park, *Introduction to Strong Interactions* (New York: Benjamin, 1966).

This book presupposes no knowledge of field theory, and develops the basic ideas of analyticity through potential theory as well as formulating some of the elementary field techniques assiciated with strong interactions.

More detailed studies of the properties of the scattering amplitude obtained in potential theory are given by

2. V. de Alfaro and T. Regge, *Non-Relativistic Potential Scattering* (New York: Benjamin, 1963).
3. A. Martin, in *Progress in Elementary Particle and Cosmic Ray Physics,* Vol. VIII, (New York: Interscience, 1965).

The mathematical techniques involved in both of these works are rather formidable, however.

Another more detailed reference is

4. M. Goldberger and K. Watson, *Collision Theory* (New York: Wiley, 1964)

wherein the patient reader can discover vast vistas of useful information on scattering processes. In particular, the unitarity equation is thoroughly explored here; the pion–nucleon dispersion relations are also obtained, with spin complications included.

A more detailed, yet very readable, review of modern developments in this field is given by

5. R. J. Eden, *High Energy Collisions of Elementary Particles* (Cambridge: University Press, 1967),

and the more advanced techniques relating unitarity and analyticity along with the methods of calculating the double spectral functions, are given in

6. R. J. Eden, P. V. Landshoff, D. I. Olive, and J. C. Polkinghorne, *The Analytic S-Matrix* (Cambridge: University Press, 1966).

The flavour of the relativistic formulation of the theory, with a detailed discussion of relativistic kinematics and physical regions, can be found in

7. R. Omnes and M. Froissart, *Mandelstam Theory and Regge Poles* (New York: Benjamin, 1963).

No reading list on this subject could fail to include the following author; the interested reader, having gotten a firm grip on the basic ideas, is therefore recommended to proceed to

8. Geoffrey F. Chew, *The Analytic S-Matrix* (New York: Benjamin, 1967);
9. Geoffrey F. Chew, *S-Matrix Theory of Strong Interactions* (New York: Benjamin, 1961).

A thoroughgoing review of the pion–nucleon dispersion relations has been given by

10. J. Hamilton, in *Strong Interactions and High Energy Physics, Proceedings of the Scottish Universities Summer School,* Ed. R. G. Moorhouse (New

York: Plenum, 1963), pp.281–369.
In the same volume, the details of the bootstrap theory are discussed by
11. F. Zachariasen, *ibid*. pp.371–409.

Part III

Regge Theory

Chapter 15

COMPLEX ANGULAR MOMENTUM

It is a bold step from real values to the energy out into the complex E-plane, but not half so adventurous as that we shall take in this third part of the book. In Part II we went from continuous real variables into a complex plane. Now we shall consider the angular momentum, a parameter physically restricted to the integers and half-integers, and move from its discrete values into a complex ℓ-plane. To make such a far-fetched transformation is a technique originally introduced by T. Regge, in potential theory, in order to solve certain problems encountered in studying the properties of the scattering amplitude in the limit $|\cos\theta| \to \infty$. From it have arisen some of the most appealing concepts in our present formulation of high-energy phenomenology, which as a whole fall into the area now known as Regge theory.

In this chapter we shall consider the significant new insights into the nature of bound states and resonances which result from thinking of the angular momentum eigenvalue ℓ labeling the partial wave amplitudes as a complex variable. To see most easily the basic idea involved, let us return to the square well potential solved in Chapter 9. The partial wave amplitude in this case was given by

$$A_\ell(k) = \frac{i}{k} \frac{kj'_\ell(ka)j_\ell(Ka) - Kj_\ell(ka)j'_\ell(Ka)}{kh^{(1)'}_\ell(ka)j_\ell(Ka) - Kh^{(1)}_\ell(ka)j'_\ell(Ka)},$$

with $k^2 = 2mE$ and $K^2 = 2m(E + V_0)$. The poles of this amplitude were associated with bound states for $E < 0$, and with resonances for $E > 0$, having angular momentum ℓ. They occur when

$$kh^{(1)'}_\ell(ka)j_\ell(Ka) - Kh^{(1)}_\ell(ka)j'_\ell(Ka) = 0. \tag{15.1}$$

In the particularly simple case $\ell = 0$, the spherical Bessel functions can be expressed trigonometrically

$$j_0(Ka) = \frac{\sin Ka}{Ka}$$

$$h_0^{(1)}(ka) = \frac{e^{ika}}{ka}$$

to yield for (15.1) the familiar equation

$$\tan ka = \frac{k}{iK} \tag{15.2}$$

locating the $\ell = 0$ bound states. By well known graphical methods, the solutions of (15.2) corresponding to real E can be explicitly obtained for any given values of V_0, a, and m. The number of bound states depends on the value of the dimensionless quantity mV_0a^2, but is certainly finite; the number of resonances for this particular potential can be shown to be infinite.

For $\ell > 0$ the situation is somewhat more complicated to write down, but the essential features are unaltered. The equation is trigonometric and leads to a finite number of bound states and an infinite set of resonances. From the known properties of the Bessel functions it can be shown that if $\ell_1 > \ell_2$, the energy of the lowest bound state with angular momentum ℓ_1 is larger than that of the lowest one for ℓ_2. (This property, and other similar ones, are, in fact, relatively independent of the detailed form of the potential.) Thus a possible situation is that shown in Fig.(15-1), where we indicate for different ℓ-values the locations of four $\ell = 0$ bound states, three with $\ell = 1$, etc.

The Bessel functions involved in (15.1) are basically defined, however, simply as solutions of the differential equation

$$\frac{1}{r}\frac{d^2}{dr^2}(rR_\ell(r)) + \left[k^2 - \frac{\ell(\ell+1)}{r^2}\right]R_\ell(r) = 0, \tag{9.7}$$

obeying stipulated boundary conditions. While symmetries arising from physics generally restrict ℓ to integer values, the mathematical properties of the solutions of (9.7) may be considered for arbitrary real or complex ℓ. Therefore we may think of $j_\ell(Ka)$ and $h_\ell^{(1)}(ka)$ as functions of a continuous variable ℓ, and regard (15.5) as an equation giving an implicit functional relationship between ℓ and E. That is, instead of fixing ℓ at an integer and looking for values of E which satisfy (15.1), suppose we reverse the procedure by fixing E *first* and finding the values of ℓ for which (15.1) is true. Ordinarily we would not expect to choose a bound

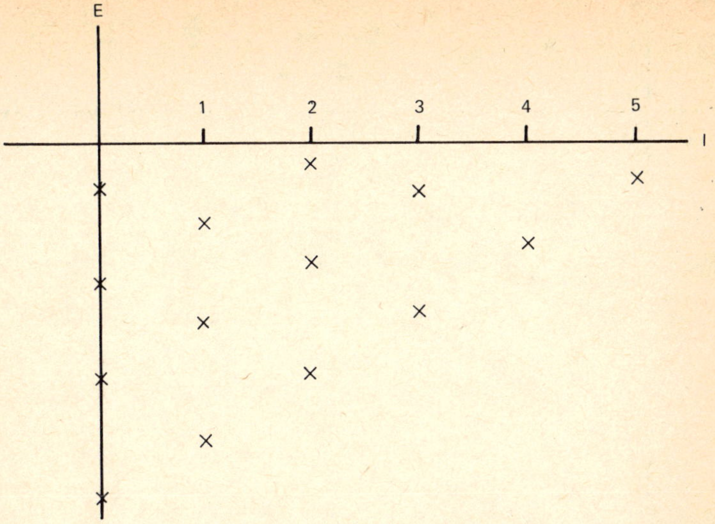

Fig.(15-1) A plot of a possible set of bound state energies against the angular momentum.

state energy for E, and consequently non-integral values α_i of ℓ would result. Since the differential equation and the boundary conditions depend analytically on ℓ, arguments reminiscent of those used in Chapter 10 assure us that a small change in E will lead to a small change in α_i. Thus carrying out the above procedure for all E will enable us to define a set of functions $\alpha_i(E)$, such that for $\ell = \alpha_i(E)$ a solution of (15.1) is obtained. These functions are known as the *Regge trajectories* of the partial wave amplitude.

The bound states in Fig.(15-1) must then correspond to the passage of the Regge trajectories through integer values, as shown in Fig.(15-2). Every bound state lies on some trajectory, and is therefore intimately associated with other bound states having different values of angular momentum but lying on that same trajectory. Consequently the pictured situation represents four Regge families of bound states, rather than fourteen independent poles.

What happens to these trajectories when the energy becomes positive? We have seen generally that resonance poles for integral ℓ must occur for complex energies, implying that the trajectories must move into the complex E-plane in order to reach integer values of ℓ. Conversely, as E becomes positive real, $\alpha_i(E)$ must become complex. If

$$\alpha_i(E_R - i\Gamma) = \ell \tag{15.3}$$

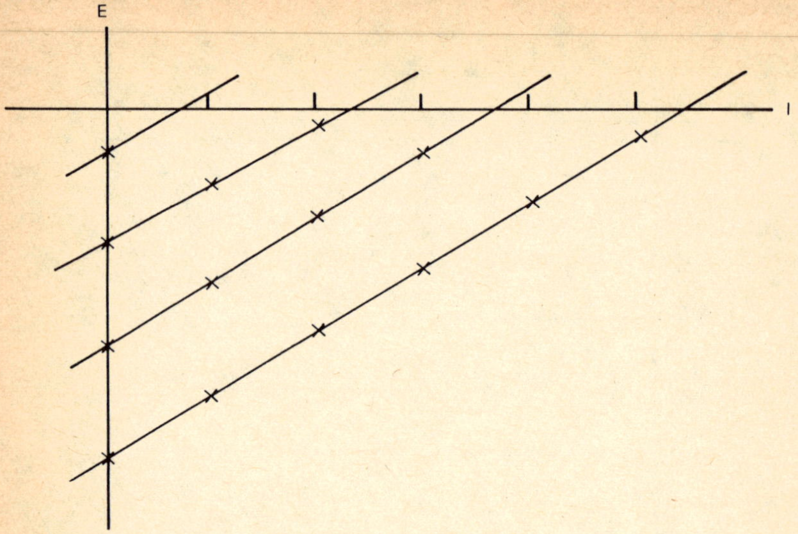

Fig.(15-2) The bound states of Fig.(15-1) grouped onto four Regge trajectories.

produces a narrow resonance with $\Gamma \ll E_R$, we may write

$$\alpha_i(E_R - i\Gamma) \approx \alpha_i(E_R) - i\Gamma\alpha_i'(E_R)$$

yielding

$$\alpha_i(E_R) \approx \ell + i\Gamma\alpha_i'(E_R). \tag{15.4}$$

In other words, when E_R is a resonant energy $\alpha_i(E_R)$ is complex, its real part giving the angular momentum of the resonance. The width of the resonance is related to the imaginary part of the trajectory there,

$$\Gamma = \frac{\text{Im}\alpha_i(E_R)}{\alpha_i'(E_R)}, \tag{15.5}$$

and, unless $\alpha_i(E)$ is varying unusually rapidly at $E = E_R$, a narrow resonance implies a small imaginary part for $\alpha_i(E_R)$. Consequently a resonance corresponds to a complex Regge trajectory $\alpha_i(E)$ passing close by an integer value.

The Sommerfeld–Watson Transformation

Thus we now wish to picture bound states and resonances as the effects of poles of the partial-wave amplitude in a complex angular momentum plane, rather than in a complex energy plane as in Part II. To be complete, we should consider the amplitude here as a function of two complex variables, E and ℓ, with singularity surfaces in a four-dimensional space as in the Mandelstam representation. For practical purposes, however, it suffices to keep the energy real and simply study analyticity in the angular momentum index. In order to avoid confusion, we shall hereafter label this quantity λ when referring to a complex variable, maintaining the usual meaning of ℓ as an integer.

It is not clear, however, that it is possible in general to continue the partial-wave amplitude meaningfully to complex values of λ. Indeed, if $a(\lambda, k)$ is such a continuation, with $a(\ell, k) = a_\ell(k)$, then we may define

$$a_\phi(\lambda, k) = a(\lambda, k) + \phi(\lambda)\sin\pi\lambda \tag{15.6}$$

for any function $\phi(\lambda)$ which is finite for $\lambda = \ell$, and it follows that also $a_\phi(\ell, k) = a_\ell(k)$. Consequently there is a rather drastic ambiguity in the continuation of the partial-wave amplitude.

We can avoid this ambiguity, however, by making an assumption on the asymptotic behavior of $a(\lambda, k)$, because of a result known as Carlson's Theorem. This theorem guarantees that if $a(\lambda, k)$ is an analytic function of λ in the region $\mathrm{Re}\lambda \geqslant N$ which takes on specified values $a_\ell(k)$ for all $\ell \geqslant N$, and if $a(\lambda, k)$ is exponentially bounded as $|\lambda| \to \infty$ with $\mathrm{Re}\lambda \geqslant N$, then $a(\lambda, k)$ is in fact unique. We shall presently show that it is possible to define a continuation $a(\lambda, k)$, either from potential theory or on the basis of the Mandelstam representation, which vanishes exponentially as $|\lambda| \to \infty$ with $\mathrm{Re}\lambda \geqslant -\frac{1}{2}$. By Carlson's theorem, the continuation is therefore unique, i.e. it is the only one which vanishes in this way.

Fortunately, this unique continuation is just the one we need in order to carry out a rather esoteric procedure, known as the Sommerfeld–Watson transformation, which converts a summation over ℓ into a contour integral in the complex λ-plane. Let us suppose that $g(\lambda)$ is an analytic function of λ in the neighborhood of some integer ℓ, and form the integral

$$G(\ell) = \frac{1}{2i}\oint_{c_\ell} \frac{g(\lambda)}{\sin\pi\lambda}\, d\lambda, \tag{15.7}$$

where c_ℓ is a closed contour surrounding the integer point $\lambda = \ell$, as shown in Fig.(15-3a). Near $\lambda = \ell$,

$$\sin \pi\lambda \approx \sin \pi\ell + \pi(\lambda - \ell) \cos \pi\ell$$

$$\approx \pi(\lambda - \ell)(-1)^\ell, \qquad (15.8)$$

so the integral has a simple pole at $\lambda = \ell$ with residue $(-1)^\ell g(\ell)/\pi$. It follows by Cauchy's theorem that

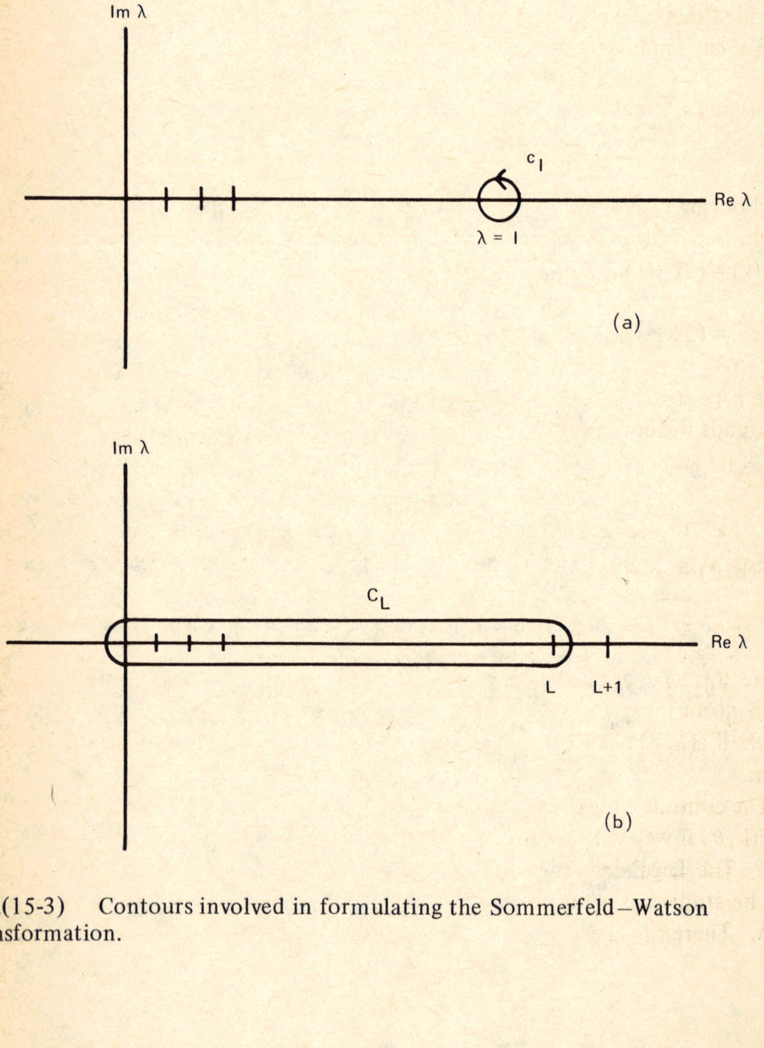

Fig.(15-3) Contours involved in formulating the Sommerfeld–Watson transformation.

$$G(\ell) = (-1)^\ell g(\ell). \tag{15.9}$$

Thus a contour integral of the form (15.7) picks out the value of $g(\lambda)$ at the integer $\lambda = \ell$. To represent a summation over ℓ, then, it is only necessary to use a sum of such contours, or equivalently a single contour enclosing all of the desired ℓ-values; if c_L is the contour shown in Fig.(15-3b),

$$\frac{1}{2i}\oint_{c_L} \frac{g(\lambda)}{\sin \pi\lambda}\, d\lambda = \sum_{\ell=0}^{L} (-1)^\ell g(\ell) \tag{15.10}$$

provided that $g(\lambda)$ is analytic within and on c_L.

We may apply this procedure to the partial wave series

$$F(k,\theta) = \sum_\ell (2\ell + 1)a_\ell(k)P_\ell(\cos\theta)$$

by choosing

$$g(\ell) = (2\ell + 1)a_\ell(k)P_\ell(\cos\theta)(-1)^\ell$$

$$= (2\ell + 1)a_\ell(k)P_\ell(-\cos\theta)$$

since $P_\ell(-\cos\theta) = (-1)^\ell P_\ell(\cos\theta)$. The summation is over all ℓ, so the contour surrounds the entire positive real λ-axis and closes at infinity, as shown in Fig. (15-4a), yielding

$$F(k,\theta) = \frac{1}{2i}\oint_c \frac{(2\lambda + 1)a(\lambda, k)P_\lambda(-\cos\theta)}{\sin \pi\lambda}\, d\lambda, \tag{15.11}$$

where $a(\lambda, k)$ denotes the continuation of $a_\ell(k)$, with $a(\ell, k) = a_\ell(k)$. Note that the contour must *not* enclose any singularities except the poles resulting from $\sin \pi\lambda$; if $a(\lambda, k)$ has Regge poles lying on the real axis, c must detour around them.

The contour c may now be deformed analytically without changing the value of $F(k,\theta)$ if we avoid all singularities of the integrand. Where are these singularities? The Legendre function $P_\lambda(-\cos\theta)$, like the Hankel functions in (15.1), can be studied as a function of λ, and it turns out to be an entire function of λ. Therefore the only singularities of the integrand (other than the poles

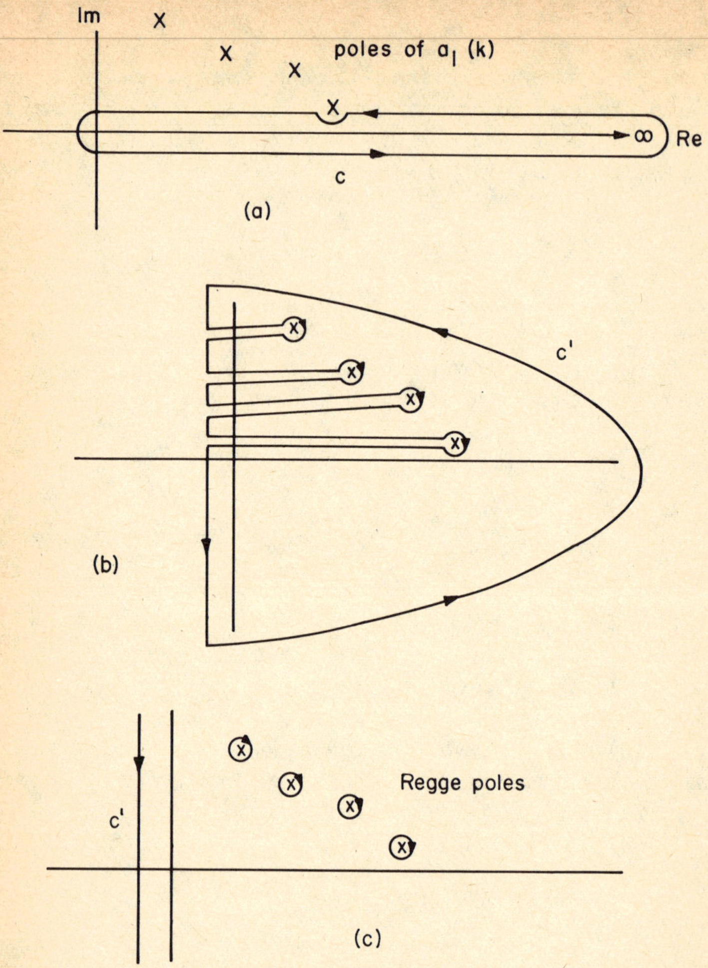

Fig.(15-4) Deformation of the contour involved in making the Sommerfeld–Watson transformation on the partial wave series.

caused by sin $\pi\lambda$) are those of a (λ, k). We have seen that there may be Regge poles of $a(\lambda, k)$ located at $\lambda = \alpha_i(k)$, and we shall neglect here the possibility of other singularities. Furthermore, if we use the unique continuation $a(\lambda, k)$ resulting from Carlson's Theorem, it may be shown (using the known asymptotic forms of $P_\lambda(-\cos\theta)$ and $\sin\pi\lambda$) that the integrand of (15.11) vanishes as $|\lambda| \to \infty$ with $\mathrm{Re}\lambda \geqslant -\frac{1}{2}$. Therefore let us deform the contour c as shown, leading to that in Fig.(15-4c), consisting of the line Reλ = $-\frac{1}{2}$, an infinite semicircle c_∞, and circles c_i excluding the Regge poles. (The "bridge" joining such a circle to c_∞ in Fig.(14-4b) makes no contribution, since its two halves cancel each other.) Consequently

$$F(k, \theta) = \frac{1}{2i}\left[\int_{-1/2-i\infty}^{-1/2+i\infty} + \int_{c_\infty} + \sum_i \oint_{c_i}\right]\frac{(2\lambda + 1)a(\lambda, k)P_\lambda(-\cos\theta)}{\sin \pi\lambda}d\lambda. \quad (15.12)$$

The integral over c_∞ vanishes, however, and the c_i simply pick up the contributions of the Regge pole terms. If these are of the form

$$a_i(\lambda, k) = \frac{1}{\pi}\frac{r_i(k)}{\lambda - \alpha_i(k)}, \quad (15.13)$$

then the clockwise integral yields

$$\frac{1}{2i}\oint_{c_i} d\lambda \frac{(2\lambda + 1)a_i(\lambda, k)P_\lambda(-\cos\theta)}{\sin \pi\lambda}$$

$$= \frac{(2\alpha_i(k) + 1)r_i(k)P_{\alpha_i(k)}(-\cos\theta)}{\sin \pi\alpha_i(k)}. \quad (15.14)$$

It follows that

$$F(k, \theta) = \sum_i \frac{(2\alpha_i(k) + 1)r_i(k)P_{\alpha_i(k)}(-\cos\theta)}{\sin \pi\alpha_i(k)} + B(k, \theta), \quad (15.15)$$

where

$$B(k, \theta) = \int_{-1/2-i\infty}^{-1/2+i\infty} d\lambda \frac{(2\lambda + 1)a(\lambda, k)P_\lambda(-\cos\theta)}{\sin \pi\lambda} \quad (15.16)$$

is known as the "background integral".

The Sommerfeld–Watson transformation thus enables us to separate the scattering amplitude into two parts – a sum over all Regge poles for which $\mathrm{Re}\,\alpha_i(k) > -\frac{1}{2}$, plus the background integral along $\mathrm{Re}\lambda = -\frac{1}{2}$. The bound states and resonances belonging to a Regge trajectory $\alpha_i(k)$ are all explicitly exhibited in the single pole term (15.14); as $\alpha_i(k)$ approaches an integer value ℓ, we have

$$\frac{(2\alpha_i(k)+1)r_i(k)P_{\alpha_i(k)}(-\cos\theta)}{\sin\pi\alpha_i(k)} \to \frac{(2\ell+1)r_i(k)P_\ell(-\cos\theta)}{\pi(\alpha_i(k)-\ell)(-1)^\ell} \tag{15.17}$$

in agreement with (15.13). The number of contributing Regge poles depends on k, of course, and as k is varied new ones may emerge from the region $\mathrm{Re}\lambda < -\frac{1}{2}$.

It behooves us therefore to inquire more deeply into the properties of the partial-wave amplitude in the complex angular momentum plane – to learn, among other things, how Regge trajectories vary with energy, and whether *all* bound states must be Regge poles. These properties were derived in detail by Regge using potential theory, and we shall summarize briefly his results. They may also be calculated on the basis of dispersion relations, as we shall see in the following chapter.

Properties of $a(\lambda, k)$ in Potential Theory

In potential theory, the properties of the partial wave amplitude extended into the complex λ-plane can be investigated by the same methods used for the complex k-plane. We define a generalized Jost function $F(\lambda, k, r)$, which is the solution of

$$\frac{d^2}{dr^2}F(\lambda,k,r) + \left(k^2 - U(r) - \frac{\lambda(\lambda+1)}{r^2}\right)F(\lambda,k,r) = 0 \tag{15.18}$$

with the boundary condition

$$F(\lambda,k,r) \approx e^{-i(kr-\lambda\pi/2)}.$$

For integral values of $\lambda = \ell$, we have

$$F(\ell,k,r) = F_\ell(k,r), \tag{15.19}$$

so that the partial wave amplitude is given by

$$a(\lambda, k) = \frac{1}{2ik}\left[\frac{F(\lambda, k, 0)}{F(\lambda, -k, 0)} - 1\right]. \tag{15.20}$$

The procedure here is somewhat more complicated than the use of Poincare's theorem in Chapter 10, but leads by similar steps to the conclusion that $F(\lambda, k, r)$ is analytic as a function of λ for $\mathrm{Re}\lambda > -\frac{1}{2}$. The behavior of $F(\lambda, k, r)$ as $|\lambda| \to \infty$, however, is much more difficult to establish; the details are given by Newton.

The result is that for sufficiently well-behaved potentials (in particular, for superpositions of Yukawa potentials such as (10.39)), the partial wave amplitude $a(\lambda, k)$ defined by (15.20) is an analytic function of λ for $\mathrm{Re}\lambda > -\frac{1}{2}$, *except* for poles occurring when $F(\lambda, -k, 0) = 0$, and that its behavior as $|\lambda| \to \infty$ in that region is satisfactory for the Sommerfeld–Watson transformation (15.14).

The locations of these poles, and the Regge trajectories $\alpha_i(E)$ which they generate as the energy is varied, can be studied by means of the Schrödinger equation. We shall merely summarize the results here, without going into the detailed arguments. To be specific, let us assume that the potential can be expressed as a superposition of Yukawas. Then for *negative* energy, the poles occurring for $\mathrm{Re}\lambda > -\frac{1}{2}$ can be shown to be simple poles lying on the real λ axis. As the energy increases through zero to positive values, the Regge trajectories will move off the real axis into the complex λ-plane, but *only* into the region $\mathrm{Im}\lambda > 0$ if $\mathrm{Re}\lambda > -\frac{1}{2}$. Furthermore, as the energy increases, the real part of $\alpha_i(E)$ must eventually turn back toward the line $\mathrm{Re}\lambda = -\frac{1}{2}$ and cross it into the left half-plane. Thus the general behavior of a Regge trajectory for a potential of

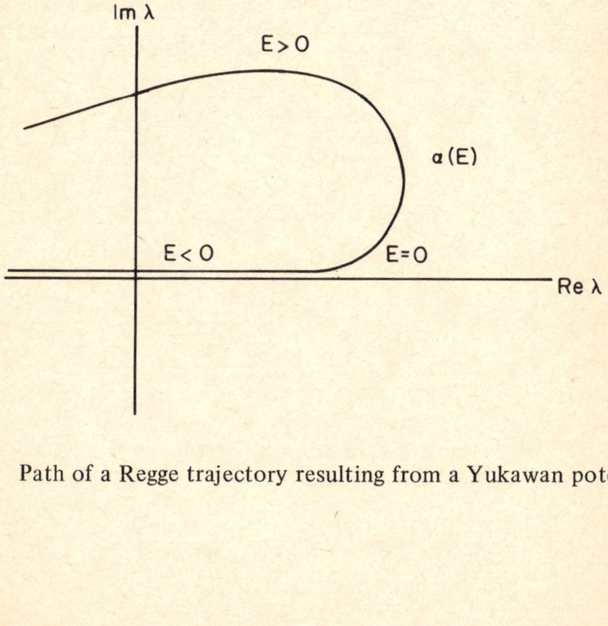

Fig.(15-5) Path of a Regge trajectory resulting from a Yukawan potential.

this type is that pictured in Fig.(15-5). The Regge trajectories of a single Yukawa potential have been explicitly calculated, and some of them are shown in Fig. (15-6). It can be proved, furthermore, that for any fixed value of the energy the number of Regge poles for such a potential is finite.

The simplicity of these results is dependent on the fact that Yukawan potentials are analytic functions of r and vanish rapidly for large r. Potentials of strictly finite range, i.e. those (such as the square well) which vanish identically for r greater than some fixed value, are drastically non-analytic, and this non-analyticity is reflected in more complicated properties in the complex λ-plane. In a similar way, poor asymptotic behavior such as that of the Coulomb potential will destroy the simplicity of the trajectories.

In fact, the Regge trajectories of the Coulomb interaction can be obtained explicitly from the known scattering amplitudes of this potential. For $V(r) = -V_0/r$, the partial-wave amplitude $a(\lambda, k)$ is

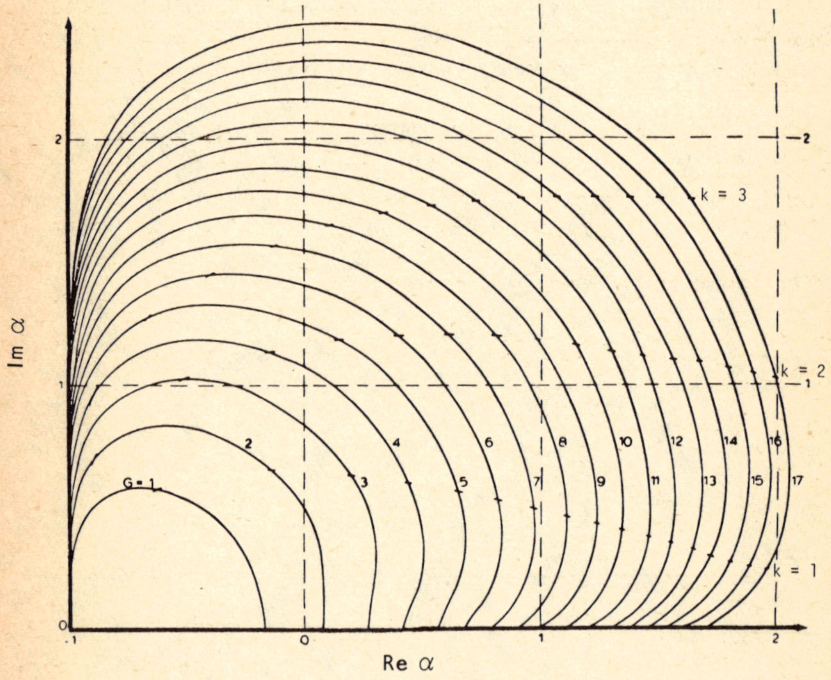

Fig.(15-6) Regge trajectories of the Yukawa potential $V(r) = -Ge^{-r}/r$, for different values of G. (From C. Lovelace and D. Masson, *Nuovo Cimento* **26**, 472 (1962).)

$$a(\lambda, k) = \frac{1}{2ik}\left[\frac{\Gamma(\lambda - iV_0/2k)}{\Gamma(\lambda + iV_0/2k)} - 1\right]. \tag{15.21}$$

The gamma function $\Gamma(z)$ is an analytic function of z having poles at the negative integers and zero, but no zeros. The ratio of two gamma functions appearing in (15.21) will therefore produce poles of $a(\lambda, k)$ whenever

$$\lambda - iV_0/2k = -N, \quad N = 1, 2, 3, \ldots$$

unless $\lambda + iV_0/2k = 2\lambda + N$ is also a negative integer; in that case the two poles cancel and $a(\lambda, k)$ is regular.

Let us consider $V_0 > 0$, an attractive Coulomb potential. For negative energy,

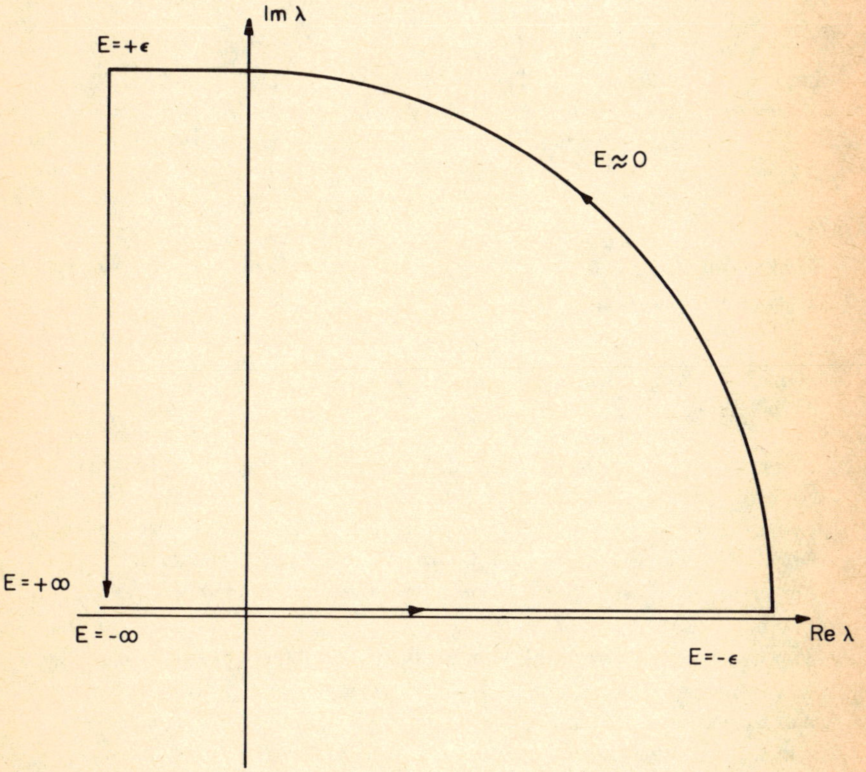

Fig.(15-7) Regge trajectory of an attractive Coulomb potential.

$k = i\kappa$ is purely positive imaginary, and the trajectory is given by

$$\alpha_N(E) = \frac{V_0}{2\kappa} - N. \tag{15.22a}$$

For $E = -\infty$, the trajectory begins at $-N$; as E increases toward zero, $\alpha_N(E)$ moves to the right, as shown in Fig.(15-7), going to $+\infty$ as $\kappa \to 0$. Thus $\alpha_N(E)$ passes through *every* positive integer, unlike the trajectory of a Yukawa potential, and produces a bound state for *every* ℓ. This infinitude of bound states results from the infinite range of V(r). As E becomes positive,

$$\alpha_N(E) = \frac{V_0}{2ik} - N \tag{15.22b}$$

moves around the λ-plane at infinity to the line $\mathrm{Re}\,\alpha_N = -N$ and heads back toward its starting point. Thus there are no Coulomb resonances.

In potential theory all bound states and resonances must be Regge poles; there cannot be any poles in the partial wave amplitudes which are independent of λ. If there were, they would have to be present for large real values of λ. But by taking λ large enough we can make the effect of the potential negligible in any region of r, and therefore the λ-independent singularities must be present in the limit of no potential, i.e. of free particles. The scattering matrix becomes unity in that limit, so it follows that there are no poles which are independent of λ. Thus in potential scattering all poles are Regge poles.

Exchange Forces and Signature

Since relativistic scattering, propagated via particle exchange, is represented to a first approximation by the Yukawa potential, the appearance of Regge poles in potential theory suggests that they may also occur there. This possibility will be investigated in the next chapter on the basis of dispersion relations. Here we may attempt, within the context of potential theory, to remedy its most significant defect – the absence of "exchange forces", or, equivalently, the failure of a single potential to describe both t and u channel exchanges in s channel scattering.

To see this problem let us consider the process $\pi^+ p \to \pi^+ p$. It may take place via ρ exchange, as shown in Fig.(15-8a), which is equivalent to a single-particle state in the t-channel reaction $\pi^+\pi^- \to \bar{p}p$. It may also occur via neutron exchange, as in Fig.(15-8b), which is a single-particle state in the u-channel $\pi^- p \to \pi^- p$. Thus there are two possible potentials, V_ρ and V_n, leading to two

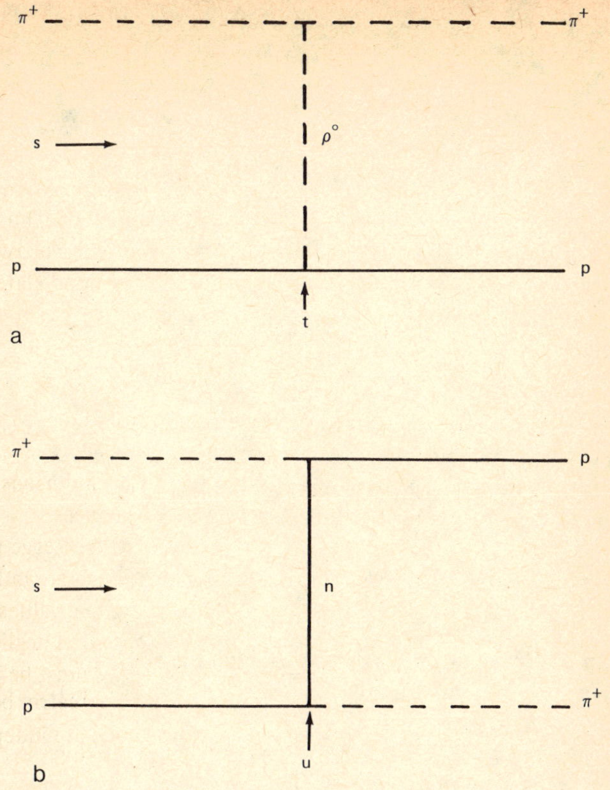

Fig.(15-8) Possible graphs for $\pi^+p \rightarrow \pi^+p$ involving (a) ρ-meson exchange and (b) neutron exchange.

scattering amplitudes. If θ is the angle through which the π^+ is scattered, then the total amplitude is

$$f(k, \theta) = f_\rho(k, \theta) + f_n(k, \pi - \theta), \tag{15.23}$$

with $f_\rho(k, \theta)$ describing the ρ-exchange process and $f_n(k, \pi - \theta)$ the "exchange process", i.e. an incoming π^+ is changed to a proton and scattered through an angle $\pi - \theta$, so that the outgoing π^+ is at an angle θ.

Since both f_ρ and f_n result from Yukawa potentials, they presumably have the corresponding Regge properties. If their partial wave expansions are

$$f_\rho(k,\theta) = \sum_\ell (2\ell+1)a_\ell^\rho(k)P_\ell(\cos\theta) \tag{15.24a}$$

$$f_n(k,\theta) = \sum_\ell (2\ell+1)a_\ell^n(k)P_\ell(\cos\theta), \tag{15.24b}$$

then since $P_\ell(\cos(\pi-\theta)) = P_\ell(-\cos\theta) = (-1)^\ell P_\ell(\cos\theta)$ we will have

$$f(k,\theta) = \sum_\ell (2\ell+1)[a_\ell^\rho(k) + (-1)^\ell a_\ell^n(k)]\,P_\ell(\cos\theta). \tag{15.25}$$

While the Sommerfeld–Watson transformation can be carried out straightforwardly on $f_\rho(k,\theta)$ or $f_n(k,\theta)$, the reversed sign of $\cos\theta$ in $f(k,\theta)$ leads, through the factor $(-1)^\ell$, to difficulty. The reason is simply that writing $(-1)^\ell = e^{-i\pi\ell}$ and extending the result to the λ-plane yields

$$a(\lambda,k) = a^\rho(\lambda,k) + e^{-i\pi\lambda}a^n(\lambda,k) \tag{15.26}$$

and the factor $e^{-i\pi\lambda}$ will diverge exponentially on the contour c_∞ in (15.12). Therefore Carlson's theorem cannot be satisfied and the Sommerfeld–Watson transformation cannot be carried out.

The innocuous factor $(-1)^\ell$ in (15.25) thus becomes quite noxious when extended to complex angular momenta, and extra work is required to remedy its problems. The solution turns out to be rather simple, however. Since the difficulty arises from the troublesome $(-1)^\ell$, let us do away with it entirely by defining the *signatured* amplitudes $f^\pm(k,\cos\theta)$,

$$f^\pm(k,\cos\theta) = \sum_\ell (2\ell+1)a_\ell^\pm(k)P_\ell(\cos\theta) \tag{15.27}$$

$$a_\ell^\pm(k) = a_\ell^\rho(k) \pm a_\ell^n(k). \tag{15.28}$$

The full physical amplitude $f(k,\theta)$ can be expressed in terms of $f^\pm(k,\cos\theta)$ as

$$f(k,\theta) = \tfrac{1}{2}(f^+(k,\cos\theta) + f^+(k,-\cos\theta))$$
$$+\ \tfrac{1}{2}(f^-(k,\cos\theta) - f^-(k,-\cos\theta)). \tag{15.29}$$

To verify this, we note that

$$\tfrac{1}{2}(f^+(k, \cos\theta) + f^+(k, -\cos\theta)) = \tfrac{1}{2}\sum_{\ell}(2\ell + 1)a_\ell^+(k)[P_\ell(\cos\theta) + P_\ell(-\cos\theta)]$$

$$= \sum_{\ell \text{ even}} (2\ell + 1)[a_\ell^\rho(k) + a_\ell^n(k)]\, P_\ell(\cos\theta), \tag{15.30a}$$

since $P_\ell(-\cos\theta) = (-1)^\ell P_\ell(\cos\theta)$, and likewise that

$$\tfrac{1}{2}(f^-(k, \cos\theta) - f^-(k, -\cos\theta)) = \sum_{\ell \text{ odd}} (2\ell + 1)[a_\ell^\rho(k) - a_\ell^n(k)]\, P_\ell(\cos\theta), \tag{15.30b}$$

reproducing thereby (15.25).

The point of this separation is that, while we cannot carry out the Sommerfeld–Watson transformation on $f(k, \theta)$, we *can* carry it out on $f^\pm(k, \cos\theta)$, since $a_\ell^\pm(k)$ both are straightforwardly continued to complex ℓ, by means of (15.28), and satisfy the necessary conditions as $|\lambda| \to \infty$. Having done so, we may *then* recombine the signatured amplitudes according to (15.29) to obtain a Regge representation of $f(k, \theta)$. Suppose, analogously to (15.15), that

$$f^\pm(k, \cos\theta) = \sum_i \frac{(2a_i^\pm(k) + 1)r_i^\pm(k)}{\sin \pi a_i^\pm(k)} P_{a_i^\pm(k)}(-\cos\theta) + B^\pm(k, \cos\theta), \tag{15.31}$$

where $a_i^\pm(k)$ are the Regge trajectories of $f^\pm(k, \cos\theta)$, $r_i^\pm(k)$ are their residues, defined as in (15.13), and $B^\pm(k, \cos\theta)$ are the background integrals. Then, since $P_a(-\cos\theta) = e^{-i\pi a}P_a(\cos\theta)$ even for non-integral a, we have

$$f(k, \theta) = \sum_i \frac{(2a_i^+ + 1)r_i(k)}{\sin \pi a_i^+}(1 + e^{-i\pi a_i^+})P_{a_i^+}(-\cos\theta)$$

$$+ \sum_i \frac{(2a_i^- + 1)r_i^-(k)}{\sin \pi a_i^-}(1 - e^{-i\pi a_i})P_{a_i^-}(-\cos\theta)$$

$$+ B(k, \theta). \tag{15.32}$$

Thus in the presence of u-channel exchange effects, the grouping of bound states and resonances onto Regge trajectories is altered by the addition of the factor $(1 + \tau e^{-i\pi\alpha_i})$, where $\tau = \pm 1$ is called the *signature* of the trajectory. When the trajectory passes through an integer value ℓ, the signature factor becomes

$$1 + \tau e^{-i\pi\ell} = 1 + \tau(-1)^{\ell}$$

which vanishes when ℓ is odd for $\tau = +1$, and when ℓ is even for $\tau = -1$. This vanishing in each case cancels the vanishing denominator $\sin \pi\alpha_i$, so that the even signature ($\tau = +1$) Regge poles produce only even-ℓ bound states and resonances in $f(k, \theta)$, and likewise only odd-ℓ states result for $\tau = -1$. Consequently the bound states and resonances lying along a given Regge trajectory will differ by *two* units of angular momentum, rather than by a single unit as would follow in potential theory from (15.15).

Factorization

In a potential theory, the Regge poles characterize the potential itself; but if we think of relativistic unitarity, some "universal" aspects of Regge poles become apparent. Let us imagine, for reasons to become clear in the next chapter, that the amplitude we are considering describes a t-channel scattering process, and write the trajectory and residue as functions of t rather than k. The particles on a trajectory contribute, as intermediate states in the unitarity equation, to all amplitudes having the same internal quantum numbers. It follows that the same Regge trajectory must appear in all such amplitudes.

For example, the ρ resonance can, as we shall shortly see, be considered a manifestation of a Regge trajectory with $I = G = 1$, $B = S = 0$, and $\tau = -1$. It appears experimentally as a resonance in the $\pi\pi$ elastic scattering amplitude. But it also will appear in the $I = 1$ nucleon–antinucleon amplitude, since that reaction also has the correct quantum numbers for the ρ. Similarly, it will appear in the $\overline{N}N \to \pi\pi$ amplitude, the $\overline{K}K$ amplitude, etc. In short the ρ pole will be present in every amplitude for which the initial and final states have the appropriate internal quantum numbers. The same will be true of any other particle lying on the same trajectory as the ρ, since it will have the same quantum numbers (except for angular momentum). It follows that *every* scattering amplitude with $I = G = 1$, $B = S = 0$ will contain the *same* Regge pole α_ρ.

It is helpful to visualize this appearance by thinking of the Regge trajectory as a single "particle", rather than several, and represent it diagrammatically by a wavy line as in Fig.(15-9a). In this way the single diagram shown represents the contributions of all particles along the trajectory α_ρ to the $\pi\pi$ scattering amplitude. When $t \approx m_\rho^2$, the amplitude will be dominated by the ρ contribution, and the residue $r_\rho^{\pi\pi}(t)$ of the ρ pole will be simply related to the square of the $\rho\pi\pi$ coupling constant $g^2_{\pi\pi\rho}$. Analogous results will hold near any other particle on the ρ trajectory, with $r_\rho^{\pi\pi}(t)$ again approximating the square of the appropriate coupling constant. In the same way, the residue $r_\rho^{\bar{N}N}(t)$ of the contribution of the ρ trajectory to the $\bar{N}N$ elastic scattering amplitude will interpolate between the squares of the coupling constants of the particles on the ρ trajectory to $\bar{N}N$, with $r_\rho^{\bar{N}N}(m_\rho^2)$ giving $g^2_{\bar{N}N\rho}$, etc. Thus the Regge residue can be thought of as the continuation of the square of the coupling constant.

Consider, however, the contribution of this trajectory to the inelastic $\bar{N}N \to \pi\pi$ amplitudes shown in Fig.(15-9c). Here the residue function $r_\rho^{\bar{N}N\to\pi\pi}(t)$ will, near $t = m_\rho^2$, approximate the *product* of the two coupling constants $g_{\pi\pi\rho}$ and $g_{\bar{N}N\rho}$. This means that near $t = m_\rho^2$, we must have

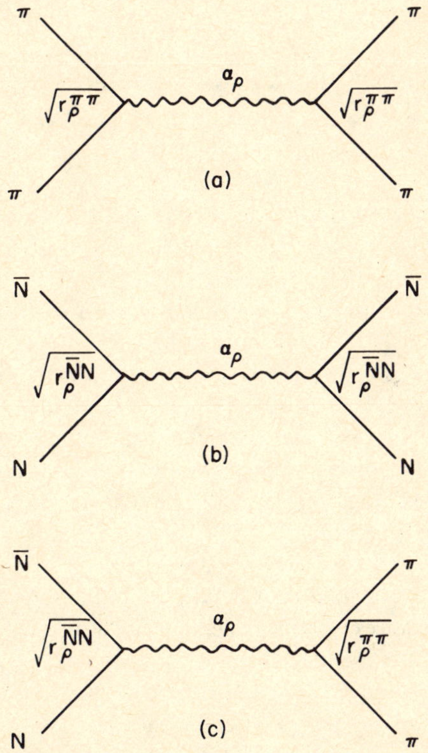

Fig.(15-9) Appearance of the trajectory α and residues in the reactions $\pi\pi \to \pi\pi$, $\bar{N}N \to \bar{N}N$, and $\bar{N}N \to \pi\pi$. The trajectories are the same, and the residues are related by factorization.

$$r_\rho^{\bar{N}N\to\pi\pi}(t) = [r_\rho^{\bar{N}N}(t)]^{1/2}[r_\rho^{\pi\pi}(t)]^{1/2}.$$

The same must, of course, be true for any other particle on the ρ trajectory, and it follows that this equation holds for all t, i.e. that the residue function can be *factorized.* For a general process AB → CD, therefore, we may write

$$r_\alpha^{AB\to CD}(t) = g_\alpha^{AB}(t) g_\alpha^{CD}(t) \tag{15.33}$$

with $g_\alpha^{AB}(t)$ and $g_\alpha^{CD}(t)$ appropriately interpolating the coupling constants of the particles on the trajectory α to AB and to CD.

Comparison with Experiment

The Regge theory therefore predicts that particles should appear in families, rather than individually, with the members of a family differing by two units of angular momentum, and the coupling constants interpolated by the residue functions. In a realistic theory, we must also take account of other quantum numbers – isospin, parity, spin, etc. – by separating out various reduced amplitudes as in Part I. If a Regge trajectory appears in one of these amplitudes, we should see particles with the same internal quantum numbers, differing only by two units in their angular momenta.

For example, let us consider $\pi\pi$ scattering. There are three isospin amplitudes, corresponding to I = 0, 1, 2. We know that the ρ appears as a resonance in the I = 1 channel with $J^P = 1^-$. If it is a Regge pole, then there should also occur other $\pi\pi$ resonances with I = 1 and $J^P = 3^-, 5^-$, etc. In fact, the g(1680) meson, with $J^P = 3^-$, seems to have exactly the right quantum numbers. Therefore we may conjecture that it is the first Regge recurrence of the ρ, i.e. that both resonances are manifestations of a single Regge trajectory $\alpha_\rho(t)$ for which $\mathrm{Re}\alpha_\rho(m_\rho^2) = 1$ and $\mathrm{Re}\alpha(m_g^2) = 3$.

The next resonance on this trajectory should have $J^P = 5^-$ and $\mathrm{Re}\alpha_\rho(M^2) = 5$, M being its mass. Unfortunately, our knowledge of the spin–parity assignments of higher meson resonances is far from complete, and it is not clear whether such a resonance exists. The functional form of $\alpha_\rho(t)$ is completely unknown, of course; if it is similar to the Yukawa trajectories of Fig.(15-5), it should eventually turn back toward the origin, possibly before reaching $\alpha_\rho = 5$. Since two points determine a straight line, however, it is appealing (although admittedly naive) to guess that

$$\alpha_\rho(s) \approx A + Bt. \tag{15.34a}$$

Solving

$$A + Bm_\rho^2 = 1$$

$$A + Bm_g^2 = 3$$

yields A = 0.47, B = 0.89 GeV^{-2}, or approximately

$$\alpha_\rho(s) \approx 0.5 + t, \tag{15.34b}$$

with t measured in GeV^2. This trajectory is shown in Fig.(15-10) as a plot of Reα versus t, usually called a Chew–Frautschi diagram. Extending it to $\alpha_\rho = 5$ predicts a $J^P = 5^-$ non-strange isovector meson with mass about 2120 MeV. Resonant structure has been observed (called the T meson) with a mass around 2200 MeV, but its quantum numbers are not yet known.

It therefore seems possible that an odd-signature Regge trajectory is present in the I = 1 $\pi\pi$ scattering amplitude. For this isospin channel, the $\tau = +1$ amplitude must vanish identically. The reason is the same as that leading to (6.45), that a

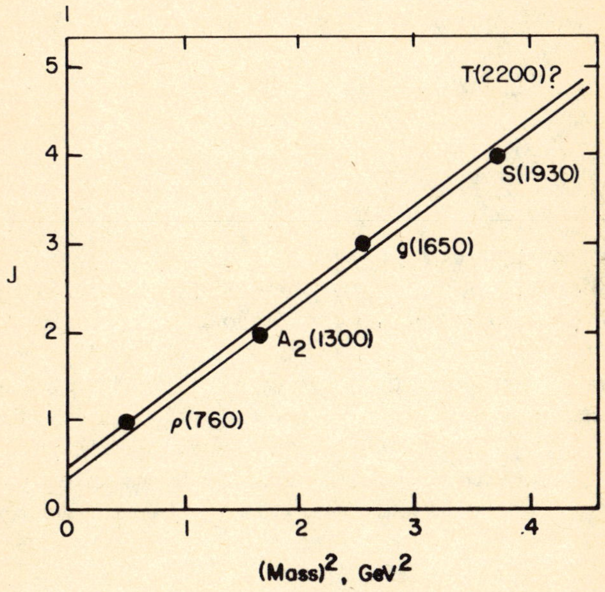

Fig.(15-10) Straight-line approximations to the ρ and A_2 trajectories, the similarity of which is evidence for exchange degeneracy.

two-pion state with I = 1 has an antisymmetric isospin wave function, and must therefore also have an antisymmetric spatial wave function. Consequently the scattering amplitude must be an odd function of $\cos\theta$.

A second example of mesonic Regge trajectories is provided by the $A_2(1310)$, an isospin 1 resonance in the $\pi\rho$ system with $J^P = 2^+$. If we assume here also a linear Regge trajectory with slope ~1 GeV^{-2}, we have

$$\alpha_{A_2}(t) \approx 0.3 + t. \tag{15.34c}$$

A spin-4 meson is predicted at a mass of about 1920 MeV, which is consistent with that of the S(1930). Here again, however, the theory cannot be thoroughly tested because of incomplete experimental knowledge of the higher meson resonances. The trajectory passes through zero for $t \approx -0.3$ GeV^2, corresponding to an imaginary mass; therefore no spin-zero particle is expected. The Regge-pole amplitude will nonetheless have a pole there, however, because of the vanishing of $\sin(\pi\alpha_{A_2}(t))$ in the denominator. Consequently it predicts the existence of a "ghost", as particles with imaginary mass are called in this context. Since ghosts are physically objectionable, the residue must vanish whenever they occur in order to cancel the pole. In this case, we might take

$$r_{A_2}(t) = \alpha_{A_2}(t) r'_{A_2}(t),$$

introducing a "ghost-killing" factor $\alpha_{A_2}(t)$, so that

$$\frac{r_{A_2}(t)}{\sin \pi\alpha_{A_2}(t)} \to \frac{r'_{A_2}(t)}{\pi}$$

yielding a finite amplitude as $\alpha_{A_2}(t) \to 0$.

It is interesting to note that the ρ trajectory and the A_2 trajectory seem almost to coincide. In the $\pi\pi$ amplitude only the ρ is allowed, but in nucleon–antinucleon scattering, both would appear in the I = 1 amplitude, differing only in their signatures. Thus it appears the resonances on these two trajectories could, in fact, be placed on a single straight-line trajectory of the type (15.34a), provided we are willing to neglect signature effects. In other words, neglecting the u-channel effects, or the exchange potential, leads to a degeneracy of opposite-signature trajectories. This effect is usually called "exchange degeneracy".

Nucleon Resonances

The spectrum of nucleon resonances may also be compared with the predictions of the Regge-pole theory if we make the appropriate generalization to include the nucleon spin. If we consider them as almost-bound states of the pion–nucleon system, then the angular momentum j of the resonance will be the combination of the orbital angular momentum ℓ of the πN system and the spin ½ of the nucleon. For given ℓ, we may have either $j = \ell + \frac{1}{2}$ or $j = \ell - \frac{1}{2}$, and correspondingly there are two partial-wave amplitudes $f_{\ell\pm}(s)$ for each value of ℓ. These two amplitudes are, of course, related to the partial wave projections of the amplitudes describing scattering with and without flipping the nucleon's spin; details of the partial-wave expansion for scattering of spin-zero and spin ½ particles are given in Appendix B. Resonances are identified with poles of the partial wave amplitudes $f_{\ell\pm}(s)$. The parity of such a resonance will be $-(-1)^\ell$, the $(-1)^\ell$ resulting from the spatial wave function in the usual way and the extra minus coming from the negative intrinsic parity of the pion.

The Sommerfeld–Watson transformation can be carried out using $f_{\ell\pm}(s)$ just as in the absence of spin, and the resulting Regge trajectories will, because of signature effects, produce resonances differing by two units of angular momentum. Trajectories with positive signature will have even ℓ and hence negative parity,

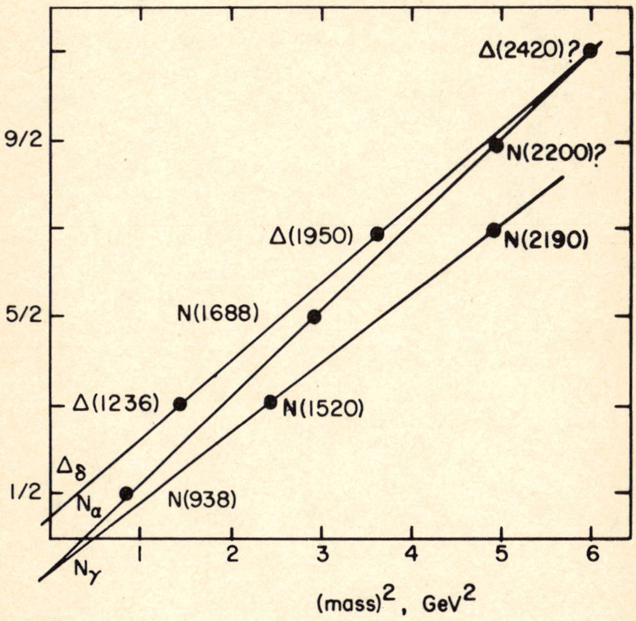

Fig.(15-11) Straight-line approximations to the N_α, N_γ, and Δ_δ trajectories.

while those with negative signature have odd ℓ and therefore positive parity. In each case the angular momentum of the resonance may be either $j = \ell \pm \frac{1}{2}$. Thus the nucleon resonances may be grouped according to their spin-parity assignments into four basic families

$$\begin{aligned}
&\alpha: \quad \tau = -, \quad && j = \ell - \tfrac{1}{2}, \quad && j^P = 1/2^+, 5/2^+, 9/2^+, \ldots \\
&\beta: \quad \tau = +, \quad && j = \ell + \tfrac{1}{2}, \quad && j^P = 1/2^-, 5/2^-, 9/2^-, \ldots \\
&\gamma: \quad \tau = +, \quad && j = \ell - \tfrac{1}{2}, \quad && j^P = 3/2^-, 7/2^-, 11/2^-, \ldots \\
&\delta: \quad \tau = -, \quad && j = \ell + \tfrac{1}{2}, \quad && j^P = 3/2^+, 7/2^+, 11/2^+, \ldots
\end{aligned} \tag{15.35}$$

The nucleon itself, for example, may be classified as belonging to the N_α trajectory. The next particle on this trajectory should have $J^P = 5/2^+$, and the N(1688) has the correct values. Furthermore, if a linear trajectory is assumed, as in (15.34), the slope of this trajectory turns out again to be about 1 GeV^{-2}, in agreement with that obtained for the mesons. A similar success is encountered for the N_γ trajectory, on which N(1520) and N(2190) can be satisfactorily classified. The $I = 3/2$ Δ_δ trajectory also seems to be realized in the decuplet Δ itself, the $\Delta(1910)$, and the $\Delta(2420)$. These three trajectories are shown in Fig.(15-11).

There are, of course, many more nucleon resonances, but until their spin-parity assignments are adequately measured it is conjectural to place them on any particular trajectory.

Chapter 16

RELATIVISTIC REGGE POLES

The presence of Regge poles in potential theory, plus the apparent existence of Regge recurrences of the ρ and A_2 mesons as well as of the nucleon and the Δ, suggest strongly that they may indeed be present in a correct relativistic scattering theory. We therefore embark in this chapter on a discussion of relativistic Regge theory, beginning with a description of how the Sommerfeld–Watson transformation may be applied to the partial-wave series obtained from the Mandelstam representation. The problem of u-channel effects necessarily arises again there, to be resolved by the concept of signature exactly as in potential theory.

The most significant aspect of relativistic Regge poles, however, and the one which has provided significant new ideas about high energy scattering, is the form of the Regge amplitude in the crossed channel. In addition to resolving a serious problem regarding its asymptotic behavior, cross-channel Regge poles lead to an elegantly simple parametrization of the scattering amplitude. Whether the consequences of this parametrization are consistent with the behavior observed experimentally is the subject of continuing research. The Regge model has become, for many physicists, a way of thinking. It is, fortunately, a very flexible model, which can be adapted to almost any turn of the data by a minor modification of some equally minor assumption. Whether it is literally true is therefore very difficult, if not impossible, to test. But even if Regge theory does not prove to be the promised land, it has certainly been a road leading in a useful direction, and we shall spend the remainder of this book exploring the countryside through which it travels.

Regge Poles from Dispersion Relations

The partial wave amplitudes resulting from the Mandelstam representation can be obtained with relative ease, as we have shown in Chapter 14. The properties of the resulting Regge trajectories, however, are less clear than in potential theory, because the double spectral functions remain unknown.

To find Regge poles we must merely make the appropriate assumptions about these unknown functions. This extra latitude is responsible for the adaptability of Regge theory; it prevents us from saying that the trajectories must have Yukawan properties, or that there are no branch points, but only poles, in the λ-plane, as in potential theory.

For future convenience, we shall consider here the t channel reaction, $t \geqslant 4m^2$, so that s and u are momentum transfer variables and are negative. Let us begin from a single dispersion relation at fixed t,

$$F(s, t, u) = F^s(s, t, u) + F^u(s, t, u)$$

$$F^s(s, t, u) = \frac{1}{\pi}\int_{4m^2}^{\infty} ds' \frac{\Delta_s(s', t, 4m^2 - s' - t)}{s' - s}$$

$$F^u(s, t, u) = \frac{1}{\pi}\int_{4m^2}^{\infty} du' \frac{\Delta_u(4m^2 - t - u', t, u')}{u' - u}. \tag{16.1}$$

(On the basis of the Mandelstam representation, we would have

$$\Delta_s(s', t, u') = \frac{1}{\pi}\int dt' \frac{[\rho_{st}(s', t', u') + \rho_{su}(s', t', u')]}{t' - t}$$

$$\Delta_u(s', t, u') = \frac{1}{\pi}\int dt' \frac{[\rho_{ut}(s', t', u') + \rho_{su}(s', t', u')]}{t' - t},$$

but these representations are not crucial to our argument.) We have seen in (14.4) how to project out the resulting partial wave amplitudes in the t channel, obtaining

$$F(s, t, u) = \sum_{\ell} (2\ell + 1) a_\ell(t) P_\ell(\cos\theta_t)$$

with

$$a_\ell(t) = a_\ell^s(t) + a_\ell^u(t) \tag{16.2a}$$

$$a_\ell^s(t) = \frac{1}{t - 4m^2} \int_{4m^2}^{\infty} ds'\Delta_s(s', t, u')Q_\ell\left(1 + \frac{2s'}{t - 4m^2}\right) \tag{16.2b}$$

$$a_\ell^u(t) = \frac{1}{t - 4m^2} \int_{4m^2}^{\infty} du'\Delta_u(s', t, u')Q_\ell\left(-1 - \frac{2u'}{t - 4m^2}\right). \tag{16.2c}$$

For the moment let us consider only $a_\ell^s(t)$. The definition (16.2b) is immediately extended to complex λ by writing

$$a^s(\lambda, t) = \frac{1}{t - 4m^2} \int_{4m^2}^{\infty} ds'\Delta_s(s', t, u')Q_\lambda\left(1 + \frac{2s'}{t - 4m^2}\right) \tag{16.3}$$

The properties of $a^s(\lambda, t)$ as a function of λ can then be deduced from the known behavior of $Q_\lambda(z)$ and that assumed for Δ_s. If, for example, we assume that $\Delta_s(s', t, u') \sim (s')^{\alpha(t)}$ for $s' > s_1$, then since

$$Q_\lambda(z) \sim z^{-\lambda-1}, \quad z \to +\infty \tag{16.4}$$

the integral has a contribution at its upper limit of the form

$$\int_{s_1}^{\infty} ds'(s')^{\alpha(t)-\lambda-1} = \frac{(s_1)^{\alpha(t)-\lambda}}{\lambda - \alpha(t)} \tag{16.5}$$

provided $\mathrm{Re}(\lambda - \alpha(t)) > 0$. If $\mathrm{Re}\lambda - \alpha(t) \leqslant 0$, the integral will diverge, and $a^s(\lambda, t)$ cannot be defined by it. But the *result* of evaluating the integral may, as in (16.5), be well defined for all λ, and in that case it is the analytic continuation of $a^s(\lambda, t)$ outside the region $\mathrm{Re}\lambda > \mathrm{Re}\alpha(t)$. In other words, this assumed asymptotic behavior will lead to a Regge pole in $a^s(\lambda, t)$.

In writing the dispersion relation (16.1), of course, we have assumed that $\Delta_\ell^s(s', t, u')$ vanishes as $s' \to \infty$, i.e. that $\alpha(t) < 0$. If this were not so, the dispersion integral would diverge, and subtractions would have to be introduced. With N subtractions, the s-cut contribution in (16.1) takes the form

$$F^s(s, t, u) = \sum_{n=0}^{N-1} c_n(t)s^n + \frac{s^N}{\pi} \int_{4m^2}^{\infty} ds' \frac{\Delta_s^{(N)}(s', t, u')}{(s')^N(s' - s)} . \tag{16.6}$$

For integer values $\lambda = \ell$, the partial wave projections of the sum in (16.6) will vanish identically for $\ell \geqslant N$. Therefore the analytic continuation of $a^s(\lambda, t)$ for $\text{Re}\lambda \geqslant N$ will result entirely from the integral term. If $\Delta_s^{(N)} \sim (s')^{N+\alpha(t)}$ as $s' \to \infty$, it again follows that $a^s(\lambda, t)$ contains a contribution of the same form as (16.5). Having obtained this Regge pole term for $\text{Re}\lambda \geqslant N$, we may continue it to $\text{Re}\lambda < N$, where it may be expected to yield the correct partial wave projections of the integral. The coefficients $c_n(t)$, however, are completely independent of the spectral function $\Delta_s^{(N)}$; therefore the contributions of the summation in (16.6) will be neglected by this continuation, and it will not reproduce exactly the true partial wave amplitude.

This matter of subtractions is, for that reason, crucial to the question whether all poles are Regge poles. If the dispersion relation exists, even with subtractions, then there is some λ large enough that the contribution of the dispersion integral to the partial wave amplitude is convergent. For t real and negative, it has been shown on fairly general grounds that two subtractions should suffice. Consequently any singularity in the λ-plane which lies in the region $\text{Re}\,\lambda > 1$ must be t-dependent, in order to satisfy this requirement when t becomes real and negative. It follows that any pole with $\text{Re}\,\lambda > 1$ must be a Regge pole, i.e. that all particles with spin greater than one are Regge recurrences. "Fixed poles", those independent of t, must have $\text{Re}\,\lambda \leqslant 1$. One is tempted by two possible conjectures at this point: that *all* poles are Regge poles, in some bootstrap sense, or that the fixed poles are truly "elementary" particles, the Regge poles being "composites" of them. The choice is between these alternatives is as yet unmade, although the present evidence seems to favor a Regge interpretation for the pion.

Signature in Relativistic Regge Theory

We must still show that the partial wave amplitude can be defined to satisfy Carlson's theorem; otherwise the Sommerfeld–Watson transformation cannot be carried out. If the integral in (16.3) (or (16.6), if there are subtractions) is convergent, the properties of $a^s(\lambda, t)$ in the λ-plane will reflect those of the Legendre function. It is known that $Q_\lambda(z)$ is an analytic function of λ except for poles at the negative integers, which are immaterial to us here, and as $|\lambda| \to \infty$ has the asymptotic form

$$Q_\lambda(z) \sim \lambda^{-1/2} \exp\left[(\lambda + \tfrac{1}{2}) \ln\{z - (z^2 - 1)^{1/2}\}\right]. \tag{16.7}$$

For $t > 4m^2$, the argument $z = 1 + 2s'/(t - 4m^2)$ occurring in (16.3) is always greater than unity, and therefore the logarithm in (16.7) is real and negative. It follows that the integrand of (16.3) vanishes exponentially as $|\lambda| \to \infty$ if $\mathrm{Re}(\lambda + \tfrac{1}{2}) > 0$, so we have

$$|a^s(\lambda, t)| \to 0, \text{ as } |\lambda| \to \infty, \ \mathrm{Re}\lambda \geqslant -\tfrac{1}{2}. \tag{16.8}$$

Consequently $a^s(\lambda, t)$ satisfies the conditions of Carlson's theorem.

For the u-channel part of the amplitude, however, problems arise. They follow from exactly the same sources we have already discussed in the preceding chapter, the existence of u-channel as well as s-channel exchanges. The root of the problem is the negative argument of the Legendre function in $a^u_\ell(t)$ as given in (16.2c). Since $Q_\ell(-z) = (-1)^\ell Q_\ell(z)$, we may rewrite this equation as

$$a^u_\ell(t) = \frac{(-1)^\ell}{t - 4m^2} \int_{4m^2}^{\infty} du' \Delta_u(s', t, u') Q_\ell\left(1 + \frac{2u'}{t - 4m^2}\right). \tag{16.9a}$$

Defining the continuation $a^u(\lambda, t)$ analogously to (16.3) then yields

$$a^u(\lambda, t) = \frac{e^{-i\pi\lambda}}{t - 4m^2} \int_{4m^2}^{\infty} du' \Delta_u(s', t, u') Q_\lambda\left(1 + \frac{2u'}{t - 4m^2}\right), \tag{16.9b}$$

and, as in (15.26), a factor $e^{-i\pi\lambda}$ arises, which causes $a^u(\lambda, t)$ to diverge as $|\lambda| \to \infty$ with $\mathrm{Im}\lambda > 0$. Therefore $a^u(\lambda, t)$ will not satisfy the necessary conditions for Carlson's theorem and the Sommerfeld–Watson transformation.

The difficulty can be avoided by projecting out the partial waves of $F^u(s, t, u)$ with respect to $(-\cos\theta_t)$ rather than $\cos\theta_t$,

$$F^u(s, t, u) = \sum_\ell (2\ell + 1) \tilde{a}^u_\ell(t) P_\ell(-\cos\theta_t). \tag{16.10}$$

Then the relevant integral defining $\tilde{a}^u_\ell(t)$ is

$$\tilde{a}^u_\ell(t) = \int_{-1}^{1} d\cos\theta\; P_\ell(-\cos\theta_t) \int_{4m^2}^{\infty} du' \frac{\Delta_u(s', t, u')}{u' - u} \tag{16.11}$$

without the $(-1)^\ell$, and we may define

$$\tilde{a}^u(\lambda, t) = -\frac{1}{t - 4m^2} \int_{4m^2}^{\infty} du' \Delta_u(s', t, u') Q_\lambda \left(1 + \frac{2u'}{t - 4m^2}\right), \tag{16.12}$$

which is entirely satisfactory as $|\lambda| \to \infty$. Combining the two partial wave series yields

$$F(s, t, u) = \sum_\ell (2\ell + 1)[a^s_\ell(t) P_\ell(\cos\theta_t) + \tilde{a}^u_\ell(t) P_\ell(-\cos\theta_t)]$$

$$= \sum_\ell (2\ell + 1)[a^s_\ell(t) + (-1)^\ell \tilde{a}^u_\ell(t)]\, P_\ell(\cos\theta_t) \tag{16.13}$$

which is equivalent to (15.25). The divergence problem is isolated in the $(-1)^\ell$, which is avoided by defining the signatured partial wave amplitudes

$$a^\pm(\lambda, t) = a^s(\lambda, t) \pm \tilde{a}^u(\lambda, t) \tag{16.14}$$

and carrying out the Sommerfeld–Watson transformation using $a^\pm(\lambda, t)$. Combining the results appropriately, as in (15.29), leads to the analogous result

$$F(s, t, u) = \sum_i \frac{(2\alpha_i^+(t) + 1) r_i^+(t)}{\sin \pi\alpha_i^+(t)} (1 + e^{-i\pi\alpha_i^+(t)}) P_{\alpha_i^+(t)}\left(-1 - \frac{2s}{t - 4m^2}\right)$$

$$+ \sum_i \frac{(2\alpha_i^-(t) + 1) r_i^-(t)}{\sin \pi\alpha_i^-(t)} (1 - e^{-i\pi\alpha_i^-(t)}) P_{\alpha_i^-(t)}\left(-1 - \frac{2s}{t - 4m^2}\right)$$

$$+ B(s, t), \tag{16.15}$$

with $\cos\theta_t$ expressed appropriately in terms of the relativistic variables.

Crossed-Channel Regge Poles

It was not until two years after Regge had formulated these ideas that the implications of Regge poles in one channel for high energy scattering in another were considered. Let us take the amplitude (16.15), obtained by continuation of the t-channel partial wave amplitudes, and investigate its form in the limit of high-energy near-forward scattering in the s channel. Then we have

$$s = 4(k^2 + m^2) \gg 4m^2$$

$$t = -2k^2(1 - \cos\theta_s) \leqslant 0, \quad |t| \ll 4m^2, \tag{16.16}$$

and the argument of the Legendre function is

$$-\cos\theta_t = -1 - \frac{2s}{t - 4m^2} \approx \frac{s}{2m^2} \tag{16.17}$$

which becomes large and positive as $s \to \infty$. The asymptotic form of the Legendre function in that limit is

$$P_{\alpha(t)}\left(-1 - \frac{2s}{t - 4m^2}\right) \approx P_{\alpha(t)}\left(\frac{s}{2m^2}\right) \approx \left(\frac{s}{2m^2}\right)^{\alpha(t)} \tag{16.18}$$

Consequently the sum of Regge pole terms

$$\sum_i \frac{(2\alpha_i(t) + 1)r_i(t)}{\sin \pi\alpha_i(t)} (1 + \tau_i e^{-i\pi\alpha_i(t)}) P_{\alpha_i(t)}(-\cos\theta_t)$$

$$\approx \sum_i \frac{(2\alpha_i(t) + 1)r_i(t)}{\sin \pi\alpha_i(t)} (1 + \tau_i e^{-i\pi\alpha_i(t)}) \left(\frac{s}{2m^2}\right)^{\alpha_i(t)} \tag{16.19}$$

will be dominated, for large enough values of s, by the contribution of that trajectory having the largest value of $\mathrm{Re}\,\alpha_i(t)$. Furthermore, the background integral B(s, t) will be of the form

$$B(s, t) \approx \int_{-1/2-i\infty}^{-1/2+i\infty} d\lambda \frac{(2\lambda + 1)a(\lambda, t)}{\sin \pi\lambda} \left(\frac{s}{2m^2}\right)^{\lambda}$$

$$= \left[\int_{-\infty}^{\infty} dx \, \frac{x a(ix - \tfrac{1}{2}, t)}{\cosh \pi x} e^{ix \ln(2m^2/s)} \right] \left(\frac{s}{2m^2} \right)^{-1/2}, \tag{16.20}$$

where we have written $\lambda = ix - \tfrac{1}{2}$ in order to extract the factor $s^{-1/2}$ explicitly. If the integral is well-behaved, it follows that $B(s, t)$ disappears like $s^{-1/2}$ as $s \to \infty$.

Therefore, if $\alpha(t)$ is the trajectory with the largest real part, and $\alpha(t) > -\tfrac{1}{2}$, the scattering amplitude is asymptotically dominated by the corresponding Regge pole,

$$F(s, t) \approx \frac{(2\alpha(t) + 1) r(t)}{\sin \pi\alpha(t)} (1 + \tau e^{-i\pi\alpha(t)}) \left(\frac{s}{2m^2} \right)^{\alpha(t)} \tag{16.21}$$

Notice that the amplitude (16.21) depends on the energy variable s *only* via the power $s^{\alpha(t)}$. If the momentum transfer variable t is held fixed, the dependence of the amplitude is given by a power law. Thus a Regge pole in the t channel produces a particularly simple energy dependence in the s channel.

The power is, moreover, dependent on the value of t. As a result of this fact we are able to avoid one of the principal difficulties of the particle exchange model, its unsatisfactory asymptotic behavior. If an s-channel interaction is propagated by exchange of a particle of spin L, the amplitude contains the corresponding t-channel bound-state term,

$$\frac{P_L(\cos \theta_t)}{t - M^2} = \frac{P_L\left(1 + \dfrac{2s}{t - 4m^2}\right)}{t - M^2} = F_L(s, t) \tag{16.22}$$

which, in the s-channel high energy region (16.16), behaves like s^L. Thus exchange of an object with spin L leads to a total cross section (remembering that we are now in the s channel)

$$\sigma_L(s) = \frac{8\pi}{\sqrt{s(s - 4m^2)}} \operatorname{Im} F_L(s, 0)$$

$$\approx c s^{L-1} \tag{16.23}$$

and as $s \to \infty$ the total cross section diverges if $L > 1$.

The asymptotic behavior of the total cross section is still open to debate, as we shall describe in Chapter 18; but they certainly are not rising like any large power of s. There are, furthermore, fairly general proofs limiting the growth of the total cross section. These proofs are far too specialized to merit detailed discussion here; the interested reader is referred to the book by Eden, who gives the following simple justification of the results. Let us assume that the target particle has a wave function that falls off exponentially, so that its probability density can be written

$$\rho(r) \propto e^{-\gamma r}$$

(at least in some asymptotic region). Then the probability of interaction at r with an incident particle of energy E will be

$$P(E, r) = p(E)e^{-\gamma r},$$

the product of the (energy-dependent) interaction probability p(E) with the density of the target. Let us now assume also that p(E) is bounded by some power of the energy,

$$|p(E)| \leqslant p_0 E^N$$

Then

$$|P(E, r)| \leqslant p_0 e^{-\gamma r + N \ln E} = p_0 e^{-\gamma (r - r_0)},$$

which implies that the interaction probability will become vanishingly small when r is sufficiently greater than

$$r_0 = \frac{N \ln E}{\gamma}.$$

The total cross section may thus be measured by

$$\sigma \leqslant \pi r_0^2 \leqslant c \ln^2 E$$

which, on expressing the energy via the relativistic variable s, becomes the *Froissart bound*

$$\sigma(s) \leqslant c' \ln^2 s. \qquad (16.24)$$

This bound can, in fact, be proven rigorously on the basis of quantum field theory.

From (16.24) it follows that a contribution of the form (16.23) is not allowed for $L > 1$. But there are certainly possible exchanges involving particles of spin greater than one – the A_2, for example, or any other higher-spin meson. If these were "elementary" particles (in the fixed-pole sense) it would therefore be necessary that they all cooperate in such a way that the divergent terms cancel each other – a rather unexpected consistency among otherwise independent states. Particles lying on a Regge trajectory, however, conspire together in just such a manner to produce a single Regge pole, replacing the offending L in (16.22) by $\alpha(s)$ in (16.21), and therefore yielding

$$\sigma(s) \approx c s^{\alpha(0)-1} \tag{16.25}$$

instead of (16.23). For (16.25), the Froissart bound will be satisfied if $\alpha(0) \leqslant 1$. This is no longer a stringent condition with respect to the spin of the exchanged particle, since that is determined by $\alpha(M^2)$ rather than $\alpha(0)$. Indeed, all of the trajectories known experimentally in the direct channel satisfy this constraint with ease. Thus the simple parametrization (16.21) obtained in Regge theory avoids the divergence problem associated with the particle exchange model; in so doing, it relates the non-increasing nature of the total cross section to the absence of high-lying trajectories and the low-mass, high-spin particles they would produce.

The Regge Amplitude

Let us now look more carefully into the properties and consequences of the scattering amplitude

$$F_R(s,t) = \frac{(2\alpha(t)+1)r(t)}{\sin \pi\alpha(t)}\left(1 + \tau e^{-i\pi\alpha(t)}\right)\left(\frac{s}{2m^2}\right)^{\alpha(t)} \tag{16.26}$$

obtained by assuming, as in (16.21), that a single Regge pole is dominant. The total cross section following from such an amplitude, as indicated in (16.25), is proportional to $s^{\alpha(0)-1}$. Asymptotically constant total cross sections are therefore obtained if the dominant trajectory saturates the Froissart bound,

$$\alpha_P(0) = 1. \tag{16.27}$$

The trajectory satisfying (16.27) is often called the Pomeranchuk trajectory, after the Russian physicist who derived some very interesting properties of the asymptotic amplitude (of which more will be said shortly).

If all total cross sections are to become asymptotically constant, the Pomeranchuk trajectory must be present in all elastic scattering amplitudes. In other words, it must have the quantum numbers of the vacuum, since otherwise we could invent a reaction in which it could not be exchanged. Consequently we expect its quantum numbers to be $I = B = S = 0, G = \tau = +1$. The positive signature avoids the production of a mass-zero particle by (16.27); the first particle on the trajectory should therefore have $J^P = 2^+$. The f(1270) and the f′(1514) mesons have the correct quantum numbers, but there is considerable question whether either of them actually belongs to the Pomeranchuk trajectory.

The factorization of the residues for the Pomeranchuk trajectory leads to relations between asymptotic cross sections. For elastic scattering in the t channel (15.33) implies that

$$r_P^{A\bar{A}}(t) r_P^{B\bar{B}}(t) = [r_P^{A\bar{A}\to B\bar{B}}(t)]^2,$$

which, when crossed to $t = 0$, yields

$$\sigma(AA)\sigma(BB) = [\sigma(AB)]^2. \tag{16.28}$$

Unfortunately, with only nucleon targets (16.28) cannot be tested experimentally; but it can be used to estimate cross sections of non-measurable processes, e.g.

$$\sigma(\pi\pi) = [\sigma(\pi N)]^2/\sigma(NN).$$

The differential cross section resulting from (16.26) is also of interest, in particular regarding its simple energy dependence. Let us write it with respect to the momentum transfer variable t,

$$\frac{d\sigma}{dt} = \frac{4\pi}{s(s-4m^2)} |F_R(s,t)|^2 \approx A(t) s^{2\alpha(t)-2} \tag{16.29}$$

for $s \gg 4m^2$. Like the total cross section, the differential cross section obeys a power law in s, the power now being $2\alpha(t) - 2$; if we look at the variation of $d\sigma/dt$ with s at a given fixed value of t, the power can be explicitly obtained. Doing this for various t will enable us to measure experimentally the trajectory $\alpha(t)$ for $t < 0$, and to compare it with the trajectories obtained in the direct channel as in the preceding chapter.

At high energy the differential cross section is, in general, in the shape of a

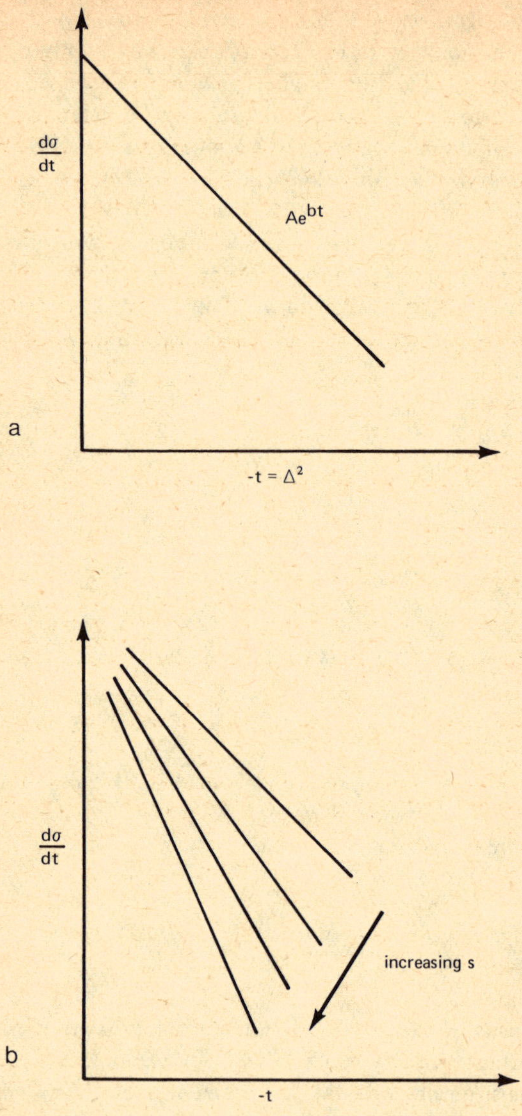

Fig.(16-1) Exponential shape of the diffraction peak, observed for many processes at high energy, showing the shrinkage predicted by Regge theory.

diffraction peak, as shown in Fig.(16-1a), which as a function of t is given (at fixed energy) by

$$\frac{d\sigma}{dt} = Be^{\beta t} = Be^{-\beta\Delta^2}. \qquad (16.30)$$

The Regge pole result (16.29) may be written in this form as

$$\frac{d\sigma}{dt} = e^{\ln A(t)+(2\alpha(t)-2)\ln s}.$$

If we now assume that, for small t, ln A(t) and the trajectory $\alpha(t)$ are approximately linear functions of t,

$$\ln A(t) \approx A_0 + A_1 t$$

$$\alpha(t) \approx \alpha_0 + \alpha_1 t \qquad (16.31)$$

then (16.28) becomes

$$\frac{d\sigma}{dt} = [A(0)s^{2\alpha(0)-2}]\, e^{(A_1+2\alpha_1 \ln s)t}. \qquad (16.32)$$

Consequently β, the slope of the diffraction peak, is given for a Regge amplitude by

$$\beta = A_1 + 2\alpha_1 \ln s. \qquad (16.33)$$

This logarithmic dependence on s will cause β to be an increasing function of s; therefore the slope of the diffraction peak will increase with energy. It follows that $d\sigma/dt$ will become more and more sharply peaked as s increases, as shown in Fig.(16.1b), if the scattering amplitude is dominated by a single Regge pole. This phenomenon is usually referred to as "the shrinkage of the diffraction peak".

A third important property of the Regge amplitude is that its phase results entirely from the signature factor

$$\zeta(t) = 1 + \tau e^{-i\pi\alpha(t)}$$

$$= 2e^{-i\pi\alpha(t)/2} \cos \tfrac{1}{2}\pi\alpha(t), \quad \tau = +$$

$$= 2ie^{-i\pi\alpha(t)/2} \sin \tfrac{1}{2}\pi\alpha(t), \quad \tau = -. \tag{16.34}$$

In the s-channel physical region, both the residue r(t) and the trajectory $\alpha(t)$ are real. This fact results from the t-channel partial wave unitarity equation. For $t \leq 0$, there are no intermediate states to contribute to (12.45), since these values of t are below all scattering thresholds. Consequently the partial-wave unitarity equation becomes

$$\text{Im}\, a_\ell(t) = 0, \quad t \leq 0, \tag{16.35}$$

and writing $a_\ell(t)$ in the form (15.13) yields

$$\text{Im} \frac{r(t)}{\ell - \alpha(t)} = 0$$

which can hold for all ℓ only if r(t) and $\alpha(t)$ are real.

It follows that all terms in $F_R(s, t)$ are real except for the signature factor (16.34), and thus that

$$F_R(s, t) = |F_R(s, t)| e^{-i\pi\alpha(t)/2}, \tag{16.36a}$$

i.e.

$$\frac{\text{Re}F_R(s, t)}{\text{Im}F_R(s, t)} = -\cot \tfrac{1}{2}\pi\alpha(t), \tag{16.36b}$$

Thus the phase of $F_R(s, t)$ is independent of the energy, and is directly related to the trajectory $\alpha(t)$. Unfortunately, the phase of the amplitude is not easy to determine, and therefore (16.36) cannot provide an accurate measurement of $\alpha(t)$.

This simplicity of the phase of a Regge amplitude has one other particularly interesting consequence which arises when spin is included in the scattering process. In pion–nucleon scattering, for example, there are two amplitudes, F and G, describing scattering without and with flip of the nucleon's spin. These are related by analytic continuation to the two amplitudes describing $\pi\pi \to \bar{N}N$, referring to parallel and antiparallel spin states for the $\bar{N}$ and N. A Regge pole in either amplitude of the latter channel will, however, appear in both F and G when crossed into the $\pi N \to \pi N$ channel. As a result, if a single Regge pole dominates, both F and G will have the same phase asymptotically, as given by (16.36). But the nucleon polarization in the $\pi N \to \pi N$ reaction is given by the form

$$P(s, t) = -2 \frac{\mathrm{Im}(F^*G)}{|F|^2 + |G|^2}.$$

If both F and G have the same phase, F^*G is real, and consequently $P = 0$. This result does not, of course, depend on the particular reaction involved; in general we conclude that if a single Regge pole is dominant, the polarization must vanish.

Thus if the Pomeranchuk pole dominates asymptotically, it leads to constant total cross sections, shrinking diffraction peaks, and vanishing polarizations. Unfortunately, the experimental data do not show this beautifully simple behavior. A more complicated Regge description, involving several poles, is required; we shall take up this question in Chapter 17.

Line Reversal and the Pomeranchuk Theorem

The phase obtained from (16.35) is the same as that predicted by the arguments of analyticity and crossing symmetry in Chapter 14, since

$$e^{-i\pi\alpha(t)/2} s^{\alpha(t)} = (e^{-i\pi/2} s)^{\alpha(t)}$$
$$= \left(\frac{s}{i}\right)^{\alpha(t)}.$$

We are considering the s channel, so the crossing symmetry arguments would apply to the s and u channels, as in (14.18). The appropriate variable is therefore given by (14.31),

$$\tfrac{1}{2}(s - u) = s + \tfrac{1}{2}t - 2m^2$$
$$\approx s.$$

Thus for even signature, $F_R(s, t)$ is a real function of (s/i), as expected for a crossing-symmetric amplitude, while for odd signature the same is true of $iF_R(s, t)$, as for a crossing-antisymmetric amplitude.

This is rather surprising, since we have not made any assumptions about the s–u crossing symmetry of the Regge amplitude; however, the use of amplitudes with definite signature automatically insures the necessary symmetries. The reason is that the signatured amplitudes in (15.30) are even or odd functions of $\cos\theta_t$, and interchanging $\cos\theta_t$ and $(-\cos\theta_t)$ is equivalent to interchanging s and u. Therefore, an even signature amplitude is automatically symmetric in s and u, while an odd signature one is antisymmetric. Specifically, let us write the amplitude without extracting the factor $e^{-i\pi\alpha(t)}$ as

$$F_R(s, t) = \frac{(2\alpha(t) + 1)r(t)}{\sin \pi\alpha(t)} [P_{\alpha(t)}(-\cos\theta_t) + \tau P_{\alpha(t)}(\cos\theta_t)]. \tag{16.37}$$

In the limit $s \to \infty$, with $t \approx 0$, we also have

$$u = 4m^2 - s - t \approx -s \to -\infty$$

so

$$\cos\theta_t = -1 - \frac{2u}{t - 4m^2} \approx \frac{u}{2m^2} \approx -\frac{s}{2m^2}.$$

Thus

$$F_R(s, t) \approx \frac{(2\alpha(t) + 1)r(t)}{\text{siu}\ \pi\alpha(t)} \left[\left(\frac{s}{2m^2}\right)^{\alpha(t)} + \tau\left(\frac{u}{2m^2}\right)^{\alpha(t)}\right] \tag{16.38}$$

and $F_R(s, t)$ is symmetric or antisymmetric with respect to crossing between s and u, according to the signature.

Equation (16.38) brings to light an interesting property of the Regge-dominated amplitude. The pole term obtained in the t channel could be crossed into the u channel as well as into the s channel. The same limiting forms will occur, except that in (16.38) we will now have $u \gg 4m^2$ and $s \ll -4m^2$. Let us denote by $\bar{F}_R(u, t)$ the amplitude obtained in this limit. Since now $s < 0$, the factor $e^{-i\pi\alpha(t)}$ should be extracted from $s^{\alpha(t)}$, and it follows that

$$\bar{F}_R(u, t) \approx \frac{(2\alpha(t) + 1)r(t)}{\sin \pi\alpha(t)} (e^{-i\pi\alpha(t)} + \tau)\left(\frac{u}{2m^2}\right)^{\alpha(t)}$$

$$= \tau \frac{(2\alpha(t) + 1)r(t)}{\sin \pi\alpha(t)} (1 + \tau e^{-i\pi\alpha(t)})\left(\frac{u}{2m^2}\right)^{\alpha(t)} \tag{16.39}$$

because $\tau^2 = 1$. Thus

$$\bar{F}_R(u, t) = \tau F_R(u, t). \tag{16.40}$$

In other words, the contribution of a t-channel Regge pole to the u-channel scattering amplitude is *identical* to its contribution to the s-channel, multiplied by the signature.

If the processes are defined as in Chapter 13, so that the t-channel reaction is $A\bar{C} \to \bar{B}D$, then $s \gg 4m^2$, $t \approx 0$ describes forward $AB \to CD$ scattering, while $u \gg 4m^2$, $t \approx 0$ describes forward $A\bar{D} \to C\bar{B}$ scattering. The corresponding graphs are shown in Fig.(16-2). The difference between the two limits can be visualized as simply reversing the direction of the bottom line, and the relation (16.40) between the contributions of the s-channel Regge pole to the two processes is often known as "line reversal".

A particularly important case of line reversal involves the choice C = A, D = B, for which the s and u channels describe respectively the elastic scattering processes $AB \to AB$ and $A\bar{B} \to A\bar{B}$, and at t = 0 the optical theorem may be applied to (16.40) to yield

$$\bar{\sigma}(u) = \tau\sigma(u). \tag{16.41}$$

In other words, the total cross sections for $AB \to AB$ and $A\bar{B} \to A\bar{B}$ are equal

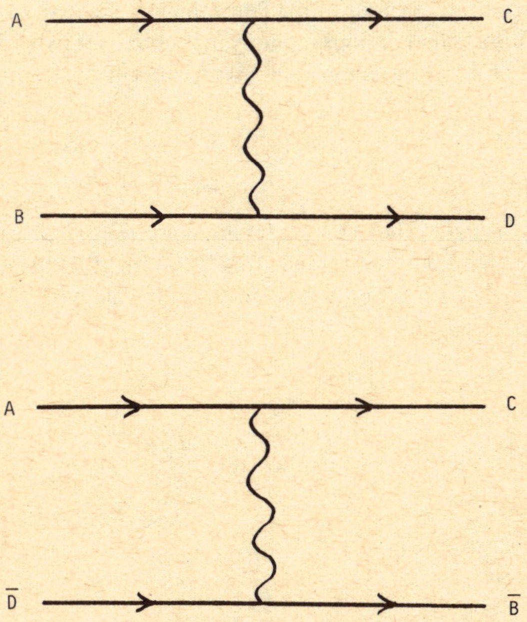

Fig.(16-2) Reactions related by line reversal.

except for the factor τ. But the total cross section is necessarily positive, so if a single pole is dominant its signature must be positive. In that case, furthermore, the total cross sections of $AB \to AB$ and $A\overline{B} \to A\overline{B}$ scattering are equal.

Thus Regge theory leads, for a single dominant pole term, to the asymptotic equality of particle–particle and particle–antiparticle scattering, which we required in (11.42a) to avoid the need of subtraction in the odd dispersion relation. That this equality should hold is the content of the Pomeranchuk theorem, which has been proved on grounds more general than Regge theory, although still not absolutely firm. When particle and antiparticle are members of the same isomultiplet, as in pion–nucleon scattering, the proof is relatively simple. By (3.23)

$$F(\pi^+p \to \pi^+p) - F(\pi^-p \to \pi^-p) = \sqrt{2}\,F(\pi^-p \to \pi^0 n),$$

so if we combine isosymmetry with the reasonable, and experimentally probable, assumption that the amplitudes of inelastic scattering processes such as charge exchange vanish with increasing energy, we have asymptotically

$$F(\pi^+p \to \pi^+p) = F(\pi^-p \to \pi^-p). \tag{16.42}$$

In a more general case, a proof of the Pomeranchuk theorem can be based on the existence of dispersion relations requiring no more than two subtractions. Let us return to (11.42a) and assume that asymptotically both σ and $\overline{\sigma}$ become constant, with

$$\overline{\sigma}(\omega) - \sigma(\omega) = c, \tag{16.43}$$

where c is some constant. Then (11.42a) is not convergent, and a subtraction is necessary. Making it at $\omega = 0$ and using (16.43), we find that the asymptotic part of the integrand makes a contribution to the integral yielding

$$\frac{\operatorname{Re} A^-(\omega, 0)}{\operatorname{Im} A^-(\omega, 0)} \approx -\frac{c \ln \omega}{\sigma}, \quad \omega \to \infty. \tag{16.44}$$

If we now make the additional assumption that

$$\frac{\operatorname{Re} A^-(\omega, 0)}{\ln \omega \operatorname{Im} A^-(\omega, 0)} \to 0 \text{ as } \omega \to \infty, \tag{16.45}$$

it follows immediately that $c = 0$, and the two total cross sections must be equal.

The assumption (16.45) is rather *ad hoc,* however; it has not been justified on any grounds other than plausibility. In Regge theory, for example, (16.35b) shows that it is indeed satisfied. But in general it need not be, and while the Pomeranchuk theorem is generally accepted because of its conceptual elegance, it remains to be proven on a fundamental basis.

Daughters

Next we turn to a consideration of some of the peculiarities which must be exhibited by Regge amplitudes in order to maintain reasonable properties in certain specialized situations. The easiest of these to discover is the existence of so-called "daughter trajectories" because of constraints on the scattering of unequal-mass particles via Regge-pole exchange.

Let us consider the kinematics of an unequal-mass scattering process, calling the s channel $AB \to CD$ as usual, where the masses involved are arbitrary and are denoted by M_A, M_B, M_C, and M_D. The physical regions, and the relation between the invariant variables and the momenta and scattering angles, become much more complicated in this situation. We are interested here particularly in how the inequality of the masses affects the expressions obtained by continuing a t-channel Regge-pole amplitude into the s channel. Therefore we must express $\cos\theta_t$ in terms of s, t, and u for unequal masses. Some calculation yields

$$\cos\theta_t = \frac{1}{2pk}\left[z + \frac{\delta}{t}\right], \tag{16.46}$$

where

$$z = \tfrac{1}{2}(s - u) = s + \tfrac{1}{2}t + \tfrac{1}{2}(M_A^2 + M_B^2 + M_C^2 + M_D^2),$$

$$\delta = \tfrac{1}{2}(M_A^2 - M_C^2)(M_B^2 - M_D^2), \tag{16.47}$$

and p and k are the CM momenta of the initial and final states for the t-channel process,

$$p^2 = \frac{1}{4t}[t - (M_A + M_C)^2]\,[t - (M_A - M_C)^2]$$

$$k^2 = \frac{1}{4t}[t - (M_B + M_D)^2]\,[t - (M_B - M_D)^2]. \tag{16.48}$$

We may note immediately from (16.46) and (16.47) that, if $M_A \neq M_C$ and $M_B \neq M_D$, the non-zero value of δ destroys some of the simplicity otherwise obtained. The interchange of s and u is no longer equivalent to that of $\cos\theta_t$ and $(-\cos\theta_t)$, differing by a term δ/t which becomes infinite as we continue toward $t = 0$. This infinity is a particularly serious problem for Regge theory, since as $t \to 0$ it leads to

$$\cos\theta_t \approx \frac{t}{\delta}\left(z + \frac{\delta}{t}\right) \to 1 \qquad (16.49)$$

for any fixed z. In other words, however large we choose s to be, there remains a region near $t = 0$ where $\cos\theta_t$ does not become large. This region lies, furthermore, within the physical region of the crossed channel and includes near-forward scattering. It follows that we cannot expect the scattering amplitude there to be given by the leading Regge pole term.

Thus the inequality of the masses destroys the direct correspondence between $P_\ell(\cos\theta_t)$ and s^ℓ which exists when $\delta = 0$. Thinking of the problem in this way makes clear a possible remedy: instead of carrying out the Sommerfeld–Watson transformation on the partial-wave summation of Legendre polynomials, we could follow a similar procedure starting directly from a *power* series in the momentum transfer s,

$$F_t(s, t) = \sum_n c_n(t) s^n. \qquad (16.50)$$

Continuing the coefficients $c_n(t)$ into the complex n-plane and finding the dominant pole at $n = \nu(t)$ then yields

$$F_t(s, t) \approx r(t) s^{\nu(t)}. \qquad (16.51)$$

This procedure is known as the Khuri transformation. The connection between Khuri poles leading to (18.23) and Regge poles has been investigated in some detail. It reveals that canceling the singularity at $t = 0$ requires the existence of families of Regge trajectories, each leading trajectory being accompanied by others, known as “daughter trajectories”, spaced below it.

The origin of this requirement is most easily understood by studying the analyticity of a Regge-pole amplitude in the unequal-mass situation. We have seen that analyticity in two variables seems to be a tenable basic assumption; but which two variables should be chosen? The connection established between analyticity and crossing symmetry in Chapter 14 indicates that the crossing-symmetric variable $z = \frac{1}{2}(s - u)$ should be one, and clearly t itself should be the other. Therefore let us construct a crossing-symmetric form of the scattering amplitude in this case. The crossing relation between the s-channel and u-channel amplitudes is

$$F_s(s + i\epsilon, t) = F_u(u - i\epsilon, t) = F_u^*(u + i\epsilon, t), \tag{16.52}$$

and if we define the amplitudes as functions of z by

$$\begin{aligned} F_s(s, t) &\equiv \mathscr{F}_s(iz, t) \\ F_u(u, t) &\equiv \mathscr{F}_u(-iz, t) \end{aligned} \tag{16.53}$$

it follows that

$$\mathscr{F}_s(iz, t) = \mathscr{F}_u^*(-iz, t). \tag{16.54}$$

Amplitudes with definite crossing symmetry can then be formulated, namely

$$\mathscr{F}_\pm(iz, t) = \mathscr{F}_s(iz, t) \pm \mathscr{F}_u(iz, t) \tag{16.55}$$

having the property that

$$\mathscr{F}_\pm^*(iz, t) = \pm\, \mathscr{F}_\pm(-iz, t), \tag{16.56}$$

implying for real t that $\mathscr{F}_+(iz, t)$ and $i\,\mathscr{F}_-(iz, t)$ are real analytic functions of the variable iz.

Now suppose that these amplitudes are obtained by continuation of a t-channel Regge pole. Then we write the s-channel amplitude in the form

$$\mathscr{F}_s(iz, t) = F_s(s, t) = R(t)P_{\alpha(t)}\left(-\frac{z + \delta/t}{2pk}\right)\left(1 + \tau e^{-i\pi\alpha(t)}\right). \tag{16.57a}$$

The u-channel term entering (16.55) has s and u interchanged, so z becomes $-z$, yielding

$$\begin{aligned} \mathscr{F}_u(iz, t) &= F_u(s, t) \\ &= R(t)P_{\alpha(t)}\left(-\frac{-z + \delta/t}{2pk}\right)\left(1 + \tau e^{+i\pi\alpha(t)}\right), \end{aligned} \tag{16.57b}$$

with a sign change in the signature because of the appearance of $(-i\epsilon)$ in (16.52). Thus obtain, with $y = iz$,

$$\mathscr{F}_{\pm}(y, t) = R(t)\left[(1 + \tau e^{-i\pi\alpha(t)})P_{\alpha(t)}\left(\frac{iy - \delta/t}{2pk}\right)\right.$$

$$\left. \pm (1 + \tau e^{i\pi\alpha(t)})P_{\alpha(t)}\left(\frac{-iy - \delta/t}{2pk}\right)\right]. \tag{16.58}$$

If $R(t)$ is real, $\mathscr{F}_{+}(y, t)$ and $i\,\mathscr{F}_{-}(y, t)$ are, as expected, real analytic functions of $y = iz$, since the Legendre function is known to be a real analytic function of its argument. At $t = 0$, however, these amplitudes clearly have a singularity, which can be isolated by expanding the Legendre functions in powers of y. Using

$$P_\nu(\xi) = \xi^\nu(1 + 0(\xi^{|\nu - 1/2| - 3/2}))$$

we find that (absorbing $(2pk)^{\alpha(t)}$ into the residue function)

$$\mathscr{F}_{\pm}(iz, t) = R'(t)\Big\{(iz)^\alpha(e^{1/2\pi\alpha} \pm e^{-1/2i\pi\alpha})(1 \pm \tau)$$

$$+ (iz)^{\alpha-1}\frac{i\delta\alpha}{t}(e^{1/2i\pi\alpha} \mp e^{-1/2i\pi\alpha})(1 \mp \tau) + \ldots\Big\} \tag{16.59}$$

with the succeeding terms containing lower powers of iz and higher powers of δ. The first term of (16.59) is just the usual Regge power law, which would be the entire result if $\delta = 0$. The second term, however, will become infinite as $t \to 0$, canceling the dominant behavior of the leading power; likewise, succeeding terms will also blow up in this case, the summation of all of them leading to a logarithmic singularity in the amplitude. If $\mathscr{F}_{\pm}(iz, t)$ is to be analytic at $t = 0$, then, we must add to it extra terms cancelling these singularities.

Specifically, we may add to the leading Regge pole term in (18.29) a daughter Regge pole,

$$F_1(s, t) = R_1(t)P_{\alpha_1(t)}(-\cos\theta_t)(1 + \tau_1 e^{-i\pi\alpha_1(t)}),$$

with residue, trajectory, and signature chosen so that at $t = 0$ the *leading* term in the expansion of $F_1(s, t)$ precisely cancels the *second* term of (18.31). This will occur if

$$\alpha_1(0) = \alpha(0) - 1 \tag{16.60a}$$

$$\lim_{t \to 0} tR_1(t) = R(0)\delta\alpha(0) \tag{16.60b}$$

$$\tau_1 = -\tau. \tag{16.60c}$$

Thus, in order to avoid a singularity at t = 0 for the unequal-mass Regge amplitude, we must introduce a "daughter trajectory" $\alpha_1(t)$ satisfying (16.60a).

If such a daughter trajectory exists, however, it should appear in all reactions where the parent trajectory $\alpha(t)$ contributes, regardless of the masses involved. Fits to the high-energy data are probably insensitive to these corrections, since the lower trajectory values make them negligible; but we should look for particles and resonances lying on the daughter trajectories as t-channel manifestations of their existence. Since (16.60a) says nothing about the variation of $\alpha_1(t)$ away from t = 0, it is impossible to predict the expected masses; indeed, there are models in which daughter trajectories have a negative slope for $t > 0$, and thus never reach positive integer values to produce particles. Alternatively, there are other models for which the daughter trajectory runs parallel to its parent. Then, for example, the daughter of the ρ would be observed as a spin-zero meson at about the ρ mass.

The second term in $F_1(s, t)$ does not cancel the third in $F_R(s, t)$; to eliminate that singularity, a "grand-daughter" $\alpha_2(t)$ is required, with $\alpha_2(0) = \alpha_1(0) - 1$, appropriate constraints on the residue, and signature $\tau_2 = \tau$. Succeeding generations must also exist to cancel the succeeding singularities. Since the Froissart bound applied to equal-mass reactions requires $\alpha(0) \leqslant 1$, only $\alpha_1(0)$ is to the right of the background integral; whether the younger generations emerge in the t-channel is a matter of conjecture.

Unequal-mass Regge poles thus hint of a proliferation of trajectories. It has been observed that daughter trajectories do occur in some model field theories, but experiment has revealed relatively little evidence for their existence; the only reasonable candidate is the $\rho'(1600)$, which could be the granddaughter of the ρ. If the entire family of trajectories is even roughly parallel, a large number of high-mass, low-spin resonances must soon come into view. It is certainly conceivable, however, that all of the offspring turn downwards, or backwards, after passing through t = 0, or that they continue upward but always decouple from physical particles because of vanishing residues. In that case they will serve no physical purpose, but merely be mathematical curiosities rescuing unequal-mass Regge theory.

Ghosts, Sense, and Nonsense

When the particles involved in a scattering process have non-zero spin, the partial wave expansion which is the starting point of Regge theory must be appropriately generalized. In particular, the Legendre polynomials are replaced by more complicated functions which arise through the appropriate representations of the rotation group in such situations. In general, the Sommerfeld–Watson transform can be applied as readily to these functions as in the simpler case we have studied. Some interesting insights do result from basically kinematical properties of these functions, however, and we shall mention these to close this chapter.

Gell-Mann was the first to notice a difficulty arising through the angular momentum coupling coefficients. Let us again consider the process $AB \to CD$ of Fig.16-2, but now assume that all four particles have non-zero spin. The vertex between two particles and the exchanged Reggeon is termed a "sense" transition if there is no spin flip between the external particles, and a "nonsense" transition if there is non-zero spin flip. The scattering amplitudes for $AB \to CD$ may then be categorized as F_{ss}, F_{sn}, F_{ns}, and F_{nn}, the subscripts describing the spin flip situation at the αAC and αBD vertices. Carrying out the Sommerfeld–Watson transform will yield amplitudes which, near $\alpha(t) = 0$, are of the form

$$F_{ss} \sim r_{ss}(t) \frac{(1 + \tau e^{-i\pi\alpha(t)})}{\sin \pi\alpha(t)} s^{\alpha(t)}$$

$$F_{ns} \sim [\alpha(t)]^{1/2} r_{ns}(t) \frac{(1 + \tau e^{-i\pi\alpha(t)})}{\sin \pi\alpha(t)} s^{\alpha(t)}$$

$$F_{nn} \sim \alpha(t) r_{nn}(t) \frac{(1 + \tau e^{-i\pi\alpha(t)})}{\sin \alpha\pi(t)} s^{\alpha(t)} \qquad (16.61)$$

(F_{sn} will be of the same form as F_{ns}).

It is the factor $[\alpha(t)]^{1/2}$ in F_{ns} which is significant. It leads to a branch point in the t plane at the value t_0 for which $\alpha(t_0) = 0$. Since there is no reason for such a singularity, we assume it must be cancelled by an appropriate factor in $r_{ns}(t)$. Because of factorization, however, we must have

$$r_{ss}(t) = g^s_{\alpha AC}(t) g^s_{\alpha BD}(t)$$

$$r_{sn}(t) = g^s_{\alpha AC}(t) g^n_{\alpha BD}(t)$$

$$r_{nn}(t) = g^n_{\alpha AC}(t) g^n_{\alpha BD}(t), \qquad (16.62)$$

where $g^s_{\alpha AC}(t)$, for example, is the coupling constant for the αAC non-flip vertex. Consequently the assumption we make on $r_{ns}(t)$ to cancel this singularity will have consequences for $r_{nn}(t)$, or $r_{ss}(t)$, or both.

What choice should be made depends on the signature. If $\tau = +$ ("right" signature) then the factor sin $\pi\alpha(t)$ will produce a pole implying a spin-zero particle. Suppose, however, that $t_0 < 0$. Then the particle would have imaginary mass. To avoid such "ghost" particles, the residue must vanish and cancel the pole.

The analyticity problem can be resolved simply by assuming $r_{ns}(t) \propto [\alpha(t)]^{1/2}$. But which of the coupling constants should contain this factor, and which should be analytic there? If we choose the sense coupling to be analytic, the ghost pole survives; but if we "choose nonsense" by taking $g^n_{\alpha AC}(t)$ analytic and $g^s_{\alpha BD} \propto [\alpha(t)]^{1/2}$, then all three amplitudes will be free of the ghost pole. If $\alpha = 0$ is a right signature point, then, the coupling must choose nonsense to kill the ghost.

If $\tau = -$, however, the "wrong signature" factor $(1 - e^{-i\pi\alpha})$ automatically dispenses with the ghost problem. Either coupling constant may then be chosen to contain the factor $[\alpha(t)]^{1/2}$; but for either choice, the spin-flip amplitudes inevitably vanish at $\alpha = 0$. The non-flip amplitude survives only if the coupling "chooses sense", taking $g^s_{\alpha AC}(t)$ analytic and $g^n_{\alpha BD} \propto [\alpha(t)]^{1/2}$.

These considerations therefore predict the vanishing of spin-flip residues at $\alpha = 0$ for odd-signature trajectories. Similar arguments can be made at other integer values of α. Also, information gained about the vertex may be applied even when all particles do not have non-zero spin; for example, considering the ρNN vertex in NN $\to$ NN in this way predicts the vanishing of the $\pi^- p \to \pi^0 n$ spin-flip amplitude because of a "wrong signature nonsense zero".

Conspiracy and Evasion

Another arcane property of Regge poles arising, like daughters, from studying the literal and detailed behavior of Regge pole amplitudes is the necessity for conspiracy or evasion. These rather treacherous-sounding terms refer to a phenomenon originating (for equal as well as unequal masses) in the description of processes involving spin. Angular momentum conservation requires that the spin-flip amplitude for pion–nucleon scattering must vanish as $t \to 0$, regardless of whether it arises from t-channel Reggeization. As it happens, in that particular case the factor $\sqrt{-t}$ arises automatically in defining the continuation of the Regge pole terms. In general, however, this fortuitous result is not obtained. The amplitude for a given spin process, which has a known behavior in the s channel as $t \to 0$, will be given by a combination of t-channel spin amplitudes which do not manifestly possess that behavior. Instead, it must be *assumed* for them, thereby yielding constraints on the Regge pole amplitudes.

The simplest physical situation in which such a problem arises is elastic proton–proton scattering. Here there are five spin amplitudes in the direct channel, which are related to five corresponding amplitudes in the t channel via the crossing matrices mentioned in Chapter 13. Carrying out the Sommerfeld–Watson transformation in the t channel and continuing to the s channel does *not*, however, automatically produce the correct behavior as $t \to 0$ for all five s-channel amplitudes. It follows that in this limit the t-channel amplitudes must cooperate in order to yield the correct vanishing behavior for spin-flip amplitudes in the s channel.

This cooperation can take place basically in either of two ways. Each amplitude will presumably contain several Regge pole terms; there may be several residue functions for each trajectory, arising from the different t-channel spin amplitudes, as well as perhaps more than a single trajectory. If two *different* trajectories combine to yield the correct t-behavior, we say that there is a *conspiracy* between these trajectories; whereas obtaining the necessary behavior in the residues of each of the trajectories is known as *evasion.* It has been shown that evasion is always possible, and therefore that conspiracy is never absolutely necessary. It is nonetheless possible, however, and if two Regge trajectories do conspire together a number of constraints on both the residues and the trajectories themselves are obtained. Experimentally, though, the case for conspiracy is far poorer than that for daughters, and the entire subject may well prove a mathematical obfuscation.

Chapter 17

REGGE PHENOMENOLOGY

Equipped with a basic knowledge of the properties of the Regge pole amplitude, let us now turn to the question of how observed high-energy phenomena may be interpreted in terms of crossed-channel Regge trajectories. As we have said, the dominance of a single trajectory leads to shrinking diffraction peaks, vanishing polarizations, and (for the Pomeranchuk trajectory) constant total cross sections. Historically, the first reaction to be compared with Regge theory was elastic proton–proton scattering at laboratory momenta going from 6 to 20 GeV. In this range the total cross section was decreasing slowly, indicating that asymptotic energies were still somewhat higher; but the differential cross section did, indeed, reveal a shrinkage of the diffraction peak! As a result, Regge theory became a very popular topic in high-energy physics.

Before long, however, pion–proton scattering was also studied in this same energy range, and here the results were less satisfactory; the total cross sections decreased by 20%, and, more seriously, the slope of the diffraction peak remained constant. Before the theory could recover from this setback came evidence that in $\bar{p}p$ and $\bar{K}p$ scattering there was (as it was then called) "antishrinkage", i.e. the slope of the diffraction peak was decreasing, over similar energies.

There followed a winter of discontent, during which the Regge theory was kept alive by a few hardy souls (notably Phillips and Rarita) who insisted that although a single Regge pole was certainly inadequate at these energies, a relatively small number of different ones might still suffice. As it became apparent that this hope was reasonable, the Regge theory burst back to life; in fact, the opposite extreme has now been reached (and hopefully passed) in which all difficulties are explained by adding a few more poles.

In this chapter we shall ignore the historical order of development and begin by discussing in some detail the Regge description of pion–nucleon scattering for incident pion momenta between 6 and 30 GeV/c. In this single example we find the greatest simplicity of all the available reactions, some of the most salient successes and the most worrisome of the difficulties. We shall then describe the extensions necessary to accommodate other measured reactions within a similar framework. Data at higher energy have called some aspects of this description into question. In order to facilitate introduction of the traditional Regge framework, we shall postpone the discussion of these data and their unresolved questions until the following chapter.

Reference will frequently be made to "fitting" the data; it is appropriate here to mention that in practice, this usually means writing the amplitude in terms of a set of variable parameters and using standard computer programs to find the values of those parameters producing the best agreement with experimental data.

Pion–Nucleon Total Cross Sections

The most direct test for Regge behavior is in the predicted power-law dependence of total cross sections. Let us, for the moment, neglect spin complications and label the three channels as

$$\begin{aligned} &s\colon p\pi^+ \to p\pi^+ \\ &t\colon p\bar{p} \to \pi^-\pi^+ \\ &u\colon p\pi^- \to p\pi^-. \end{aligned} \tag{17.1}$$

Elastic $\pi^\pm p$ scattering will then be described by the continuations into the s and u channels of the Regge pole terms obtained in the t-channel process $p\bar{p} \to \pi^-\pi^+$. There are two isospin amplitudes for the t channel, corresponding to the I = 0 and I = 1 nucleon–antinucleon states. Decomposing as in Chapter 3 leads to

$$F(p\bar{p} \to \pi^-\pi^+) = \tfrac{1}{2}F_1(N\bar{N} \to \pi\pi) + \frac{1}{\sqrt{6}}F_0(N\bar{N} \to \pi\pi), \tag{17.2}$$

where $F_I(N\bar{N} \to \pi\pi)$ is the reduced amplitude for isospin I. The Regge trajectories producing I = 1 bound states and resonances, such as the ρ, appear as poles in F_1, while those for I = 0 are correspondingly in F_0. We have seen already how the symmetries of the $\pi\pi$ isospin wave functions restrict the values of the signature τ; only odd-ℓ resonances can decay to I = 1 $\pi\pi$ combinations, and only even-ℓ to I = 0. Thus F_1 can contain only $\tau = -1$ Regge poles, and F_0 only $\tau = +1$.

It follows then from the line reversal symmetry (16.40) that F_0 will make identical contributions to the s and u channel processes, while those of F_1 will be equal in magnitude but opposite in sign. This result is, in fact, independent of Regge theory in this particular case, deriving directly from the symmetries of the $\pi\pi$ isospin amplitudes. If we absorb the coefficients in (17.2) by writing $f_1(s, t) = \tfrac{1}{2}F_1$ and $f_0(s, t) = F_0/\sqrt{6}$, then the continuation of $F(p\bar{p} \to \pi^-\pi^+)$ into the s channel yields

$$F_{\pi^+p}(s, t) = f_0(s, t) + f_1(s, t), \tag{17.3a}$$

while continuation into the u channel, using (16.40), leads to

$$F_{\pi^-p}(u, t) = f_0(u, t) - f_1(u, t). \tag{17.3b}$$

It should be kept in mind just how the amplitudes $F_{\pi^\pm p}$ are related by crossing symmetry: there exists a single function F(s, t, u) for which F(s, t, u) = $F_{\pi^+p}(s, t) = F_{\pi^-p}(u, t)$, becoming the physical $\pi^\pm p$ scattering amplitude in the appropriate physical region.

Now let us write the π^-p amplitude as a function of s, with $s \gg (m + \mu)^2$, rather than u,

$$F_{\pi^-p}(s, t) = f_0(s, t) - f_1(s, t). \tag{17.4}$$

That is, we now *forsake* the labeling of channels given in (17.1) and use s to label the square of the total energy of the π^-p system. Then the contributions of f_0 and f_1 may be separated into

$$f_0(s, t) = \tfrac{1}{2}[F_{\pi^+p}(s, t) + F_{\pi^-p}(s, t)]$$

$$f_1(s, t) = \tfrac{1}{2}[F_{\pi^+p}(s, t) - F_{\pi^-p}(s, t)], \tag{17.5}$$

and we may compare these amplitudes with experimental data. The optical theorem yields the total cross sections

$$\sigma_0(s) = \frac{4\pi}{kW} \operatorname{Im} f_0(s, 0) = \tfrac{1}{2}[\sigma_{\pi^+p}(s) + \sigma_{\pi^-p}(s)]$$

$$\sigma_1(s) = \frac{4\pi}{kW} \operatorname{Im} f_1(s, 0) = \tfrac{1}{2}[\sigma_{\pi^+p}(s) - \sigma_{\pi^-p}(s)] \tag{17.6}$$

which, if a single pole is indeed dominant, should have the power-law dependence (16.25).

In Fig.(17-1) are shown the πp total cross sections from 6 to 30 GeV/c laboratory pion momentum. (Data at higher energies will be dealt with later.) Let us consider first the I = 1 contribution $\sigma_1(s)$, the behavior of which is indicated in Fig.(17-2a). Although the experimental errors introduce some uncertainty, the difference of total cross sections does seem to be decreasing satisfactorily. Fitting these data with a power-law dependence such as

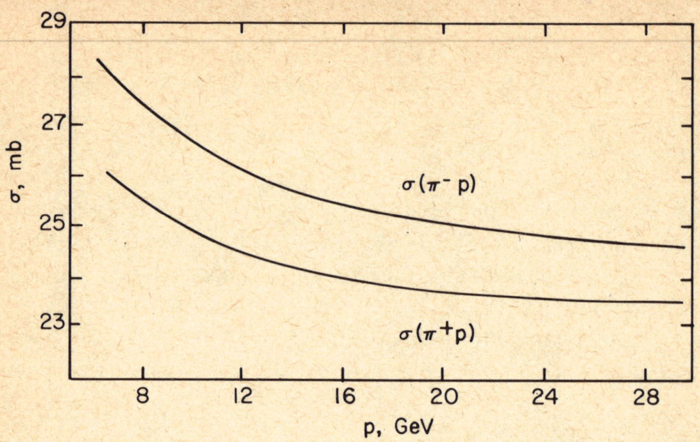

Fig.(17-1) Curves showing the approximate variation of the pion–nucleon total cross sections. (From the data of K. J. Foley *et al.*, *Phys. Rev. Letters* **14**, 189 (1967).)

$$\sigma_1(s) \approx c_1 s^{\alpha_1(0)-1} \tag{17.7}$$

yields an intercept of the order $\alpha_1(0) \sim 0.5$, the exact value depending on what other data (e.g. differential cross sections, polarizations, etc.) have been fitted simultaneously. Significantly, the intercept $\alpha_1(0)$ is entirely consistent with the extrapolation of the linear ρ trajectory (15.34b), which yields $\alpha_\rho(0) = 0.47$.

For the I = 0 total cross section σ_0, however, the dominance of a single trajectory is less clear. The data show, as in Fig.(17-2b), that $\sigma_0(s)$ decreases substantially in this energy range. While it is certainly conceivable that $\sigma_0(s)$ could be fitted using a single pole term, to do so would require an intercept less than unity, and the total cross sections would consequently be predicted to vanish asymptotically. In order to maintain asymptotically constant total cross sections, we must assume that $\sigma_0(s)$ has contributions from two distinct trajectories, one of which is the Pomeranchuk trajectory $\alpha_P(t)$ (which, for brevity, is often called simply the Pomeron). Thus we write

$$\begin{aligned}\sigma_0(s) &= \sigma_0(\infty) s^{\alpha_P(0)-1} + c_0 s^{\alpha_{P'}(0)-1} \\ &= \sigma_0(\infty) + c_0 s^{\alpha_{P'}(0)-1}\end{aligned} \tag{17.8}$$

assuming $\alpha_P(0) = 1$. The second trajectory is customarily called the P′, to indicate that it has the same quantum numbers as the P – namely those of the vacuum.

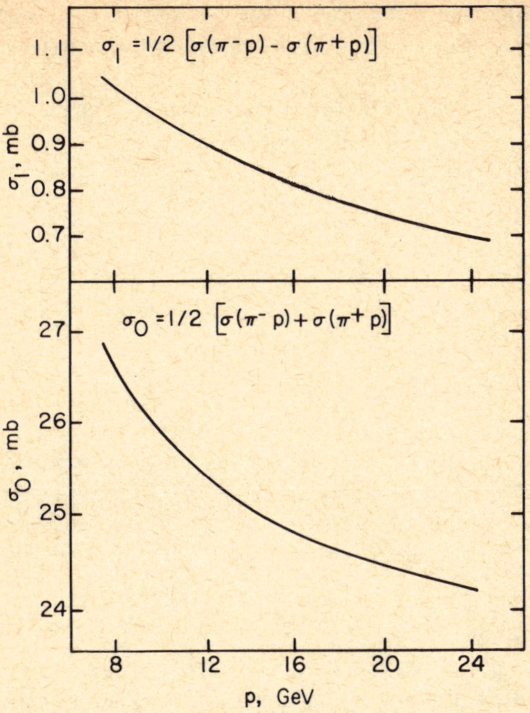

Fig.(17-2) Curves showing the approximate variation of the difference and the sum of the pion–nucleon total cross sections shown in Fig.(7-1).

Fitting (17.8) to the data shows that the exact values of c_0 and $\alpha_{P'}(0)$, as well as of the asymptote $\sigma_0(\infty)$, are rather elusive; that is, they may vary considerably and yet cooperate to maintain a good fit. Typical values are $\alpha_{P'}(0) \sim \frac{1}{2}$ and $\sigma_0(\infty) \sim 19$ mb, with about 5 mb of total cross section due to P′.

Thus it appears that, insofar as the total cross sections are concerned, pion–nucleon scattering cannot be described by a single Regge pole. These energies are simply not high enough for that elegant simplicity to have set in; or, as it has been punningly put, we have not yet reached "asymptopia". Instead, at least three trajectories are required – two with I = 0 and one, apparently the ρ, with I = 1.

Charge Exchange and the ρ Trajectory

It is possible, however, to isolate the contribution of the I = 1 amplitude by considering the charge-exchange process $\pi^- p \to \pi^\circ n$ rather than elastic scattering, and to study in this way a process which may be dominated by a single pole. If this process is the s channel, then the t channel is $p\bar{n} \to \pi^+\pi^\circ$, for which there is a single I = 1 isospin amplitude (i.e. a charged meson must be exchanged). Instead of (17.2), we have

$$F(p\bar{n} \to \pi^+\pi^\circ) = \frac{1}{\sqrt{2}} F_1(N\bar{N} \to \pi\pi),$$

yielding in the crossed channel

$$F_{\pi^- p \to \pi^\circ n}(s, t) = \sqrt{2}\, f_1(s, t) \tag{17.9}$$

which, of course, agrees with the isospin equality

$$F_{\pi^- p \to \pi^\circ n} = \frac{1}{\sqrt{2}}(F_{\pi^+ p} - F_{\pi^- p}).$$

It turns out experimentally that in the charge exchange reaction significant contributions arise from the spin-flip amplitude. A detailed treatment of the Regge description of this process thus provides the best possible example of one-pole phenomenology. We define the two spin amplitudes F(s, t) and G(s, t) as in Appendix B, so that the differential cross section and polarization are given by

$$\frac{d\sigma}{dt}(s, t) = \frac{\pi}{k^2 W^2}(|F(s, t)|^2 + |G(s, t)|^2) \tag{17.10a}$$

$$P(s, t) = -2\, \frac{\mathrm{Im}(F^*(s, t) G(s, t))}{|F(s, t)|^2 + |G(s, t)|^2}. \tag{17.10b}$$

Both F(s, t) and G(s, t) can be obtained by continuation of the Regge pole amplitudes for the crossed channel $p\bar{n} \to \pi^+\pi^\circ$. There are two of these, corresponding to singlet and triplet spin orientations for the $p\bar{n}$ combination. The details of the Reggeization procedure are rather obscure, and we shall only give the results here.

Assuming that the ρ trajectory is completely dominant, the non-flip amplitude F(s, t) is given, analogously to the spinless case, by

$$F(s, t) = F_\rho(s, t) = \beta_\rho(t) \frac{\zeta_\rho(t)}{\sin \pi\alpha_\rho(t)} \left(\frac{s}{s_0}\right)^{\alpha_\rho(t)} \tag{17.11}$$

where $\zeta_\rho(t) = (1 - e^{-i\pi\alpha_\rho(t)})$ is the signature factor and we have absorbed all other residue terms into the real function $\beta_\rho(t)$. A normalization constant s_0 has been extracted so that the quantity raised to the power $\alpha_\rho(t)$ will be dimensionless; from the continuation of $\cos\theta_t$, we would have $s_0 = 2m\mu$, but since $s_0^{-\alpha_\rho(t)}$ is only a function of t and could therefore be included with $\beta_\rho(t)$, this choice is not crucial, and $s_0 = 1\ \text{GeV}^2$ is often chosen for simplicity.

The corresponding parametrization of G(s, t) is

$$G(s, t) = G_\rho(s, t) = \sqrt{-t}\,\gamma_\rho(t) \frac{\zeta_\rho(t)}{\sin \pi\alpha_\rho(t)} \alpha_\rho(t) \left(\frac{s}{s_0}\right)^{\alpha_\rho(t)}, \tag{17.12}$$

containing a factor $\sqrt{-t}$ arising from spin kinematics, reflecting the fact that for forward scattering (t = 0) the spin-flip process is forbidden by conservation of angular momentum. In addition, (17.12) contains a real residue $\gamma_\rho(t)$, the signature $\zeta_\rho(t)$, and an extra factor $\alpha_\rho(t)$. This last factor arises because of the sense–nonsense mechanism described in the preceding chapter, which requires the ρNN vertex to vanish as $\alpha_\rho \to 0$. A simpler heuristic explanation involves noticing that the crossed-channel amplitude from which G(s, t) is obtained has a partial wave expansion involving the angular dependence $\sin\theta_t P'_\ell(\cos\theta_t)$ rather than simply $P_\ell(\cos\theta_t)$. When continued to large s, $t \leqslant 0$, this becomes

$$\sqrt{1 - z^2}\, P'_\alpha(z) \approx iz \frac{d}{dz}(z^\alpha) = i\alpha z^\alpha$$

yielding a factor $\alpha(t)$ because of the differentiation.

It follows from (17.11) and (17.12) that, for large s, the differential cross section is

$$\frac{d\sigma}{dt} = \frac{\pi}{k^2W^2} \frac{|\zeta_\rho(t)|^2}{\sin^2 \pi\alpha_0(t)} \left\{|\beta_\rho(t)|^2 - t|\alpha_\rho(t)|^2|\gamma_\rho(t)|^2\right\} \left(\frac{s}{s_0}\right)^{2\alpha_\rho(t)}$$

$$\approx B(t) \left(\frac{s}{s_0}\right)^{2\alpha_\rho(t)-2} \tag{17.13}$$

Thus shrinkage is predicted for the differential cross section just as in the absence of spin. By studying the variation of (17.13) with s at fixed t, we can obtain experimentally the value of $\alpha_\rho(t)$ as in (16.29). The results of such a measurement at several different values of t are shown in Fig.(17-3), along with the extrapolation of the linear trajectory (15.34b). While it appears that the trajectory is becoming somewhat flatter in the negative t region, the two are certainly not incompatible. In fact, the ρ trajectory obtained for the charge-exchange differential cross section may be parametrized linearly with the approximate result

$$\alpha_\rho(t) \approx ½ + t \qquad (17.14)$$

in excellent agreement with the linear trajectory obtained using the ρ and g masses.

The shape of $d\sigma/dt$, as a function of t, provides information on the residue functions $\beta_\rho(t)$ and $\gamma_\rho(t)$. In general, for charge exchange the differential cross section differs from the simple diffraction peak (16.30) in that there is a dip in the forward direction, typically as shown in Fig.(17-4). It is this dip which is

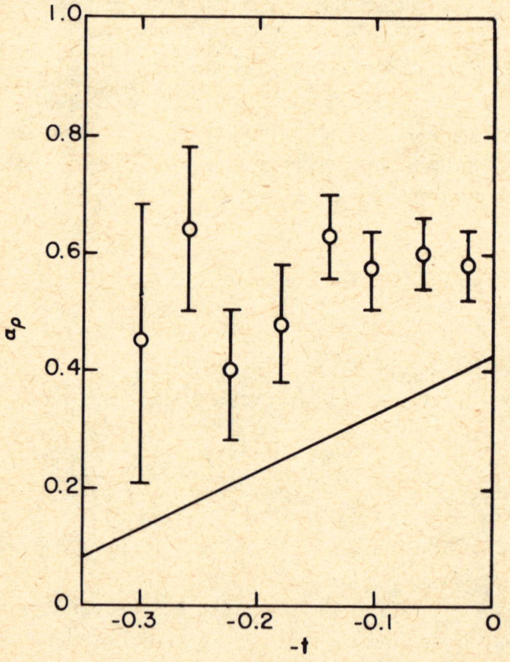

Fig.(17-3) Values of $\alpha_\rho(t)$ for $t < 0$ obtained from the pion–nucleon charge exchange process, along with the straight-line trajectory (15.34b).

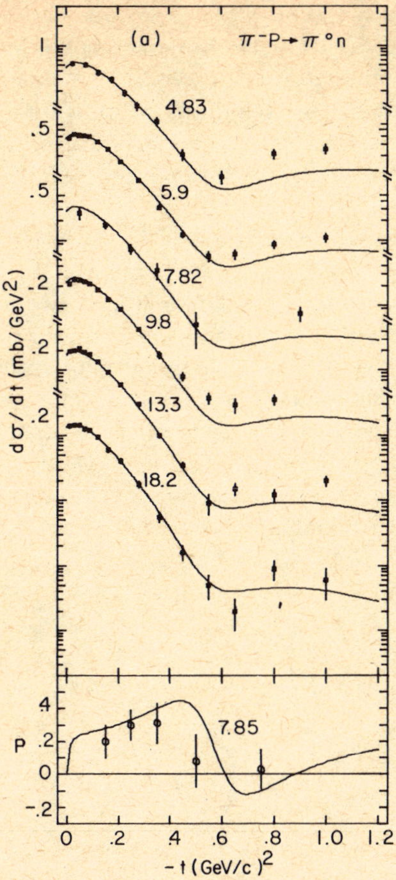

Fig.(17-4) Differential cross section and polarization for pion–nucleon charge exchange, showing the structure typical of this reaction. The curves are a fit to the data. (From S. E. Egli, D. W. Duke, and N. W. Dean, *Phys. Rev.* **D9**, 1365 (1974).)

interpreted as evidence for significant contributions from spin-flip, since G(s, t) vanishes as $t \to 0$. Although it is, of course, possible that $\beta_\rho(t)$ could be entirely responsible for this structure, it seems more likely that both $\beta_\rho(t)$ and $\gamma_\rho(t)$ are smooth functions of t, with the dip resulting from the factor t in (17.13a) which eliminates the spin-flip contribution as $t \to 0$.

If that is the case, the data should also show the vanishing of G(s, t) where $a_\rho(t) = 0$. According to (17.14), this should occur for $t \approx -0.5\ \mathrm{GeV}^2$; and there is, in fact, a second dip in the differential cross section very near that region.

The fact that $d\sigma/dt$ does not vanish at this point indicates that the ρNN vertex "chooses sense" to yield $P_\rho(t) \neq 0$; the vanishing factor is in the nonsense coupling, and this is a "wrong-signature nonsense zero." Whether the factor $\alpha_\rho(t)$ in (17.12) is actually responsible for this dip, however, is not entirely clear; there are other reactions in which a similar phenomenon is expected but not observed, and there are various other explanations of the dip not involving the vanishing of the trajectory.

Finally, let us turn to the polarization. If the amplitudes (17.11) and (17.12) are correct, then P(s, t) must vanish, as shown in Chapter 16. Experimentally, this prediction is contradicted; the polarization is positive and as large as about 25%, as shown for a typical value of laboratory momentum. It follows inescapably that the ρ trajectory, despite its successes regarding the differential cross section and the difference of total cross sections, is not sufficient to describe completely the charge-exchange amplitudes. Another term of some sort, having a different phase from that of the ρ, is needed.

Various alterations of F(s, t) and G(s, t) have been proposed in order to account for this failure. The simplest of these is the addition of a second Regge pole term, usually called ρ', with $\alpha_{\rho'}(t) < \alpha_\rho(t)$. Writing

$$F(s,t) = \beta_\rho(t)\frac{\zeta_\rho(t)}{\sin \pi\alpha_\rho(t)}\left(\frac{s}{s_0}\right)^{\alpha_\rho(t)} + \beta_{\rho'}(t)\frac{\zeta_{\rho'}(t)}{\sin \pi\alpha_{\rho'}(t)}\left(\frac{s}{s_0}\right)^{\alpha_{\rho'}(t)}$$

$$G(s,t) = \sqrt{-t}\left\{\gamma_\rho(t)\frac{\zeta_\rho(t)}{\sin \pi\alpha_\rho(t)}\alpha_\rho(t)\left(\frac{s}{s_0}\right)^{\alpha_\rho(t)} + \gamma_{\rho'}(t)\frac{\zeta_{\rho'}(t)}{\sin \pi\alpha_{\rho'}(t)}\alpha_{\rho'}(t)\left(\frac{s}{s_0}\right)^{\alpha_{\rho'}(t)}\right\} \tag{17.15}$$

leads to more complicated generalizations of (17.13); since the ρ pole alone was capable of reproducing $d\sigma/dt$ satisfactorily, the contributions of ρ' to it must be small. The polarization resulting from (17.15) arises only from interference between ρ and ρ', and is given by

$$P(s,t) = \sqrt{-t}\, p(t)\left(\frac{s}{s_0}\right)^{\alpha_{\rho'}-\alpha_\xi} \sin \tfrac{1}{2}\pi(\alpha_\rho - \alpha_{\rho'}) \tag{17.16}$$

where p(t) is a combination of the residue functions and we have neglected the ρ' contribution to the denominator of (17.10b). The polarization predicted by (17.16) cannot reproduce the experimental data as a function of t unless $\alpha_\rho - \alpha_{\rho'} \gtrsim \frac{1}{2}$; but in that case the energy dependence $s^{\alpha_\rho - \alpha_{\rho'}}$ would be too rapid. Consequently a second trajectory $\alpha_{\rho'}(t)$ cannot explain this polarization.

Elastic Differential Cross Sections and Polarizations

The elastic $\pi^{\pm}p$ scattering amplitudes are much more complicated functions than for charge exchange, since each of them consists of at least three Regge pole terms P, P′, and ρ. We may write them conveniently as

$$F_{\pi^{\pm}p}(s,t) = F_P(s,t) + F_{P'}(s,t) \pm F_{\rho}(s,t)$$

$$G_{\pi^{\pm}p}(s,t) = G_P(s,t) + G_{P'}(s,t) \pm G_{\rho}(s,t), \qquad (17.17)$$

where $F_{\alpha}(s, t)$ and $G_{\alpha}(s, t)$ denote the contributions of each trajectory as in (17.11) and (17.12). The presence of so many terms leads to considerable freedom in adapting the predictions of (17.17) to the experimental data.

It may be observed in Fig.(17-5), however, that the differential cross sections for both π^+p and π^-p show no indication of a dip in the forward direction. We may therefore assume that the contributions of GP(s, t) and $G_{P'}(s, t)$ are small relative to the corresponding non-flip amplitudes $F_P(s, t)$ and $F_{P'}(s, t)$; in fact, comparing the elastic and charge-exchange data shows that they are also much smaller than $G_{\rho}(s, t)$. By contrast, $F_{\rho}(s, t)$ is much smaller than $F_P(s, t)$, as shown by the small difference of total cross sections.

The polarizations should therefore be dominated by the interference between $(F_P(s, t) + F_{P'}(s, t))$ and $G_{\rho}(s, t)$,

$$P_{\pi^{\pm}p}(s,t) \approx -2\,\frac{\mathrm{Im}\{[F_P(s,t) + F_{P'}(s,t)]^*[\pm G_{\rho}(s,t)]\}}{|F(s,t)|^2 + |G(s,t)|^2} \qquad (17.18a)$$

i.e.

$$P_{\pi^{\pm}p}(s,t) \approx -P_{\pi^-p}(s,t). \qquad (17.18b)$$

Because $G_{\rho}(s, t)$ contributes with opposite signs in the two reactions, we thus expect that $P_{\pi^+p}(s, t)$ and $P_{\pi^-p}(s, t)$ are opposite in sign but approximately equal in magnitude. The experimental data, shown for lab momentum 10 GeV/c in Fig.(17-5), bear out this expectation. Furthermore, if the dip in the charge-exchange differential cross section at $t \approx -0.6\ \mathrm{GeV}^2$ is caused by the vanishing of $G_{\rho}(s, t)$, then there should be a corresponding zero in the polarization. In fact, it is easily shown that this zero should be a quadratic one. The relevant t-dependence in (17.12) is in the factors

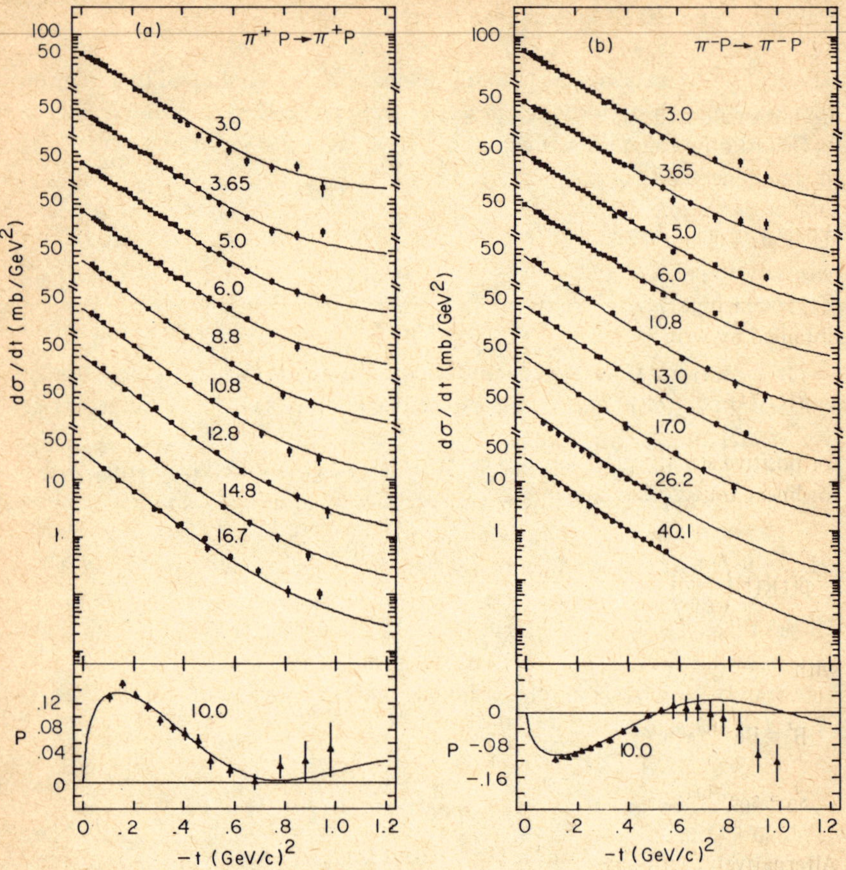

Fig.(17-5) Plots analogous to Fig.(17-4) for elastic pion–nucleon scattering, from the same source.

$$\frac{a_\rho(t)\zeta_\rho(t)}{\sin \pi a_\rho(t)} = \frac{i a_\rho e^{-1/2 i\pi a_\rho} \sin \tfrac{1}{2}\pi a_\rho}{\sin \pi a_\rho}$$

while the $(P + P')$ contribution is basically structureless in t. Furthermore, the latter contributes principally through $F_P(s, t)$ and is therefore essentially purely imaginary. Consequently the polarization is proportional to the real part of $G_\rho(s, t)$, and near $a_\rho(t) \to 0$

$$\mathrm{Im}F_P^*G_\rho \propto \frac{\alpha_\rho \sin^2 \tfrac{1}{2}\pi\alpha_\rho}{\sin \pi\alpha_\rho} \sim \alpha_\rho^2$$

produces a double zero. The data show exactly this behavior.

Quantitative calculations have been carried out using the three-pole model to fit simultaneously a large amount of pion–nucleon scattering data, including total cross sections and the phase of the forward scattering amplitude as well as elastic and charge-exchange differential cross sections and polarizations. It is necessary to parametrize somehow the t-dependence of the residues $\beta(t)$ and $\gamma(t)$ for each pole; a simple form containing the correct phase, for example, is obtained by writing

$$\beta(t)\zeta(t) = Be^{bt}e^{-i\pi\alpha(t)/2} \tag{17.19a}$$

for positive signature, adding an extra factor i if $\tau = -1$. This choice really amounts, for linear trajectories, to a redefinition of s_0, since then we have

$$\beta(t)\zeta(t)\left(\frac{s}{s_0}\right)^{\alpha_0+\alpha_1 t} = B'\left(\frac{s}{is_0'}\right)^{\alpha_0+\alpha_1 t} \tag{17.19b}$$

with

$$B' = Be^{-b\alpha_0/\alpha_1}$$

$$s_0' = s_0e^{-b/\alpha_1}.$$

Alternatively, various known terms in $\beta(t)$, such as the factor $(2\alpha(t) + 1)$, may be written explicitly in the parametrization. The details of the resulting fits will vary accordingly, of course, but in general they are quite satisfactory with the exception of the charge-exchange polarization problem mentioned earlier. Typical values for the two vacuum trajectories are

$$\alpha_P(t) = 1 + 0.3t$$

$$\alpha_{P'}(t) \approx 0.5 + t \tag{17.20}$$

The Pomeranchuk trajectory obtained in this way usually turns out to be rather flat, its small slope accounting for the non-shrinkage of the elastic diffraction peaks.

Regge Parametrization of Other Processes

A similar analysis can be applied to any two-particle reaction. There is one other process, $\pi^- p \to \eta^\circ n$, in which it appears that only a single trajectory may be involved. The crossed channel in this case is $p\bar{n} \to \pi^+\eta^\circ$, so the allowed trajectories have the internal quantum numbers of a $\pi^+\eta^\circ$ bound state – specifically, $I = 1$, $B = S = 0$, and $G = -1$. The negative G-parity rules out the ρ trajectory; in fact, of the known Regge trajectories, only the A_2 is allowed. (An odd signature trajectory with the same quantum numbers would also be allowed, but no evidence exists in the t channel for particles belonging to such a trajectory.) This process has been studied experimentally at several energies (Fig.(17-6)). It is quite similar to the $\pi^- p \to \pi^\circ n$, both in the shape of $d\sigma/dt$ and in the non-vanishing polarization. The measured values of $\alpha_{A_2}(t)$ require the trajectory to flatten out as compared with the linear extrapolation (15.34c) even more than for the ρ, as shown in Fig.(17-7).

More accessible to experiment are the elastic and charge-exchange reactions involving the other available incident particles. For example, both $K^\pm p$ and $K^\pm n$ elastic scattering have been studied, as have the charge exchanges $K^- p \to \bar{K}^\circ n$ and, to a more limited extent, $K^+ n \to K^\circ p$. For $K^- p$ elastic scattering, we analyze the crossed channel process $p\bar{p} \to K^+K^-$, in which both $I = 0$ and $I = 1$ states can be formed. In contrast to the pion–nucleon crossed channel, there is no connection between the isospin and the signature, since K^+ and K^- belong to different isomultiplets. Thus we expect to find both positive and negative signature trajectories contributing in each isospin channel. In addition to P, P′, and ρ, contributions may arise from poles with $I = 1$, $\tau = +$, such as the A_2; and with $I = 0$, $\tau = -$, with corresponds to the trajectory on which we might classify the ω meson. Thus a "minimal" Regge parametrization of $K^- p$ scattering will involve at least five poles. Denoting the contribution of each pole by $F_\alpha(s, t)$, as in (17.11), we may write

$$F_{K^-p}(s, t) = F_P + F_{P'} + F_\rho + F_\omega + F_{A_2} \tag{17.21a}$$

(the P, P′, and ρ contributions have different residues here, of course, from those in (17.17)).

An easy isospin calculation shows that the $K^- n$ elastic scattering amplitude differs from that of $K^- p$ only in the sign of the $I = 1$ exchange contribution, so (17.21a) is accompanied by

$$F_{K^-n}(s, t) = F_P + F_{P'} - F_\rho + F_\omega - F_{A_2}\,. \tag{17.21b}$$

Line reversal symmetry, on the other hand, relates both of these amplitudes to

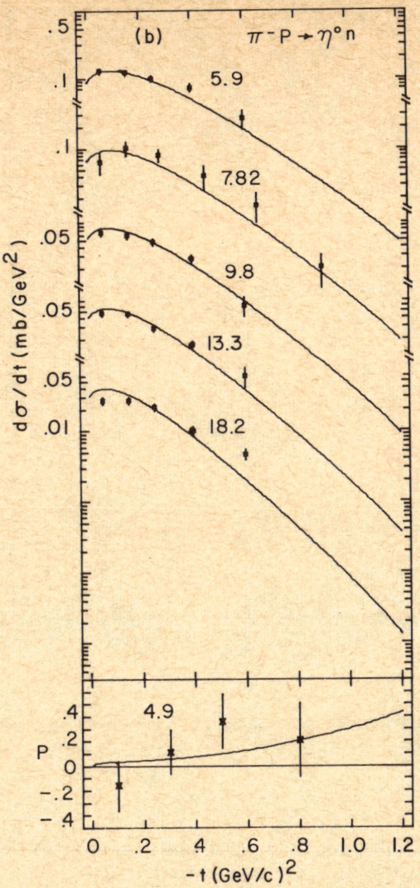

Fig.(17-6) Differential cross sections and polarization for $\pi^- p \to \eta^\circ n$, from the same source as Fig.(17-4).

those for K^+p and K^+n scattering, as in (17.4); changing K^- to K^+ requires us only to change the sign of the odd signature pole terms, yielding

$$F_{K^+p} = F_P + F_{P'} - F_\rho - F_\omega + F_{A_2} \tag{17.21c}$$

$$F_{K^+n} = F_P + F_{P'} + F_\rho - F_\omega - F_{A_2}. \tag{17.21d}$$

The $K^\pm$-nucleon total cross sections are shown in Fig.(17-8). At first glance,

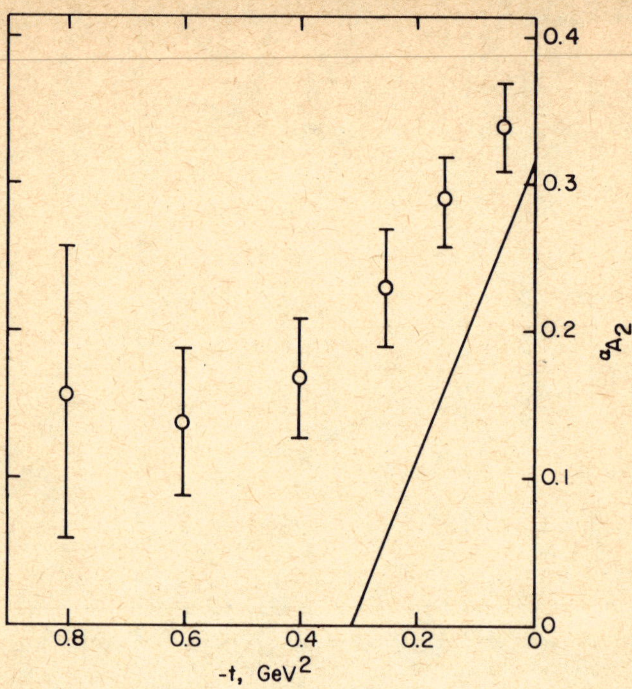

Fig.(17-7) The A_2 trajectory, as measured in $\pi^- p \to \eta^\circ n$, compared with the straight line trajectory.

the constancy of the K^+ data might suggest that we have been overcautious; the Pomeranchuk pole could explain the sum of these cross sections without P′ or ω being needed. If that were so, however, the line reversal symmetry would require identical constant values for the sum of the K^- cross sections, which are clearly not observed. In fact, we may invert equations (17.21) to obtain the contributions of each of the four types of pole,

$$F_P + F_{P'} = \tfrac{1}{4}[F_{K^-p} + F_{K^-n} + F_{K^+p} + F_{K^+n}]$$

$$F_\rho = \tfrac{1}{4}[F_{K^-p} - F_{K^-n} - F_{K^+p} + F_{K^+n}]$$

$$F_\omega = \tfrac{1}{4}[F_{K^-p} + F_{K^-n} - F_{K^+p} - F_{K^+n}]$$

$$F_{A_2} = \tfrac{1}{4}[F_{K^-p} - F_{K^-n} + F_{K^+p} - F_{K^+n}], \qquad (17.22)$$

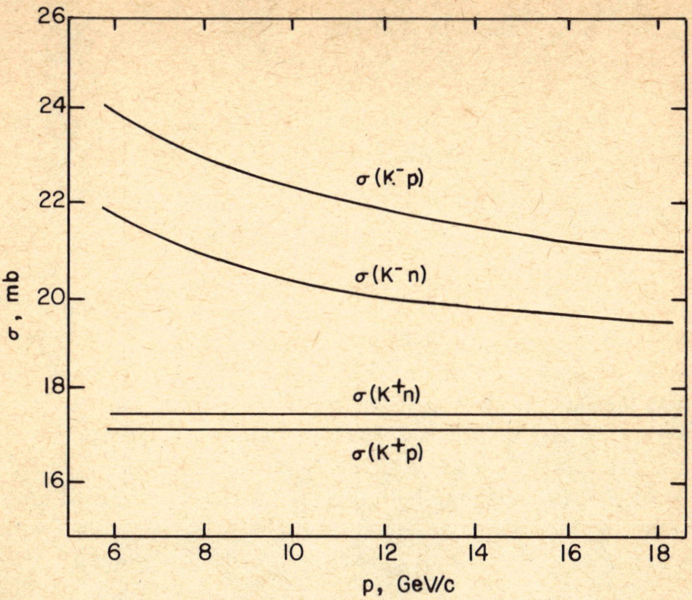

Fig.(17-8) Curves showing the approximate variation of the kaon–nucleon total cross sections. (From the data of W. Galbraith *et al.*, *Phys. Rev.* **138**, B913 (1965).)

and the data show clearly that none of these can be neglected. Thus the constant K^+p and K^+n total cross sections result from the mutual cancellation, rather than the absence, of the non-Pomeranchuk contributions. Specifically, both can be constant only if the P' pole term is canceled by the ω, and the ρ by the A_2, throughout the entire energy range, i.e.

$$F_{P'}(s, t) \approx F_{\omega}(s, t) \tag{17.23a}$$

$$F_{\rho}(s, t) \approx F_{A_2}(s, t). \tag{17.23b}$$

These equalities can hold only if the trajectories involved are equal and the residues opposite in sign. The identity of the ρ and A_2 trajectories has already been mentioned in Chapter 15 as an example of exchange degeneracy; by the same token, (17.23a) implies that two $I = 0$ trajectories with opposite signature are equal, i.e. that the P' and the ω are also exchange degenerate. Furthermore, if exchange-degenerate partners are indeed a manifestation of the insignificance of the u-channel contribution to the amplitude, we would expect them to have the same residues.

By using parametrizations similar to (17.19) for the amplitudes (17.21) and for the charge-exchange amplitudes

$$F_{K^-p\to\bar{K}^\circ n}(s,t) = 2F_\rho + 2F_{A_2}$$

$$F_{K^+n\to K^\circ p}(s,t) = -2F_\rho + 2F_{A_2}, \qquad (17.24)$$

plus similar forms for the spin flip amplitudes, satisfactory fits to the available kaon–nucleon data in this range can be obtained. The exact details will vary, of course, with the data used and with the form used to parametrize the residues, but the general features are relatively insensitive to such variations. The P, P′, and ρ trajectories resulting from (17.22) generally agree with those found in pion–nucleon scattering; in some cases both types of data have been fitted simultaneously, so agreement is guaranteed. The A_2 trajectory, likewise, is consistent with that obtained from $\pi^-p \to \eta^\circ n$, and the ω passes reasonably close to the ω when extrapolated to $a_\omega(t) = 1$.

The consideration of nucleon–nucleon and antinucleon–nucleon scattering is exactly isomorphic to that of the kaon–nucleon system. The same intermediate states are allowed in the t channel, and we may write analogously to (17.21) and (17.24)

$$F_{\bar{p}p}(s,t) = F_P + F_{P'} + F_\rho + F_\omega + F_{A_2}$$

$$F_{\bar{p}n}(s,t) = F_P + F_{P'} - F_\rho + F_\omega - F_{A_2}$$

$$F_{pp}(s,t) = F_P + F_{P'} - F_\rho - F_\omega + F_{A_2}$$

$$F_{pn}(s,t) = F_P + F_{P'} + F_\rho - F_\omega - F_{A_2}$$

$$F_{\bar{p}p\to\bar{n}n}(s,t) = 2F_\rho + 2F_{A_2}$$

$$F_{pn\to np}(s,t) = -2F_\rho + 2F_{A_2} \qquad (17.25)$$

with the appropriate redefinition of the residue functions. The total cross sections for these processes, pictured in Fig.(17-9), again show evidence for $\rho - A_2$ and $P' - \omega$ exchange degeneracy in the nucleon–nucleon total cross sections, which are not absolutely constant here but decrease much less rapidly than the other two. Detailed fits to these interactions are more difficult than for the

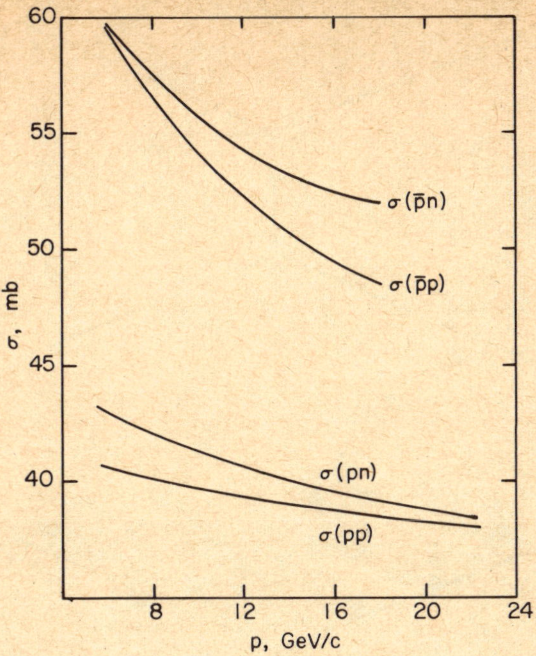

Fig.(17-9) Curves showing the approximate variation of the nucleon–nucleon and antinucleon–nucleon total cross sections. (From the data of W. Galbraith *et al., Phys. Rev.* **138**, B913 (1965).)

meson–nucleon processes because of the fact that (for *each* pole) five amplitudes are required to describe the scattering of two spin-½ particles, as compared with only two involved in (17.10). The results, however, are consistent with those obtained for the πN and KN reactions.

Finally, we may mention here one application of baryonic Regge trajectories, the description of backward pion–nucleon scattering. If the s channel reaction is $p\pi^\pm \to p\pi^\pm$, then backward scattering is dominated by the u-channel Regge trajectories arising from $p\pi^+ \to \pi^+p$. Thus backward π^+p scattering should be described at high energy by the dominant Regge poles arising in π^-p scattering, and vice versa As shown in Fig.(17-10), either $I = \frac{1}{2}$ or $I = \frac{3}{2}$ baryon trajectories can contribute to $\pi^+p \to p\pi^+$; but only $I = \frac{3}{2}$ is allowed for $\pi^-p \to p\pi^-$, since two units of charge must be exchanged. In terms of known trajectories, then, we expect the former to be dominated by the N_α and the Δ_δ trajectories, the latter only by Δ_δ. Experimentally it is observed that the differential cross section for $\pi^+p \to p\pi^+$ is in fact considerably larger than that for $\pi^-p \to p\pi^-$, indicating that the N contributions

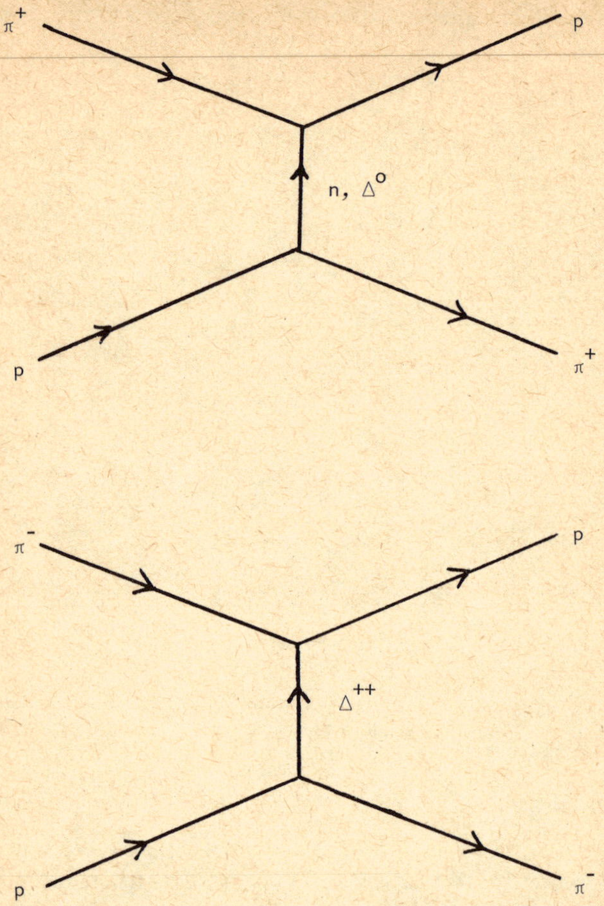

Fig.(17-10) Allowed exchanges for $\pi^{\pm}p \rightarrow p\pi^{\pm}$.

are quite significant. Furthermore, it can be argued that the N_α amplitude might vanish because of a "ghost-killing" effect when $\alpha_N(u) = -\frac{1}{2}$. This occurs when $u \approx -0.2$, and a dip in the differential cross section is observed in precisely that region (Fig.(17-11)).

The techniques which have been presented here can be applied to any number of other processes. Strangeness exchange, for instance, can be parametrized using Regge trajectories which may be compared with those on which we would classify strange particles and resonances. Other inelastic processes may be studied;

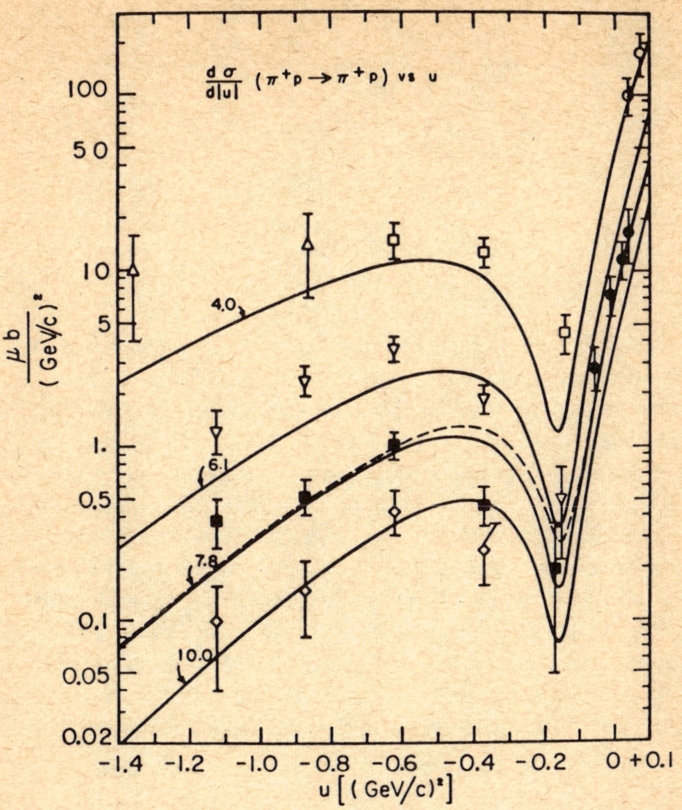

Fig.(17-11) Differential cross section for backward π^+p scattering, $\pi^+p \rightarrow p\pi^+$. (From C. B. Chiu and J. D. Stack, *Phys. Rev.* **153**, 1575 (1967).)

for example, $\pi N \rightarrow \omega\Delta$ should be dominated by the ρ trajectory, while both ρ and A_2 could contribute to $KN \rightarrow K^*\Delta$. A particularly interesting type of reaction is one which is inelastic but may involve Pomeranchuk exchange, such as $pp \rightarrow pN_\alpha(1688)$; the amplitude for such processes depends on the energy only through the shrinkage of its diffraction peak. The partial cross sections for these reactions will therefore decrease only logarithmically with energy, while all other inelastic reactions will disappear much more quickly.

Chapter 18

REGGE CUTS

What we have presented in the preceding chapter may be regarded as "basic" Regge phenomenology, in the sense that since about 1964 it has become a traditional tool in the high energy physicist's workshop. Its only significant failure in the 5–30 GeV/c range of incident momentum, apart from losing the elegance of the single-pole theory, was the non-vanishing polarization observed in pion–nucleon charge exchange; and it was hoped that some minor modification such as a ρ' pole would remedy that discrepancy. In 1969, however, the first data at higher energies became available from the Russian accelerator at Serpukhov, and with them appeared some clouds on the Regge horizon. With subsequent data from the intersecting storage rings (ISR) at CERN, near Geneva, and from the Fermi National Accelerator Laboratory (FNAL), near Chicago, these clouds erupted into a substantial storm. In this chapter we shall consider the nature of the difficulties in these higher-energy data and indicate some possible directions for their resolution.

Problems at Higher Energies

The meson–nucleon total cross sections in the Serpukhov energy range are shown in Fig.(18-1a). It is the unexpected constancy of $\sigma(\pi^{\pm}N)$ and $\sigma(K^{\pm}N)$ which posed problems for Regge theory. The fits obtained at lower energies, extended into this region, fall considerably below the Serpukhov data as they decrease toward an asymptote.

The difficulties go deeper, however, than the need for readjusting an asymptote. The K^-N total cross sections, in particular, appear to be leveling off to a constant value which is several millibarns *larger* than the already-constant K^+N data. If this behavior is interpreted as indicating different asymptotic total cross sections for K^+N and K^-N, then the Pomeranchuk theorem must be violated. To violate this rather basic theorem within the structural confines of Regge theory requires considerable imagination, and most of the forms proposed for this purpose seemed clumsy and contrived. The emphasis consequently developed around

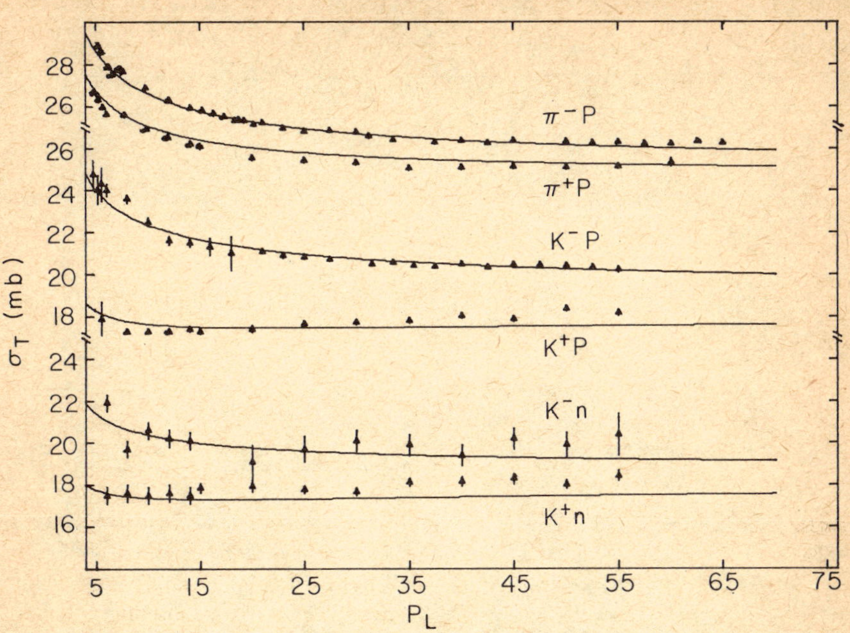

Fig.(18-1a) The meson–nucleon total cross sections measured up through the Serpukhov energy range. The solid lines are fits using a Regge cut model. (From S. E. Egli, D. W. Duke, and N. W. Dean, *Phys. Rev.* **D9**, 1365 (1974).)

methods of altering the approach to asymptopia so that it can accommodate the Serpukhov data.

The simplest way to do this was indicated by the Serpukhov K^+N total cross sections, which – although certainly not beyond experimental errors – seemed to show a slight *increase.* Measurements of the proton–proton total cross section at drastically higher energies gave a similar result for that reaction – an apparent slight rise in σ(pp). Data from FNAL have more recently begun to fill in the gap between Serpukhov and ISR energies, leading to the situation shown in Fig.(18-1b). Experimental normalizations may differ, of course, and the increases seen are not large compared with experimental uncertainties; but at present all available data (including also some cosmic ray data at superhigh energies) seem to confirm the same phenomenon. The total cross section is *increasing* with energy.

But is it increasing toward an asymptote? Or is it increasing without bound? The Froissart bound (16.24) limits the asymptotic growth of the total cross section according to

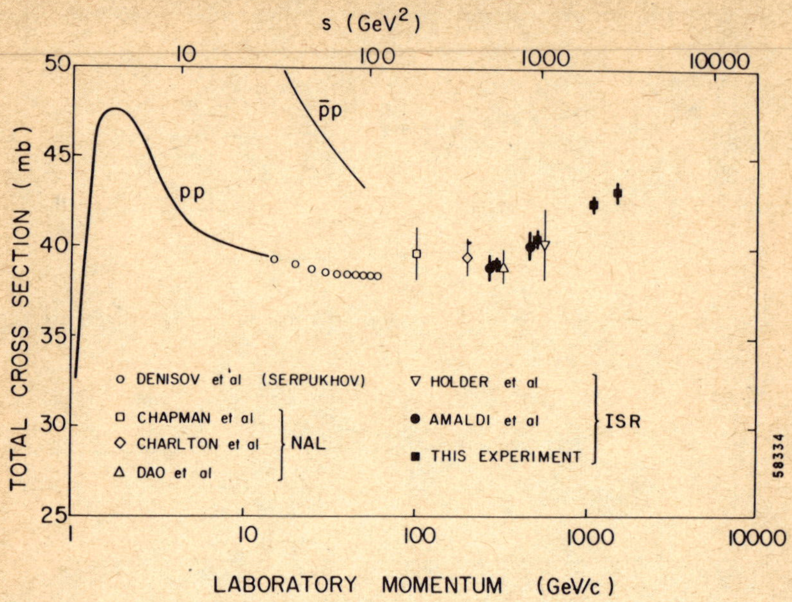

Fig.(18-1b) Measurements of the proton–proton total cross section including the Serpukhov, NAL, and ISR energy ranges. (From U. Amaldi *et al., Phys. Lett,* **44B**, 112 (1973).)

$$\sigma(s) \leqslant c \ln^2 s. \tag{16.24}$$

If the Pomeron is actually a Regge pole, therefore, it must have $\alpha_P(0) = 1$, and the total cross section must be asymptotically constant. The alternatives for producing a rise in $\sigma(s)$ are therefore crucially dependent on the nature of this singularity. If we maintain it as a pole term of the usual form, secondary terms must be included which *subtract* from it but vanish with increasing energy, yielding an increase toward the asymptote. Alternatively, we may dispense with the assumption that there is an asymptote, and let the term we have been calling the Pomeranchuk contribution have an energy dependence different from that of a simple Regge pole. The latter choice is not unreasonable, since it has been found difficult to associate any direct-channel resonances with the Pomeranchuk trajectory. Furthermore it seems esthetically unlikely that nature would choose a behavior less than the most extreme possible; since it is not possible to prove any limit better than the Froissart bound, one should be surprised if that bound is not attained.

Regge Cuts

The simplest way to generate correction terms allowing the Pomeron contribution to rise toward its asymptote is via the effects of a branch point in the angular momentum plane. Such "Regge cut" terms arise naturally in several models, and had been used as a possible explanation of the non-vanishing $\pi^- p \to \pi^\circ n$ polarization. Indeed, the use of such models for pion–nucleon interactions (by the author) and for nucleon–nucleon interactions (by Frautschi and Margolis) had led to predictions of rising total cross sections even before the Serpukhov measurements were made. To begin this chapter, therefore, we shall elaborate briefly upon the consequences of having cuts as well as poles in the angular momentum plane.

Regge cuts can arise in the Sommerfeld–Watson transformation if, for example, the spectral function in (16.3) has the asymptotic form

$$\Delta_t(s, t') \sim (t')^{\alpha_c(s)} (\ln t')^{\beta(s)}$$

yielding

$$\int_{t_1}^{\infty} dt' (t')^{\alpha_c(s) - \lambda} (\ln t')^{\beta(s)} \sim \frac{1}{(\lambda - \alpha_c(s))^{1+\beta(s)}}$$

instead of (16.4). Then the singularity at $\lambda = \alpha_c(s)$ becomes a branch point, rather than a simple pole, and the deformation of the contour must include a detour around this branch point and its attached cut, as shown in Fig.(18-2). There results in the amplitude a term of the form

$$F_{cut}(s, t) = \int^{\alpha_c(s)} d\lambda \, \frac{(2\lambda + 1)\Delta a(\lambda, t)}{\sin \pi\lambda} [P_\lambda(-\cos\theta_t) + \tau P_\lambda(\cos\theta_t)] ; \qquad (18.4)$$

in the crossed channel, (18.4) becomes

$$F_{cut}(s, t) \approx \int^{\alpha_c(s)} d\lambda \, \frac{(2\lambda + 1)\Delta a(\lambda, t)}{\sin \pi\lambda} \left(\frac{s}{2m^2}\right)^{\lambda} [1 + \tau e^{-i\pi\lambda}] , \qquad (18.5)$$

so that a Regge cut yields an integral over powers of s, rather than a simple power law.

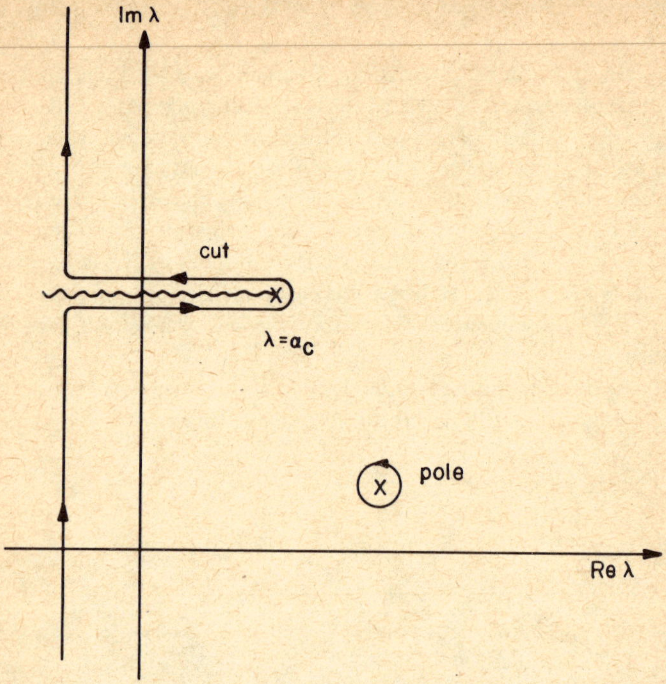

Fig.(18-2) Effect of a branch point on the Sommerfeld–Watson transformation contour

A more physically intuitive argument providing some justification for Regge cuts follows from using unitarity with the assumption that the amplitude contains a Regge pole. Elastic unitarity, in the form

$$\mathrm{Im}F(s, t) \propto \frac{1}{s^2}\int d\Omega F^*(s, t_1)F(s, t_2),$$

where t_1 and t_2 describe intermediate momentum transfers as indicated in Fig. (18-3), then contains a contribution of the form

$$\mathrm{Im}F(s, t) \propto \frac{1}{s^2}\int d\Omega r^*(t_1)\left(\frac{s}{s_0}\right)^{\alpha(t_1)} r(t_2)\left(\frac{s}{s_0}\right)^{\alpha(t_2)}. \tag{18.6}$$

It is not necessary to carry out the integral in (18.6) to see that it involves a superposition of powers of s analogous to (18.5). Regge cuts were first proposed on this basis by Amati, Fubini, and Stanghellini; later work by Mandelstam showed that, although their original conjecture was incorrect, more complicated diagrams than Fig.(18-3) do lead to branch points in the λ plane. The details of these arguments are discussed exhaustively in the book by Eden, Landshoff, Olive, and Polkinghorne.

It can be shown that the branch point arising from the iteration, as in (18.6), of two linear Regge trajectories

$$\alpha_1(t) = \alpha_1 + \alpha_1' t$$

$$\alpha_2(t) = \alpha_2 + \alpha_2' t$$

is given by

$$\alpha_c(t) = \alpha_1 + \alpha_2 - 1 + \frac{\alpha_1' \alpha_2'}{\alpha_1' + \alpha_2'} t. \tag{18.7}$$

Furthermore, it is reasonable to assume that the cut can be made to run to the left from $\alpha_c(t)$, as shown in Fig.(18-2); if so, its contribution will be dominated, as $s \to \infty$, by the part of the contour nearest $\alpha_c(t)$, where the power of s is largest. Therefore we can obtain an estimate for the asymptotic form of the Regge cut amplitude by writing

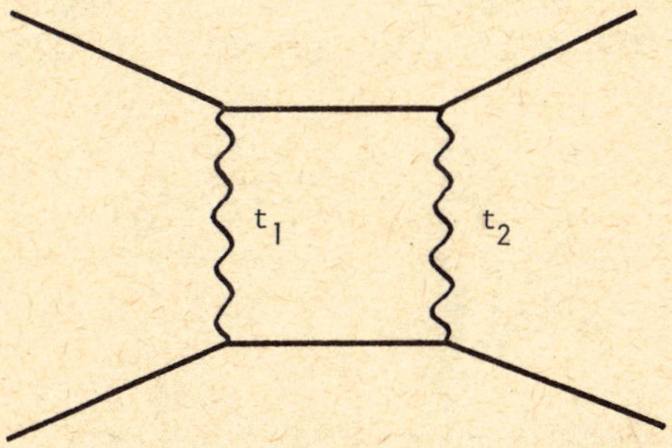

Fig.(18-3) Double Regge exchange graph which, via unitarity, appears to yield a Regge cut term.

$$(1 + \tau e^{-i\pi\lambda}) \frac{(2\lambda + 1)\Delta a(\lambda, t)}{\sin \pi\lambda} = (\lambda - \alpha_c(t))^\nu g(\lambda, t) e^{-i\pi\lambda/2} \tag{18.8}$$

with $g(\alpha_c(t), t)$ finite for $\lambda \to \alpha_c(t)$ and either real or imaginary according to $\tau = \pm 1$. It follows that

$$F_{cut}(s, t) \sim \frac{g(\alpha_c(t), t)}{\left[\ln\left(\frac{s}{is_0}\right)\right]^{\nu+1}} \left(\frac{s}{is_0}\right)^{\alpha_c(t)}; \tag{18.9}$$

the superposition of powers of s is almost equivalent to a pole at $\alpha_c(t)$, modified only by a power of $\ln (s/is_0)$ in the denominator.

Thus a reasonable simple form to assume for a scattering amplitude dominated by the Pomeranchuk pole, plus the branch point obtained as in (18.7) using

$$\alpha_1(t) = \alpha_2(t) = 1 + \alpha'_P t,$$

is given in the form (17.19b) by

$$F(s, t) = -C\left(\frac{s}{is_0}\right)^{1+\alpha'_P t} + \frac{D}{\left[\ln\left(\frac{s}{is_0}\right)\right]^{\beta}} \left(\frac{s}{is_0}\right)^{1+1/2\alpha'_P t}, \tag{18.10}$$

with C and D real constants. Although this amplitude is certainly not sufficient to describe available data, it possesses the principal features to be expected in the presence of a Regge cut, many of which can be seen in experimental results.

For example, the total cross section resulting from (18.10) when $s \gg s_0$ is

$$\sigma(s) = \frac{8\pi}{s_0}\left[C - \frac{D}{\left[\ln\left(\frac{s}{s_0}\right)\right]^{\beta}}\right]. \tag{18.11}$$

As $s \to \infty$, $\sigma(s)$ becomes a constant; but it will do so very slowly, because the logarithmic variation of the cut contribution will be very small unless β is quite large. The sign of D can be determined in various models, and it seems almost certain that D will be positive, i.e. that the cut contribution will *subtract* from the asymptotic cross section. If that is the case, then a slow, logarithmic rise in the total cross section toward a constant asymptote should be observed, as indicated in Fig.(18-4).

The rise of the proton–proton total cross section observed at FNAL and ISR, however, seems to be somewhat faster than (18.11) indicates. Most of the models incorporating simple Regge cuts are rather hard-pressed to generate sufficient curvature. At present it is unclear whether this difficulty reflects incorrect formulation of the cut term or a more fundamental problem in the nature of the Pomeron itself. For example, it is possible that the Pomeranchuk term should explicitly contain asymptotic behavior reaching the Froissart bound, such as

$$F_P(s, t) = -c \left(\frac{s}{is_0}\right)^{1+\alpha_P' t} \ln^2\left(\frac{s}{is_0}\right)$$

rather than the usual Regge pole form. Behavior of this type has been derived in quantum field theory by summing certain sets of Feynman graphs, but it is not yet clear whether that approach is the correct one.

The differential cross section resulting from (18.10) is given for large s by

$$\frac{d\sigma}{dt} = 4\pi\ C^2 \left\{ \left(\frac{s}{s_0}\right)^{2\alpha_P' t} - \frac{2CD}{\left[\ln\left(\frac{s}{s_0}\right)\right]^{\beta}} \cos(\tfrac{1}{4}\pi\alpha_P' t)\left(\frac{s}{s_0}\right)^{3/2\alpha_P' t} + \frac{D^2}{\left[\ln\left(\frac{s}{s_0}\right)\right]^{2\beta}} \left(\frac{s}{s_0}\right)^{\alpha_P' t} \right\}. \tag{18.12}$$

In addition to the diffraction peaks resulting from the pole and cut terms, there is an extra term arising from interference between the two. If D is positive, this term represents destructive interference and leads to a dip in the differential cross section, as shown in Fig.(18-5). We have already commented in Chapter 17 that such a dip is observed in many reactions, and can often be associated with the vanishing of the trajectory (because it multiplies the spin flip amplitude). Interference between pole and cut terms thus provides an alternative explanation of this phenomenon. The appearance of the double zero in elastic πp polarization, however, is difficult to reconcile with a pole-cut interference model, since the latter (even when extended to include spin) produces a single zero; the sign of the amplitude changes as the cut term becomes dominant. At present it seems clear

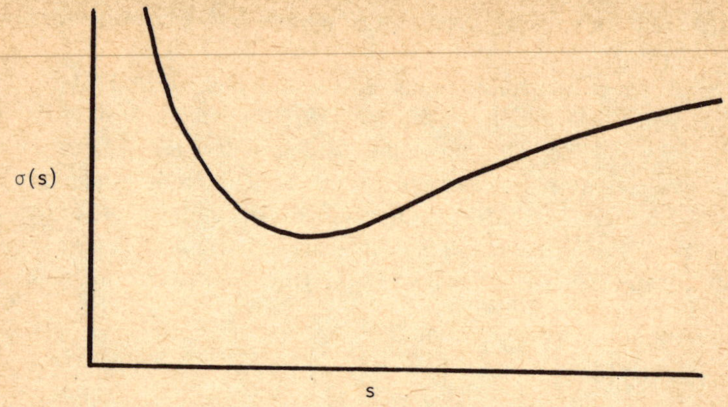

Fig.(18-4) The rise of the total cross section toward its asymptotic value resulting from the presence of a subtractive Regge cut.

that scattering amplitudes in fact possess both pole-cut interference effects and wrong-signature nonsense zeros.

Another feature of the differential cross section (18.12) is that calculation of the trajectory by studying the shrinkage of the diffraction peak is no longer as

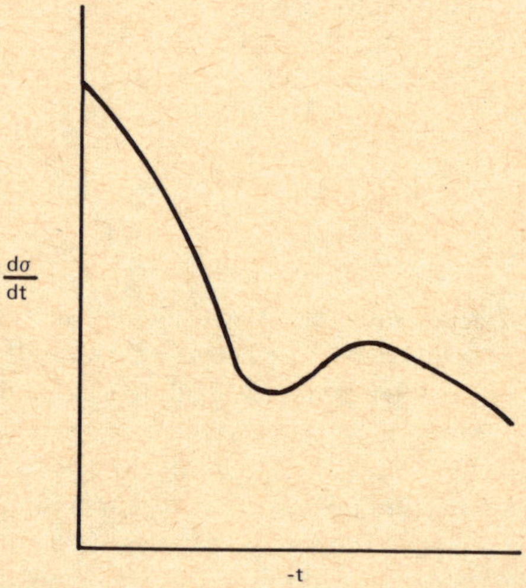

Fig.(18-5) Structure appearing in the differential cross section as a result of pole-cut interference.

simple as in (16.32). The subtractive contribution of the interference term will exaggerate the steepness of the peak, and the different s-dependences of the three terms will complicate the behavior of its slope. For $t < 0$, (18.7) shows that in general $\alpha_c(t) > \alpha_P(t)$. Therefore the cut contribution, at fixed t, will fall off less rapidly with increasing s than that of the pole. It follows that a measurement of the trajectory by means of the diffraction peak's shrinkage in this region is likely to be influenced by the slower variation of the cut term, and therefore yield a *higher* value for $\alpha(s)$ in (16.29) than the actual trajectory. The same conclusion is reached regardless of the identity of the two trajectories generating the cut.

In $\pi^- p \to \pi^\circ n$, for example, we might assume that the amplitude contains some contribution from a Regge cut generated by combining the P trajectory with the ρ, as indicated in Fig.(18-6). The appropriate amplitude is then of the form

$$F_{\pi^- p \to \pi^\circ n}(s, t) = i \left[C_\rho \left(\frac{s}{is_0} \right)^{\alpha_\rho + \alpha_\rho' t} - \frac{D_\rho}{\left[\ln \left(\frac{s}{is_0} \right) \right]^3} \left(\frac{s}{is_0} \right)^{\alpha_\rho + \alpha_c' t} \right] \tag{18.13}$$

where

$$\alpha_c' = \frac{\alpha_P' \alpha_\rho'}{\alpha_P' + \alpha_\rho'} < \alpha_\rho' .$$

The influence of α_c' will cause measured values of $\alpha(t)$ obtained via (16.29) to lie higher than the correct values. Curiously, the data shown in Fig.(17-3) reveal exactly that behavior as compared with the extrapolated straight-line trajectory. A similar result was also observed for the A_2 trajectory in Fig.(17-8).

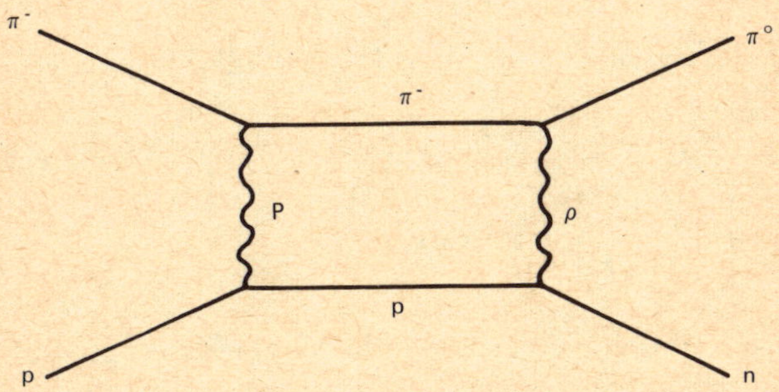

Fig.(18-6) Cut term contributing to $\pi^- p \to \pi^\circ n$.

Finally we may note, as intimated already in Chapter 17, that the presence of a Regge cut leads to a non-vanishing polarization. A spin-flip amplitude adjunct to (18.13) will be of the form

$$G_{\pi^- p \to \pi^\circ n}(s, t) = i\sqrt{-t}\left[C'_\rho\left(\frac{s}{is_o}\right)^{a_\rho + a'_\rho t} - \frac{D'_\rho}{\left[\ln\left(\frac{s}{is_0}\right)\right]^\beta}\left(\frac{s}{is_0}\right)^{a_\rho + a'_c t}\right]. \tag{18.14}$$

The pole contributions are, as before, in phase with each other and produce no polarization. Similarly, the cut term in (18.13) is in phase with its analog in (18.14); the polarization results entirely from pole-cut interference,

$$P(s, t) = -2\,\frac{\mathrm{Im}(F^*_{\mathrm{pole}}G_{\mathrm{cut}} + F^*_{\mathrm{cut}}G_{\mathrm{pole}})}{|F|^2 + |G|^2}. \tag{18.15}$$

We have

$$F^*_{\mathrm{pole}}G_{\mathrm{cut}} = \sqrt{-t}\,C_\rho\left(\frac{is}{s_0}\right)^{a_\rho + a'_\rho t} D'_\rho\left(\frac{s}{is_0}\right)^{a_\rho + a'_c t}\left[\ln\left(\frac{s}{is_0}\right)\right]^{-\beta},$$

so for $s \gg s_0$

$$\mathrm{Im}F^*_{\mathrm{pole}}G_{\mathrm{cut}} = \frac{\sqrt{-t}\,C_\rho D'_\rho \sin(\tfrac{1}{2}(a'_\rho - a'_c)t)}{\left[\ln\left(\frac{s}{s_0}\right)\right]^\beta}\left(\frac{s}{s_0}\right)^{2a_\rho + (a'_\rho + a'_c)t}$$

A similar result is obtained for the second term of (18.15), yielding

$$P(s, t) = \frac{2\sqrt{-t}\,(C_\rho D'_\rho - C'_\rho D_\rho)\sin(\tfrac{1}{2}\pi(a'_\rho - a'_c)t)}{\left[\ln\left(\frac{s}{s_0}\right)\right]^\beta\left\{|F|^2 + |G|^2\right\}}\left(\frac{s}{s_0}\right)^{2a_\rho + (a'_\rho + a'_c)t}. \tag{18.16}$$

To determine the asymptotic behavior of P(s, t) with s, it must be realized that for $t < 0$, the cut contribution asymptotically dominates the differential cross section, because $a_c(t) > a_\rho(t)$. Therefore the denominator of (18.15) becomes proportional asymptotically to $(s/s_0)^{2(a_\rho + a'_c t)}/[\ln(s/s_0)]^{2\beta}$, yielding

$$P(s, t) \sim \sqrt{-t}\, p(t) \left[\ln \left(\frac{s}{s_0} \right) \right]^{\beta} \left(\frac{s}{s_0} \right)^{(\alpha'_\rho - \alpha'_c)t} \tag{18.17}$$

Since $\alpha'_\rho > \alpha'_c$, the power to which s is raised in (18.17) is negative; but, being proportional to t, it will be quite small in the region of the diffraction peak, in contrast to the fixed power occurring in (17.16). Thus the polarization resulting from a Regge cut is expected to disappear as the energy increases much less rapidly than that obtained using a second ρ' pole. The data clearly indicate a preference for this slower decrease.

As in the case of the total cross section, however, there are indications that the simple cut correction in (18.13) is not sufficient to account for all the data. In particular, cut-induced breaking of line-reversal symmetry disagrees drastically with experimental data. The details of this disagreement are too technical to discuss here, but they indicate the need for some modification of the phase of the cut term. According to the phase-energy relations developed in Chapter 14, this would also require a modification of the energy dependence.

In summary, then, Regge cut terms arising from branch points in the angular momentum plane possess several properties which are required by the higher-energy data. They produce a total cross section rising toward an asymptote, a dip in the differential cross section, and non-vanishing polarizations which are nearly independent of the energy. But they do not reproduce the observed behavior as accurately as had originally been hoped. The theoretical justification of Regge cuts, and the precise form of their energy dependence, are thus subjects for continuing research.

Chapter 19

DUALITY

In the preceding chapters we have seen that the concept of Regge poles serves a dual purpose. At low energies, the resonance structure of the amplitude may be cogently described by means of the recurrences lying on a few basic direct-channel Regge trajectories; at high energies, the diffraction peak and its behavior are satisfactorily interpreted, for the most part, by the exchange of a small number of crossed-channel Regge trajectories. Regge phenomenology is thus pertinent at both ends of the energy spectrum.

Attempts to exploit this dual structure have led to some rather intriguing observations, the password for which is "duality." Its origin can be found in a simple – indeed, almost philosophical – question: how do we write a scattering amplitude possessing both of these dual aspects of Reggeology? Such an amplitude should provide both the low-energy and high-energy limits correctly, as well as interpolate correctly the "intermediate energy" region between them.

The most obvious answer to this question – an incorrect one, as we shall see – is simply to add two terms together,

$$F = F_d + F_c, \qquad (19.1)$$

with F_d containing the direct-channel Regge trajectories and F_c the crossed-channel Regge terms. This "interference model," so-called because its first use was in explaining the $\pi^- p \rightarrow \pi^\circ n$ polarization via interference between F_d and F_c, assumes that F_d does not contribute at high energies, and that the crossed-channel term F_c does not resonate at low energies. The first assumption is certainly plausible, since (9.48) shows that the total cross section of a resonance vanishes with increasing energy. Indeed, if the direct channel trajectories turn back toward the left half of the ℓ plane as in potential theory, only the existence of an infinite number of them can prevent F_d from vanishing at high energy. The second assumption is more difficult to evaluate, however, because it involves taking the direct-channel partial wave projection of a crossed channel Regge pole.

If, as usual, the s and t channels are the direct and crossed channels, respectively, we must study

$$f_\ell^R(s) = \int_{-1}^{1} d\cos\theta_s F_c(s,t);$$

taking the Regge parametrization in the form

$$F_c(s,t) = \beta(t)P_{\alpha(t)}(-\cos\theta_t),$$

this becomes (assuming equal masses for simplicity)

$$f_\ell^R(s) = \frac{2}{4m^2 - s} \int_0^{4m^2-s} dt\beta(t)P_{\alpha(t)}\left(-1 - \frac{2s}{t-4m^2}\right)P_\ell\left(1 + \frac{2t}{s-4m^2}\right). \quad (19.2)$$

In order to evaluate (19.2), the residue and trajectory functions $\beta(t)$ and $\alpha(t)$ must be known for $0 \geqslant t \geqslant 4m^2 - s$. In the resonance region, however, the value of $s - 4m^2$, and consequently the range of t required, is quite small. Parametrizations of $\beta(t)$ and $\alpha(t)$ obtained from high-energy fits of the diffraction peak may therefore be used.

At first glance, it seems unlikely that smooth residues and trajectories such as those in (16.31) could produce a resonant pole in (19.2). Surprisingly, however, calculations carried out by C. Schmid in this way showed that the Argand diagrams of $f_\ell^R(s)$ do indeed resonate! What is more, the resulting curves, depending on the exact parametrizations of $\beta(t)$ and $\alpha(t)$, could be made to agree quite well with experiment. An example of this agreement is shown in Fig.(19-1).

Thus it appears that the Regge term F_c not only resonates but in fact seems to "know" quite accurately about the low-energy resonance spectrum. In that case, much of the content of F_d is contained in F_c, so the amplitude (19.1) must be incorrect; it will involve double-counting of the direct-channel resonances. Some other method must be used in order to write an amplitude that is dual.

Finite Energy Sum Rules

The ability of the crossed-channel Regge asymptote to reproduce the low-energy spectrum becomes less surprising if we recall the analyticity of the amplitude as a function of the energy. In fact, dispersion relations provide a relation between the amplitude at *any* energy and its singularities at *all* energies. The high-energy

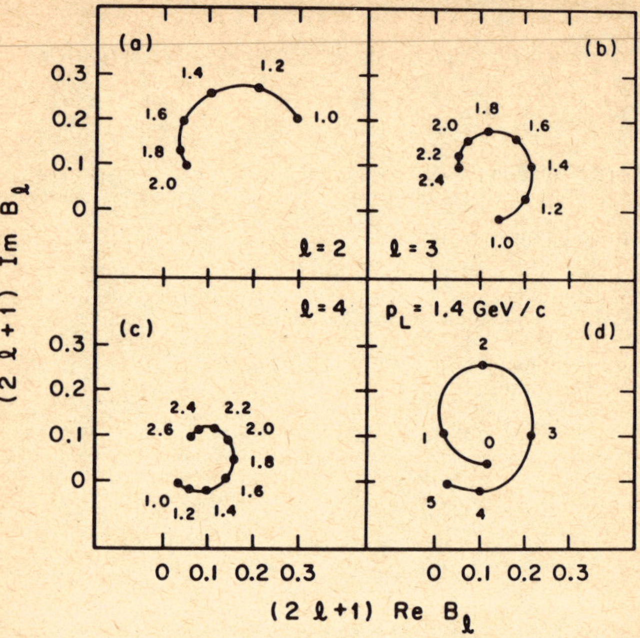

Fig.(19-1) Argand diagrams for the s-channel partial wave projections of the t-channel ρ trajectory amplitude for $\pi^- p \to \pi^0 n$. The loops correspond to the N(1920), N(2190), and N(2420) resonances. (From C. Schmid, *Phys. Rev. Letters* **20**, 689 (1968).)

limit is thereby automatically and intimately connected with the resonance and bound state poles.

Assuming Regge behavior at large s allows us to evaluate exactly the contributions of integrals in this region to dispersion relations. The need for expanding contours to infinity disappears, and a class of *finite-energy sum rules* can be obtained. In order to formulate these ideas more exactly, let us consider the particularly simple case of a crossing-symmetric amplitude dominated asymptotically by a single Regge pole. Since crossing symmetry implies even signature, we may write the asymptotic amplitude in the form

$$A(\nu + i\epsilon, t) = \frac{\beta(t)}{\sin \pi\alpha(t)} \left(\frac{\nu}{i} \right)^{\alpha(t)}, \tag{19.3}$$

where the residue $\beta(t)$ is real and the symmetric variable $\nu = (s - 2m^2)/2m$ is used.

This Regge form is assumed to be valid for real $\nu \geqslant \nu_0$, ν_0 being large, and infinitesimal ϵ, i.e. in the physical region of the ν-plane. Crossing symmetry implies that

$$A(-\nu - i\epsilon, t) = A(\nu + i\epsilon, t) \tag{19.4}$$

and analyticity equates the discontinuity across the cut to the imaginary part of the amplitude in the usual way.

Under these conditions, it can be proved that the asymptotic form (19.3) must be valid for all ν satisfying $|\nu| > \nu_0$, and not merely on the real axis. Then we may write a dispersion relation for $A(\nu, t)$, using (19.3) for the parts of the contour in the asymptotic region. Let us neglect any bound-states poles in $A(\nu, t)$, so that it is analytic in the ν-plane except for threshold branch points producing cuts running from $\pm m$ to $\pm\infty$. The contour shown in Fig.(19-2) then encloses no singularities, so the integral of $\nu^n A(\nu, t)$ around it must vanish for any non-negative integer n,

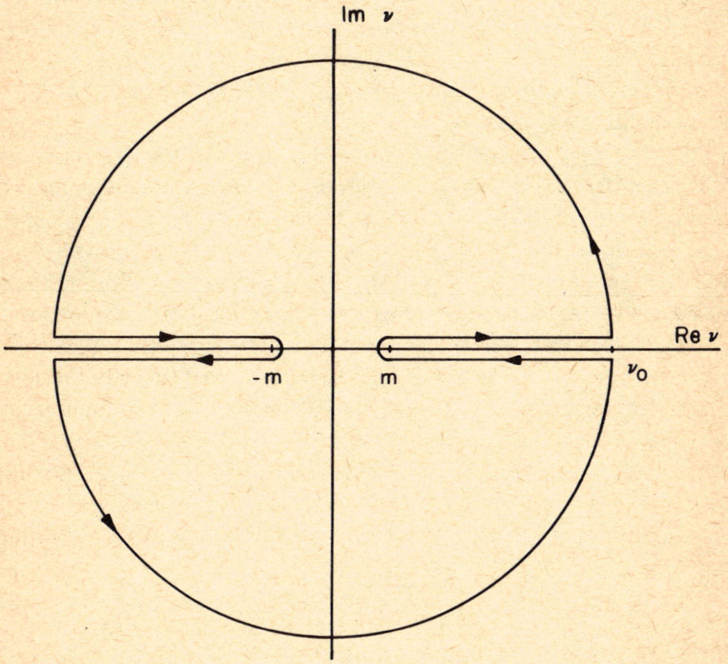

Fig.(19-2) Contour used for developing finite energy sum rules.

$$\oint_c d\nu\nu^n A(\nu, t) = 0. \tag{19.5}$$

Using the asymptotic form (19.3), however, we may evaluate exactly the integral around the large semicircles, and the integrals over the cuts may be combined using (19.4). It follows that

$$\int_m^{\nu_0} d\nu\nu^n \mathrm{Im}A(\nu + i\epsilon, t) = \frac{\beta(t)\nu_0^{n+\alpha(t)+1}}{n + \alpha(t) + 1} \tag{19.6}$$

provided n is odd; for even n, the integral (19.5) vanishes trivially because of the symmetry of $A(\nu, t)$.

A similar procedure can be followed using an amplitude which is antisymmetric under crossing and therefore is dominated by an odd-signature Regge pole. The result in that case is identical with (19.6) but holds for even n, while (19.5) vanishes trivially for odd n. Also, only trivial modifications are required to include bound states or a sum of Regge poles.

The finite energy sum rule (19.6) reveals a rather remarkable property of the scattering amplitude. The right-hand side contains only crossed-channel Regge pole terms, which we might have thought to be independent of the direct-channel resonances or their trajectories. Yet we see clearly that the Regge asymptote must be equal to a fixed-t integral over the low-energy region, and therefore the residue and trajectory functions know about the resonances. In other words, the continuations of the t-channel coupling constants and trajectories are determined by the s-channel resonance structure. Since the latter are presumably direct-channel Regge recurrences, a strong correlation between trajectories in crossed channels seems required.

Further insight into the implications of (19.6) can be gained by defining

$$A^R(\nu, t) = \frac{\beta(t)}{\sin \pi\alpha(t)}(m^2 - \nu^2)^{\alpha(t)/2}\ , \tag{19.7}$$

which has both the analytic structure and the asymptotic behavior assumed for $A(\nu, t)$. For $|\nu| > \nu_0$, therefore, the difference between the two is of order $\nu^{\alpha(t)-2}$, and it follows that

$$\int_m^{\nu_0} d\nu\nu^n \mathrm{Im}[A(\nu + i\epsilon, t) - A^R(\nu + i\epsilon, t)] \propto \nu_0^{n+\alpha(t)-1}. \tag{19.8}$$

If $n + \alpha(t) < 1$, and ν_0 is large, (19.8) must be small. But the integral can be small only if

$$\int_m^{\nu_0} d\nu\nu^n \mathrm{Im}A(\nu + i\epsilon, t) \approx \int_m^{\nu_0} d\nu\nu^n \mathrm{Im}A^R(\nu + i\epsilon, t),$$

i.e. if the Regge amplitude is an "average" of $A(\nu, t)$; in other words, $\mathrm{Im}[A(\nu + i\epsilon, t) - A^R(\nu + i\epsilon, t)]$ must average to near zero between m and ν_0. Consequently we cannot identify $(A - A^R)$ as the direct channel resonances, as implied by (19.1). A resonance would always make a positive contribution to (19.8), and the sum of such contributions would not be small. Instead, A^R must pass *through* the resonance spectrum, as shown in Fig.(19-3), so that the sign of $\mathrm{Im}(A - A^R)$ fluctuates.

This statement can be made even stronger if we use the arbitrariness of n. Suppose first that we could choose A^R to reproduce exactly, to *all* orders of ν, the asymptotic form of A. Then the right-hand side of (19.8) would vanish pre-

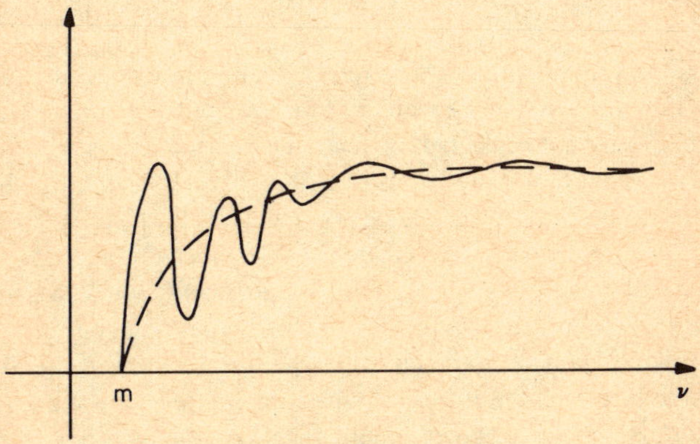

Fig.(19-3) Oscillations of Im A (solid curve) about Im A^R (dashed curve) implied by the finite energy sum rule (19.8).

cisely for any value of n. But the integral can vanish for *all* n only if the integrand does, i.e. if

$$\mathrm{Im}A(\nu + i\epsilon, t) \equiv \mathrm{Im}A^{R}(\nu + i\epsilon, t).$$

A less exact agreement between A and A^R asymptotically will change the identity to an approximation, but the basic result will be the same; the more accurately we parametrize the *high*-energy limit, the more closely that parametrization must reproduce the *low*-energy amplitude.

A Dual Amplitude

Finite energy sum rules thus provide a natural explanation of the agreement between the partial wave projections of a good Regge fit and the observed resonance spectrum. To write the amplitude as a sum of direct and crossed channel terms, as in (19.1), is incorrect. Instead, it is more appropriate to assume that a single term contains everything – Regge behavior at high energy, resonances at low energy, and some smooth transition between the two. Regge poles and resonances are then complementary; they are different ways of approximating the same function. At high energy only a few Regge terms are needed, whereas a great many resonances would be required to sum to the correct behavior. Conversely, the low energy region is simply described by a small number of resonances, but would require very many Regge pole terms. A single function which would correctly produce both of these limits could appropriately be called a *dual* amplitude.

To formulate a more precise definition of duality, however, we must know more about how the two complementary limits are obtained. Specifically, we must ask what terms are neglected. Is there, behind the resonances, a nonresonant background that nonetheless contributes, via (19.6), to the high energy region? Is there, underneath the Regge asymptote, a "negligible" term without which the true resonance behavior cannot be reproduced? Dolen, Horn, and Schmid, in the first extensive application of finite-energy sum rules, made the simplest possible conjecture – that both of these types of terms were indeed absent. In treating pion–nucleon charge exchange, they assumed that the integral in (19.5) could be calculated using *only* the contributions of the known nucleon resonances. Only the ρ trajectory should appear on the right-hand side, and by evaluating the integral at various values of t a prediction for the trajectory and residue functions $\beta(t)$ and $\alpha(t)$ was obtained and compared with the corresponding high-energy parametrizations. Although only low-energy data were used as input, the finite energy sum rules proved quite satisfactory in predicting the high-energy behavior of the charge-exchange amplitude.

The success of that calculation shows two things. First, it serves as verification of the analytic structure assumed in writing (19.5). As we have remarked previously, however, analyticity is rather sacrosanct; if the calculation had failed, we would blame our parametrization rather than the dispersion relation. Thus the second point made by the Dolen–Horn–Schmid result is the viability of an amplitude containing only resonances at low energy and a single Regge asymptote.

Can a dual amplitude, so defined, be applied to other processes? A problem arises immediately when we recall that certain elastic scattering processes have "exotic" quantum numbers, for which resonances do not exist. For example, K^+–nucleon scattering could produce only $S = +1$ direct-channel resonances, which experimentally seem absent. Likewise, proton–proton scattering does not produce resonances. And since no $I = 2$ mesons have been observed, we feel confident that the $\pi^+\pi^+$ elastic scattering amplitude would also be free of resonances. In general, we see that for elastic scattering in exotic channels there are no resonances to contribute to the integral in (19.6). Yet the amplitude is certainly not zero for these processes in the resonance region; and the Pomeranchuk trajectory should dominate the asymptotic behavior, so the right-hand side of (19.6) is also non-zero.

It is important to note, however, that in both of the exotic channels for which we have experimental data, the contributions of exchange-degenerate pairs of Regge trajectories seem to cancel, leaving only the Pomeranchuk pole dominant. We have already seen this effect in Chapter 17, where the constancy of the K^+–nucleon and nucleon–nucleon total cross sections was attributed to cancellations between ρ and A_2 and between ω and P' contributions. This correlation between the absence of direct-channel resonances and the absence of non-Pomeranchuk crossed-channel trajectories now becomes very significant, for it suggests that only the Pomeranchuk contribution to the amplitude is not dual. If we subtract away this "diffractive background" term, then, the ideas of duality may be applicable to the remainder.

Let us therefore suppose that the amplitude can be separated in this manner, into

$$A = A_P + A_{res}, \tag{19.9}$$

where A_{res} contains *all* resonances (or, equivalently, the direct-channel Regge trajectories) while the non-resonant background is entirely in A_P. Duality now assumes that in a finite energy sum rule A_P will correspond precisely with the Pomeranchuk trajectory, while A_{res} will reproduce all other trajectories in the high-energy limit. The generalization of (19.6) then becomes

$$\int_m^{\nu_0} d\nu\nu^n \mathrm{Im}A_P(\nu + i\epsilon, t) = \frac{\beta_P(t)\nu_0^{n+\alpha_P(t)+1}}{n + \alpha_P(t) + 1} \tag{19.10}$$

for the Pomeranchuk term and

$$\int_m^{\nu_0} d\nu\nu^n \mathrm{Im}A_{res}(\nu + i\epsilon, t) = \sum_{j \neq P} \frac{\beta_j(t)\nu_0^{n+\alpha_j(t)+1}}{n + \alpha_j(t) + 1} \tag{19.11}$$

for the dual part of the amplitude, the sum including all other contributing trajectories.

Some very interesting results can now be obtained by disregarding the crossing symmetry required in deriving (19.11). For example, consider the process $K^+n \to K^\circ p$, which has no direct channel resonances by proceeds via exchange of the ρ and A_2 trajectories. It follows immediately that the left-hand side of (19.11) vanishes, while the right-hand side contains two terms; consequently these two terms must cancel each other. This is possible only if

$$\alpha_\rho(t) = \alpha_{A_2}(t)$$

$$\beta_{\rho KN}(t) + \beta_{A_2 KN}(t) = 0 \tag{19.12}$$

implying exact exchange degeneracy for ρ and A_2. In a similar manner, one can consider $\pi^+\pi^+ \to \pi^+\pi^+$, which has no direct-channel resonances; the contributing trajectories are P′ and ρ. It follows that P′ and ρ must also be exchange degenerate. Likewise, analyzing $K^+K^+ \to K^+K^+$ shows that the ω trajectory must be exchange degenerate with the other three. Experimentally, all of these results seem very nearly correct.

A much simpler way of arriving at the same conclusions, which does not require forsaking the fundamental crossing symmetry assumption, may be mentioned here. Consider first the $\pi\pi$ scattering processes. Three isospin channels are available to the $\pi\pi$ system, and we have assumed that the dual part of the I = 2 amplitude vanishes in both the direct and crossed channels. The isospin amplitudes in crossed channels are simply related, however, via an isospin crossing matrix; if $A_I^{s,t}$ denotes the isospin-I amplitude in the s or t channel, one finds in particular that

$$A_2^s = \frac{1}{6}A_2^t + \frac{1}{2}A_1^t + \frac{1}{3}A_0^t.$$

If $A_2^t = A_2^s = 0$, then necessarily

$$A_1^t = -\frac{2}{3}A_0^t \,. \tag{19.13}$$

The ρ-exchange contribution is contained in A_1^t, and the P′ in A_0^t; the relation (19.13) is identical with the exchange degeneracy of (19.12).

A similar analysis can be applied in other reactions involving exotic channels, often with equally pleasing results. Unfortunately, however, there are also some serious failures. For example, the processes $K^+\pi^+ \to \pi^+K^+$ and $K^+\pi^- \to \pi^-K^+$, related by crossing, have isospin $\frac{3}{2}$ and $\frac{1}{2}$ amplitudes satisfying

$$A_{3/2}^s = \frac{1}{3}A_{3/2}^t + \frac{2}{3}A_{1/2}^t.$$

If $A_{3/2}^s = A_{3/2}^t = 0$, it follows also that $A_{1/2}^t = 0$, forbidding the existence of K* resonances. The same sort of difficulty arises in $\Delta\Delta \to \Delta\Delta$ and $\Delta\bar{\Delta} \to \bar{\Delta}\Delta$. Whether it implies a failure of duality, or can be resolved by such methods as adding Regge cuts, is not yet clear.

Dual Bootstraps and the Veneziano Amplitude

One of the most interesting processes to which duality has been applied is $\pi\pi \to \pi\omega$, for which all three crossed channels are identical. Only the ρ trajectory can contribute, so it should appear both in the high-energy limit and in the direct-channel spectrum, yielding for (19.11)

$$\int_m^{\nu_0} d\nu\nu^n \mathrm{Im}A_\rho(\nu + i\epsilon, t) = \frac{\beta_\rho(t)\nu_0^{n+\alpha_\rho(t)+1}}{n + \alpha_\rho(t) + 1}, \tag{19.14}$$

where A_ρ contains the resonances on the ρ trajectory. (We are neglecting here for simplicity the $\pi - \omega$ mass difference.) Equation (19.14) immediately suggests a bootstrap scheme, in which one parametrizes the *direct*-channel ρ trajectory, carries out the integration, and requires the result to reproduce the *crossed-channel* trajectory.

An immediate question arises regarding how high the cutoff energy ν_0 should be chosen, since that choice will certainly affect the result. If the integral is carried out over a region in which only the ρ pole itself contributes, it turns out that the self-consistent trajectory obtained is quite satisfactory. If g(1680), the first recurrence of the ρ, is also included, good results are again obtained. But including a third particle on the ρ trajectory under the integral leads to discrepancies, which become worse as ν_0 is raised still higher. The reason for this failure is simply that by extending the region of integration to include more and more direct channel resonances, we are effectively improving our parametrization of the low-energy amplitude. Consequently the result is reproducing the actual asymptotic behavior, rather than merely its leading term as in (19.14). In other words, we are bootstrapping a much more complicated function which necessarily includes daughter trajectories and other lower-order terms on the right-hand side of (19.14). The full, correct amplitude should indeed be self-consistent; but the resonances and the Regge asymptote are different limits of that more complicated function, and one cannot be expected to bootstrap the other completely.

In order to construct a more reasonable facsimile of the correct amplitude we must invent a function which is crossing symmetric, yields the correct asymptotic behavior, and has all of the poles implied by the ρ trajectory in all channels. But what poles should we expect from this trajectory? We have seen that the amplitude can be considered as direct channel trajectories or as crossed channel trajectories, but not simultaneously as both. Suppose we consider its high-energy behavior in terms of trajectories in the *direct* channel, rather than the crossed. The contribution of a single resonance falls off, according to (9.48), much more rapidly with energy than the crossed-channel Regge pole term does. The latter provides the true asymptotic form; but it cannot be obtained from a single resonance, or even from a finite set of them. Instead, an *infinite* set of direct channel resonances must be used in order to sum to a crossed-channel Regge term. Consequently it appears that the direct-channel trajectories must continue to rise indefinitely, rather than turning back as in potential theory.

An amplitude using a linearly rising trajectory was invented by G. Veneziano which fulfills all of the above conditions remarkably well. It is given by

$$V(s, t, u) = B(s, t) + B(s, u) + B(t, u) \tag{19.15}$$

where B is essentially a *beta* function,

$$B(s, t) = \frac{\Gamma(1 - \alpha(s))\Gamma(1 - \alpha(t))}{\Gamma(1 - \alpha(s) - \alpha(t))}; \tag{19.16}$$

$\Gamma(z)$ is the usual gamma function, and the trajectory is assumed to be real and linear,

$$\alpha(s) = \alpha_0 + \alpha_1 s. \tag{19.17}$$

Because $\Gamma(z)$ has a simple pole whenever z becomes a negative integer or zero, B(s, t) will have poles whenever $\alpha(s)$ or $\alpha(t)$ passes through a positive integer value, just as a single Regge pole term does. The residue of an s-channel pole, occurring for $\alpha(s) = N$, is

$$R_N(t) = \frac{(-1)^N}{N!} \frac{\Gamma(1 - \alpha(t))}{\Gamma(1 - N - \alpha(t))}; \tag{19.18}$$

if now $\alpha(t)$ approaches a positive integer, $R_N(t)$ does not have a pole, since the denominator also becomes infinite. Double poles, which would allow s-channel residues to have t-singularities, are consequently absent.

In addition to a satisfactory resonance structure, the Veneziano amplitude has the desired asymptotic behavior *except* on the positive real axes, where the infinite sets of bound state poles exist. This result can be obtained using the Stirling approximation

$$\Gamma(z) \underset{|z| \to \infty}{\to} \sqrt{2\pi}\, e^{-z} z^{z-1/2},$$

which is valid throughout the z-plane except for the negative real axis. In the limit of large $|s|$ and small negative t, for example, both $(1 - \alpha(s))$ and $(1 - \alpha(s) - \alpha(t))$ become large, and it follows that

$$B(s, t) \approx \Gamma(1 - \alpha(t))[e(1 - \alpha(s))]^{\alpha(t)}$$

which, using (19.17) and $\Gamma(z)\Gamma(1 - z) = \pi \csc \pi z$, is simply the Regge-like form

$$B(s, t) \approx \frac{\pi}{\Gamma(\alpha(t)) \sin \pi\alpha(t)} (e\alpha_1)^{\alpha(t)} e^{-i\pi\alpha(t)} s^{\alpha(t)}. \tag{19.19a}$$

A similar treatment of the other two terms of (19.15) shows that as $|s| \to \infty$

$$B(t, u) \approx \frac{\pi}{\Gamma(\alpha(t)) \sin \pi\alpha(t)} (e\alpha_1)^{\alpha(t)} s^{\alpha(t)} \tag{19.19b}$$

and

$$B(s, u) \to 0. \tag{19.19c}$$

Thus the limiting form of V(s, t, u) is

$$V(s, t, u) \approx \frac{\pi}{\Gamma(\alpha(t)) \sin \pi\alpha(t)} (1 + e^{-i\pi\alpha(t)}) s^{\alpha(t)}, \tag{19.20}$$

in essentially exact agreement with (16.21).

The Veneziano formula (19.15) has a great many interesting and unexpected features. For example, one notes immediately the appearance of $\Gamma(\alpha(t))$ in the residues of (19.20), which will cancel the pole introduced by $\sin \pi\alpha(t)$ when $\alpha(t)$ is a negative integer; a "ghost-killer" is built into the amplitude. As a result, the signature factor will produce zeroes whenever it vanishes, an effect already noted in connection with (16.61).

It is also possible to use the Veneziano amplitude for processes which are not fully crossing symmetric, by merely using the appropriate trajectories in each term. In particular, processes with an exotic channel will require only a single B function. For example, if the s channel is $\pi^+\pi^+ \to \pi^+\pi^+$, which should contain no poles, we may write

$$V_{\pi\pi}(s, t, u) = \frac{\Gamma(1 - \alpha(t))\Gamma(1 - \alpha(u))}{\Gamma(1 - \alpha(t) - \alpha(u))}. \tag{19.21}$$

This function has poles only in the t and u channels, describing $\pi^+\pi^- \to \pi^\pm\pi^\mp$. The signature factor in (19.20) arises only from combining *two* terms, however, so (19.21) automatically implies ρ-P$'$ exchange degeneracy; in fact, the mechanism responsible is essentially that of (19.13).

On the other hand, a definite disadvantage of the Veneziano formula is that it does not satisfy the basic requirement of unitarity; it has only bound state poles, and no threshold branch points. And although we have constructed the amplitude to have an infinite set of resonances, it turns out that we have gotten more of them than we bargained for. The reason is that the residue (19.18), when expressed in terms of $\cos\theta$ rather than t, is not a single Legendre polynomial. Instead it is a sum of N + 1 of them, producing parallel daughter trajectories creating particles of all spins up to N.

We shall not go into further detail on the Veneziano amplitude. Much work on it has been carried out, in particular regarding its extension to multiparticle processes. It appears at present that, like the original Regge theory itself, it is qualitatively reasonable but quantitatively far less than perfect in its simplest form.

Appendix B

SPIN COMPLICATIONS

When the particles involved in a scattering process have non-zero spin, the larger number of degrees of freedom is reflected in a correspondingly larger number of independent scattering amplitudes. The transition matrix is then indeed a matrix, the elements of which are transition amplitudes between the possible spin configurations of the initial and final states. Although arguments such as time reversal, parity, etc., can often be used to relate various amplitudes, the number of independent ones which remain is large enough to be cumbersome in any but the simplest cases. The quantities which can be measured experimentally (except for total cross sections) involve bilinear combinations of these amplitudes – often very complicated ones – and the experimental determination of the amplitudes themselves is consequently quite difficult. Even in the spinless case, the amplitude can be determined completely only in the forward direction, where the optical theorem applies; at non-zero angles, $d\sigma/d\Omega = |f|^2$ measures only the magnitude of f, never its phase. The same result is true for arbitrary spins. Except in the forward direction, only the *relative* phases of different amplitudes can be measured; an overall phase is inevitably lost.

A thorough discussion of the treatment of spin in scattering processes would carry us far beyond both the scope and the tenor of this book. A properly relativistic consideration of high energy processes requires, in fact, a revision of our basic language, since spin itself becomes unsatisfactory as a description of the state of a particle; instead, it is necessary to consider eigenstates of helicity, the component of spin parallel to a particle's momentum. The points which are important to our treatment, however, can be adequately shown much more simply. For this reason we shall restrict our attention to the scattering of a non-relativistic pseudoscalar meson by a spin-½ target – in other words, to pion–nucleon or kaon–nucleon scattering.

We must thus consider the two spin states (eigenstates of S_z) for a target nucleon, and it is convenient to choose the direction of the incident beam as the Z-axis. The transition matrix then has four elements, $T_{m_f m_i}$, with $m = \pm\frac{1}{2}$ labeling the initial and final spin states:

$$T = \begin{pmatrix} T_{++} & T_{+-} \\ T_{-+} & T_{--} \end{pmatrix}.$$

A more illuminating expression for T, however, results from making use of the fact that any operator in a spin-½ space can be written as a linear combination of the unit matrix and the three Pauli spin matrices $\boldsymbol{\sigma}_i = 2S_i$, i.e.

$$T = F + iG\hat{n}\cdot\boldsymbol{\sigma}. \tag{B.1}$$

We may clearly choose $\hat{n}$ to be a unit vector. Because of rotational invariance, $\hat{n}$ may not depend on any vector except those occurring physically in the scattering process. Hence we are restricted to only three possibilities; the incident beam momentum vector **k**, the momentum $\mathbf{k}'$ of the scattered meson, and the cross product $\mathbf{k} \times \mathbf{k}'$ of the two. No other vectors have been defined by the problem. Because of parity conservation, however, T must be a scalar operator; and since $\boldsymbol{\sigma}$ is an axial vector operator, the only satisfactory choice for $\hat{n}$ is another axial vector, in order for $\hat{n} \cdot \boldsymbol{\sigma}$ to be scalar rather than pseudoscalar. Therefore we must take

$$\hat{n} = \frac{\mathbf{k} \times \mathbf{k}'}{|\mathbf{k} \times \mathbf{k}'|}; \tag{B.2}$$

no other choice will simultaneously satisfy the requirements of rotational invariance and parity conservation.

For forward scattering, $\mathbf{k}' = \mathbf{k}$, the cross product $\mathbf{k} \times \mathbf{k}'$ vanishes, and it follows from (B.1) and (B.2) that $G = 0$ in the forward direction. This result merely reflects the conservation of angular momentum; both the initial and final meson states have $L_z = 0$, so the nucleon's spin must be unchanged in order to conserve J_z. Since $\hat{n}$ is perpendicular to the beam, $n \cdot \boldsymbol{\sigma} = n_x \sigma_x + n_y \sigma_y$ is an operator which changes the nucleon spin state,

$$\hat{n} \cdot \boldsymbol{\sigma} = \begin{pmatrix} 0 & n_x - in_y \\ n_x + in_y & 0 \end{pmatrix}.$$

Thus G is the probability amplitude for spin-flip processes and vanishes in the forward direction.

Consequently the description of meson–nucleon scattering involves two independent amplitudes, F and G, which are functions only of momentum and of the scattering angle θ; because the process has overall axial symmetry, they cannot depend on the azimuthal angle ϕ. Two complex functions correspond to four real quantities to be measured at each value of k and θ, but, as we have pointed out, the absolute phase cannot be determined for $\theta \neq 0$. Thus there are three independently measurable quantities, $|F|$, $|G|$, and their relative phase, which must be obtained in order to know T as completely as possible. In order to consider measuring these quantities, we must analyze a scattering process and describe carefully the initial and final states.

Polarization

For simplicity, let us think of the nucleon as a fixed scattering center, neglecting its recoil. The initial state may then be written simply as the product of an incident plane wave $|\mathbf{k}\rangle$ and the nucleon spin state,

$$|\Psi_i\rangle = |\mathbf{k}\rangle|i\rangle = e^{i\mathbf{k}\cdot\mathbf{r}}|i\rangle$$

with $|i\rangle$ some combination of the two spin states,

$$|i\rangle = a|½, ½\rangle + b|½, ½\rangle = \begin{pmatrix} a \\ b \end{pmatrix}. \tag{B.3}$$

normalized to $|a|^2 + |b|^2 = 1$. A more convenient description of this state is by means of its *polarization vector*

$$\mathbf{P} = \frac{\langle i|\boldsymbol{\sigma}|i\rangle}{\langle i|i\rangle}; \tag{B.4}$$

for the state (B.3), we have

$$P_x = 2\mathrm{Re}\, a^*b$$

$$P_y = 2\, \mathrm{Im}\, a^*b$$

$$P_z = |a|^2 - |b|^2. \tag{B.5}$$

Using the normalization condition shows that $\mathbf{P}^2 = 1$. The interpretation of the polarization vector follows from the observation that $|i\rangle$ is an eigenstate of $\mathbf{P} \cdot \boldsymbol{\sigma}$:

$$\mathbf{P} \cdot \boldsymbol{\sigma}\binom{a}{b} = (|a|^2 + |b|^2)\begin{pmatrix} a \\ b \end{pmatrix} = \begin{pmatrix} a \\ b \end{pmatrix}.$$

In other words, had we chosen the z-axis along $\mathbf{P}$ rather than along $\mathbf{k}$, $|i\rangle$ would have been a pure spin state. Thus $\mathbf{P}$ is effectively the "direction" of the spin.

Similarly, the final state will contain a superposition of the incident state and a scattered state,

$$|\Psi_f\rangle = |\Psi_i\rangle + |\Psi_{sc}\rangle,$$

where $|\Psi_{sc}\rangle$ is assumed (as in Chapter 9) to contain only outgoing spherical waves. Thus we write

$$|\Psi_{sc}\rangle = \frac{e^{ikr}}{r}\,|f\rangle \tag{B.6}$$

with a final spin state

$$|f\rangle = a'(\theta, \phi)|½. ½\rangle + b'(\theta, \phi)|½, -½\rangle = \begin{pmatrix} a' \\ b' \end{pmatrix}. \tag{B.7}$$

The coefficients $a'(\theta, \phi)$ and $b'(\theta, \phi)$ in (B.7) are the scattering amplitudes for transitions to the corresponding spin states. In terms of them, the differential cross sections for scattering to final states with spin ±½ are, respectively

$$\frac{d\sigma_+}{d\Omega} = |a'|^2$$

$$\frac{d\sigma_-}{d\Omega} = |b'|^2. \tag{B.8}$$

The net differential cross section is the sum over both available final states,

$$\frac{d\sigma}{d\Omega} = \frac{d\sigma_+}{d\Omega} + \frac{d\sigma_-}{d\Omega} = |a'|^2 + |b'|^2 = \langle f|f\rangle, \tag{B.9}$$

which is just the norm of the final spin state. Correspondingly, the polarization vector of the final state is

$$\mathbf{P}' = \frac{\langle f|\boldsymbol{\sigma}|f\rangle}{\langle f|f\rangle}. \tag{B.10}$$

Now let us express these quantities in terms of the scattering amplitudes F and G defined above. We have as a definition, omitting the spatial states,

$$\begin{pmatrix} a' \\ b' \end{pmatrix} = T\begin{pmatrix} a \\ b \end{pmatrix} = (F + iG\hat{n}\cdot\boldsymbol{\sigma})\begin{pmatrix} a \\ b \end{pmatrix}. \tag{B.11}$$

Taking advantage of the fact that $\hat{n}$ has no z-component, we may write

$$T = F + iG(n_x\sigma_x + n_y\sigma_y) = \begin{pmatrix} F & in^*G \\ inG & F \end{pmatrix}$$

where $n = n_x + in_y$. Then (B.11) becomes

$$\begin{pmatrix} a' \\ b' \end{pmatrix} = \begin{pmatrix} Fa + in^*Gb \\ inGa + Fb \end{pmatrix}.$$

Calculating the net differential cross section (B.9) then yields, after a little algebra,

$$\frac{d\sigma}{d\Omega} = |F|^2 + |G|^2 - 2\hat{n}\cdot\mathbf{P}\,\mathrm{Im}(F^*G). \tag{B.12}$$

The polarization of the final state is more complicated in this general situation; the result is

$$\mathbf{P}' = \frac{\{(|F|^2 - |G|^2)\mathbf{P} + 2\hat{n}(\hat{n}\cdot\mathbf{P})|G|^2 - 2\hat{n}\times\mathbf{P}\,\mathrm{Re}(F^*G) - 2\hat{n}\,\mathrm{Im}(F^*G)\}}{d\sigma/d\Omega}. \tag{B.13}$$

Since $\mathbf{P}'^2 = 1$, only two of the components of $\mathbf{P}'$ are independent; these, plus the differential cross section (B.12), provide in principle the three measurable quantities mentioned earlier.

Experimentally, of course, targets do not consist of single-particle states, and one must speak therefore of measurements averaged over the entire ensemble of nucleons involved. In the absence of any external field, the polarizations $\mathbf{P}_i$ of the individual nucleons will be randomly oriented, and the average polarization

$$\mathbf{P}_{av} = \frac{1}{N} \sum_{i=1}^{N} \mathbf{P}_i$$

will vanish. The resulting average values observed for $d\sigma/d\Omega$ and $\mathbf{P}'$ will therefore be simply

$$\frac{d\sigma}{d\Omega_0} = |F|^2 + |G|^2$$

$$\mathbf{P}' = -2 \frac{\operatorname{Im}(F^*G)}{|F|^2 + |G|^2} \hat{n}. \qquad \text{(B.14)}$$

It should be noticed in particular here that, even with an initially unpolarized target, the scattering process produces some polarization in the final state unless F and G have the same phase. The magnitude of this effect,

$$P = -\frac{2\operatorname{Im}(F^*G)}{|F|^2 + |G|^2}, \qquad \text{(B.15)}$$

is known as the polarization parameter, and clearly satisfies the relation $-1 \leqslant P \leqslant 1$.

Thus, in principle, scattering from an unpolarized target can provide at a given energy and angle, measurements of $(|F|^2 + |G|^2)$ and of $\operatorname{Im}(F^*G)$. In practice, however, it is rather difficult to measure P in this way, since every scattering involves its own value of $\hat{n}$, and the corresponding $\mathbf{P}'$ must be determined for each event separately. In any case, these two measurements are not sufficient to determine F and G. If we use instead a polarized target, having a net polarization $\mathbf{P}_{av}$, the resulting average differential cross section and polarization will be given by (B.12) and (B.13) with $\mathbf{P}$ replaced by $\mathbf{P}_{av}$. The differential cross section is simply

$$\frac{d\sigma}{d\Omega} = |F|^2 + |G|^2 - 2\mathbf{P}_{av} \cdot \hat{n} \operatorname{Im} F^*G$$

$$= \frac{d\sigma}{d\Omega_0} (1 + (\mathbf{P}_{av} \cdot \hat{n})P) \qquad \text{(B.16)}$$

A measurement of the polarization parameter P can therefore be obtained by choosing $\mathbf{P}_{av}$ normal to the beam and comparing the differential cross sections for different directions of $\hat{n}$. If, for example, $\mathbf{P}_{av}$ is vertically upward, scattering to the right will have $\hat{n} \cdot \mathbf{P}_{av} > 0$, while scattering to the left yields $\hat{n} \cdot \mathbf{P}_{av} < 0$. The difference between the two, averaged over all angles in each hemisphere, yields

$$\frac{d\sigma_r/d\Omega - d\sigma_\ell/d\Omega}{d\sigma_r/d\Omega + d\sigma_\ell/d\Omega} = \frac{2}{\pi} |\mathbf{P}_{av}| P. \tag{B.17}$$

In order to determine F and G, we still need a third measurement in addition to $d\sigma/d\Omega$ and P. Clearly it must be obtained from the final polarization, and it is consequently difficult to measure. In fact, it has become traditional to define two quantities called the Wolfenstein parameters

$$\begin{aligned} R &= R_0 \cos\theta + A_0 \sin\theta \\ A &= -R_0 \sin\theta + A_0 \cos\theta \end{aligned} \tag{B.18}$$

where

$$\begin{aligned} R_0 &= \frac{2\mathrm{Re}(F^*G)}{|F|^2 + |G|^2}, \\ A_0 &= \frac{|F|^2 - |G|^2}{|F|^2 + |G|^2}; \end{aligned} \tag{B.19}$$

these are not independent, since $P^2 + R^2 + A^2 = 1$. Knowing $d\sigma/d\Omega_0$, P, R, and A, one can determine the magnitudes of F and G and their relative phase α, finding

$$\begin{aligned} |F|^2 &= \frac{1}{2} \frac{d\sigma}{d\Omega_0} (1 + A) \\ |G|^2 &= \frac{1}{2} \frac{d\sigma}{d\Omega_0} (1 - A) \\ \sin\alpha &= P/\sqrt{1 - A^2}. \end{aligned} \tag{B.20}$$

Partial Wave Analysis

The overall angular momentum of meson-nucleon system, in terms of which we must identify the spins of nucleon resonances, is a combination of orbital angular momentum with the nucleon's spin. The resulting partial wave expansion appropriate to spin-0 spin-½ scattering is simplified considerably if we choose the initial state to involve a single spin eigenvalue; the generalization of the results to arbitrary states is relatively straightforward. Let us therefore assume that a = 1 and b = 0 in (B.3). The initial state of the entire system may then be written as

$$|\Psi_i\rangle = \frac{1}{\sqrt{2\pi}} \quad e^{ikz}\,|\tfrac{1}{2}, \tfrac{1}{2}\rangle \tag{B.21}$$

and is an eigenfunction of $J_z = L_z + S_z$ with eigenvalue +½. The scattered wave function, according to (B.6), is asymptotically

$$|\Psi_f\rangle \sim \frac{1}{\sqrt{2\pi}} \frac{e^{ikr}}{r} \quad (F\,|\tfrac{1}{2}, \tfrac{1}{2}\rangle + inG|\tfrac{1}{2}, -\tfrac{1}{2}\rangle), \tag{B.22}$$

and since the scattering situation is axially symmetric, $|\Psi_f\rangle$ must also be purely $J_z = \tfrac{1}{2}$. It follows immediately that F and nG must be expressible as sums of eigenstates of L_z containing only m = 0 and m = 1, respectively, i.e.

$$FG = \sum_\ell \sqrt{\frac{1}{2\ell + 1}} \quad F_\ell|\ell, 0\rangle$$

$$nG = \sum_\ell \sqrt{\frac{\ell(\ell + 1)}{2\ell + 1}} \quad G_\ell|\ell, 1\rangle \tag{B.23}$$

where, in the usual way, we have the spherical harmonics as the spatial representation of the eigenfunctions

$$|\ell, 0\rangle = Y_{\ell 0}(\theta, \phi) = \sqrt{\frac{2\ell + 1}{4\pi}} \quad P_\ell(\cos\theta)$$

$$|\ell, 1\rangle = Y_{\ell 1}(\theta, \phi) = \sqrt{\frac{2\ell + 1}{4\pi\ell(\ell + 1)}} \quad P_\ell^1(\cos\theta)e^{i\phi} \tag{B.24}$$

Since $n = n_x + in_y = e^{i\phi}$, these equations are equivalent to

$$F = \frac{1}{\sqrt{4\pi}} \sum_\ell F_\ell P_\ell(\cos\theta)$$

$$G = \frac{1}{\sqrt{4\pi}} \sum_\ell G_\ell P_\ell^1(\cos\theta). \tag{8.25}$$

The scattered state is given in terms of F_ℓ and G_ℓ by

$$|\Psi_f\rangle = \frac{1}{\sqrt{2\pi}} \frac{e^{ikr}}{r} \sum_{\ell} \frac{1}{\sqrt{2\ell+1}} \left\{ F_\ell |\ell, 0\rangle|\tfrac{1}{2}, \tfrac{1}{2}\rangle + i\sqrt{\ell(\ell+1)}\, G_\ell |\ell, 1\rangle|\tfrac{1}{2}, -\tfrac{1}{2}\rangle \right\}. \tag{B.26}$$

If we wish to speak of resonance formation, of course, it is more appropriate to consider an expansion of $|\Psi_f\rangle$ in eigenfunctions of the total angular momentum $\mathbf{J} = \mathbf{L} + \mathbf{S}$. The total angular momentum j can be either $\ell \pm \tfrac{1}{2}$; the two possibilities correspond to the existence of two scattering amplitudes for each ℓ, as in (B.26). If we write

$$|\Psi_f\rangle \sim \frac{1}{\sqrt{2\pi}} \frac{e^{ikr}}{r} \sum_{\ell} \left\{ \sqrt{\ell+1}\, f_{\ell+} |\ell + \tfrac{1}{2}, \tfrac{1}{2}\rangle - \sqrt{\ell}\, f_{\ell-} |\ell - \tfrac{1}{2}, \tfrac{1}{2}\rangle \right\} \tag{B.27}$$

and use the Clebsch-Gordan coefficients in

$$|\ell + \tfrac{1}{2}, \tfrac{1}{2}\rangle = \sqrt{\frac{\ell+1}{2\ell+1}}\, |\ell, 0\rangle|\tfrac{1}{2}, \tfrac{1}{2}\rangle + \sqrt{\frac{\ell}{2\ell+1}}\, |\ell, 1\rangle|\tfrac{1}{2}, -\tfrac{1}{2}\rangle$$

$$|\ell - \tfrac{1}{2}, \tfrac{1}{2}\rangle = -\sqrt{\frac{\ell}{2\ell+1}}\, |\ell, 0\rangle|\tfrac{1}{2}, \tfrac{1}{2}\rangle + \sqrt{\frac{\ell+1}{2\ell+1}}\, |\ell, 1\rangle|\tfrac{1}{2}, -\tfrac{1}{2}\rangle$$

then it follows that

$$|\Psi_f\rangle = \frac{1}{\sqrt{2\pi}} \frac{e^{ikr}}{r} \sum_{\ell} \frac{1}{\sqrt{2\ell+1}} \left\{ [(\ell+1) f_{\ell+} + \ell f_{\ell-}]\, |\ell, 0\rangle|\tfrac{1}{2}, \tfrac{1}{2}\rangle + \sqrt{\ell(\ell+1)}\, [f_{\ell+} - f_{\ell-}]\, |\ell, 1\rangle|\tfrac{1}{2}, -\tfrac{1}{2}\rangle \right\} \tag{B.28}$$

Comparing (B.28) with (B.24) shows that

$$F_\ell = (\ell+1) f_{\ell+} + \ell f_{\ell-}$$

$$iG_\ell = f_{\ell+} - f_{\ell-} \tag{B.29}$$

A similar procedure must be applied in reactions involving higher spins; the analysis becomes much more complex in all such cases.

Reading List for Regge Theory

There are many books and articles which discuss the principles and practice of Regge theory. A particularly good treatment on approximately the same level we have attempted here is that of

1. B. E. Y. Svensson, *High Energy Phenomenology and Regge Poles,* Vol.II, Proceedings of the 1967 CERN School of Physics (CERN Publication 67–24),

which reviews experimental data in some detail. An excellent review of the theory from a somewhat more sophisticated viewpoint is given by

2. R. J. Eden, reference 5 of the reading list for Part II.

The most complete and exhaustive treatment of the entire field, however, including spin effects and all other complications, is

3. P. D. B. Collins and E. J. Squires, *Regge Poles in Particle Physics* (Berlin: Springer, 1968).

The student who masters this book can, with some pride, consider himself an expert on the subject.

Thorough discussions of the principles of Regge theory in potential scattering, including the necessary proofs of convergence, etc., are

4. R. G. Newton, *The Complex J-Plane: Complex Angular Momentum in Non-Relativistic Quantum Theory* (New York: Benjamin, 1964),
5. V. de Alfaro and T. Regge, reference 2 of the Part II reading list.

A more detailed study of Regge poles in relativistic scattering theory has been given by

6. R. Oehme, in Strong Interactions and *High Energy Physics, Proceedings of the Scottish Universities Summer School,* Ed. R. G. Moorhouse (New York: Plenum, 1963).

Proofs of the presence of Regge poles in sums of Feynman diagrams can be found in

7. R. J. Eden *et al.,* reference 6 of the Part II reading list.

The question of the existence of Regge cuts is also discussed there.

An excellent review of the phenomology associated with Regge poles, plus numerous other topics, has been given by

8. Vernon D. Barger and David B. Cline, *Phenomenological Models of High Energy Scattering* (New York: Benjamin, 1969).

These are but a few of the abundant works on Regge theory. Further reading can be found in the reference lists given in many of them, or by consulting review articles such as

9. E. Leader, *Rev. Mod. Phys.* 38, 476 (1966).
10. G. Hite, *Rev. Mod. Phys.* 41, 669 (1969).

A good review of the development of duality has been given by

11. M. Kugler, *Duality*, lectures at the IX Internationale Universitatswochen fur Kernphysik, Schladming, Austria, 1970.

Postscript

When this book was first written, total cross sections were becoming asymptotically constant, the A_2 meson was curiously "split", and quark searches were being vigorously pursued.

As it was being prepared for the printer, it was revised because cross sections were rising, the A_2 meson was a single bump again, and quark searches were still being relentlessly pursued.

As the galley proofs are being corrected, total cross sections are still rising. The A_2 is still a normal meson. Quark searches proceed, with diminishing hope of success. And the world of particle physics has been electrified by the discovery of a phenomenally new type of particle appearing first in e^+e^- collisions.

Can a text stop here, treating only these "old-fashioned" topics – some as much as fifteen years old! – and not include at least a brief introduction to the still-spawning world of the $J/\psi(3105)$ and its relatives? This text will do so, in the belief that whatever the future reveals about these or other new particles, the basic ideas of SU(3), analyticity, and Regge poles will still provide the basic language for describing strong interactions.

Plus ça change . . .

Index

Angular momentum 21
Antishrinkage 319
Asymptopia 323
A_2 trajectory 290, 332, 360

Barger–Rubin relation 96, 115, 123
Baryon 1
  in Sakata model 71
  in SU(3) 75
  in SU(6) 108
Bootstrap 260, 361
Born approximation 6, 165, 191, 240
Bound state 143, 151
Branch point 182, 216, 343

Casimir invariants 29
  of SU(3) 57, 64
Charge conjugation 14
Clebsch–Gordan coefficients
  of O(3) and SU(2) 26
  of SU(3) 69
Commutation relations
  of O(3) 26
  of SU(2) 37
  of SU(3) 52
  of SU(4) 100
  of SU(6) 105
Conspiracy 317
Coupling constant 11
  and factorization 288
  in SU(3) 84
  pion–nucleon 93
C parity 15
Crossed-channel Regge poles 299, 352
Crossing 192, 216, 227, 232, 307
Cut 182, 343

D/f ratio 91, 110
Dalitz plot 237
Daughters 311
Decay of unstable particle
  in SU(2) 41
  in SU(3) 88
  in SU(6) 109
  rate 4

Differential cross section 5, 145, 305, 325, 347
Diffractive background 359
Discontinuity 183
Dispersion relations 190, 293
  double 246
  for NN → NN 221
  for πN → πN 218, 243
Double spectral function 248
Dynamics 143

Eightfold way 72
Elasticity 211
Electromagnetic properties
  in SU(3) 84
  in quark model 117
Evasion 317
Exchange degeneracy 290, 335, 364
Exotic 359

Factorization 286, 303, 316
Feynman diagrams 10
Finite-energy sum rule 353
Fixed pole 290
Form factor 110, 122
Froissart bound 301, 341

Gell-Mann–Okubo mass formula 81
Generators
  infinitesimal 29
  of O(3) 21, 29
  of SU(2) 36
  of SU(3) 50
  of SU(4) 97
  of SU(6) 103
Ghost 290, 315, 364
G parity 39

Hadrons 2
Hypercharge 14

Interference, pole-cut 348
Interference model 352

J/ψ particle 373

Johnson–Treiman relation 115
Jost function 167, 278

Kinematics 143

Leptons 1
Lie group 19
Line reversal 309, 320, 332

Magnetic moment
in quark model 119
in SU(3) 86
Mandelstam representation 248, 253, 294
Mass relations
in quark model 117
in SU(3) 76
in SU(4) 103
in SU(6) 113
Mesons 1
in Sakata model 70
in SU(3) 74
in SU(6) 108
pseudoscalar 70
tensor 72
vector 71
Mixing
ω–ϕ 71, 83, 117
f–f′ 84

N/D method 256
Nonsense 316, 325
Nucleons 12
Nucleon trajectories 291, 338

O(3) 19
Optical theorem 149, 206, 213

P′ trajectory 322
Parity
charge 15
G 39
intrinsic 15
of pion 16
operator 15
Partial wave dispersion relations 183, 253
Partial wave expansion 148
Phase of amplitude 262, 306
Phase shifts 148
Phase space 160
Physical region 231
Physical sheet 182

Polarization 307, 324, 328, 330, 350, 366
Pomeranchuk theorem 310, 340
Pomeranchuk trajectory 302, 322
Pomeron 322

Quarks 74

ρ trajectory 288, 323, 360
Rank of group 29
Real analytic function 186
Relativistic variables s, t, u 225
Representation, matrix 24
conjugate 31, 39
elementary 47
irreducible 25
of O(3) 24
of permutation group 129
of SU(2) 37
of SU(3) 59
of SU(4) 97
of SU(6) 103
regular 31
Representation space 30
Resonance 153
lifetime 156
mass 158
width 156, 271
Root vectors 30
Rotation group 19

Sakata model 65
Scalar operator 25
Scattering amplitude 4
forward 188, 201
Regge pole form 302, 346
s channel 231
Sense 316, 325
Shrinkage of diffraction peak 305, 325, 348
Signature 282, 296, 324
S matrix 163, 203
Sommerfeld–Watson transformation 273, 343
Spin-flip amplitude 325, 366
Strangeness 13
Structure constants 29
of SU(3) 56
Subtracted dispersion relation 200, 296

t channel 231

Tensor operator 26, 80
Threshold 215
Total cross section 149, 186, 209, 300, 320, 346
Transition rate 4
Triality 75
Two-particle scattering 5
 in quark model 121
 in SU(2) 43
 in SU(3) 93
 in SU(6) 114
 potential theoretic 145
 relativistic 225

u channel 231
Unitarity 149, 211
Unphysical sheet 182

Vector operator 26
Veneziano amplitude 362
Virtual particles 7

Wigner–Eckart theorem 28, 42
Wrong-signature nonsense zeros 317, 348

Young diagram 130
Yukawa potential 11, 176, 189, 280